T0255546

Universitext

Universitext

Series Editors

Sheldon Axler
San Francisco State University, San Francisco, CA, USA

Carles Casacuberta
Universitat de Barcelona, Barcelona, Spain

John Greenlees
University of Warwick, Coventry, UK

Angus MacIntyre
Queen Mary University of London, London, UK

Kenneth Ribet
University of California, Berkeley, CA, USA

Claude Sabbah
École Polytechnique, CNRS, Université Paris-Saclay, Palaiseau, France

Endre Süli
University of Oxford, Oxford, UK

Wojbor A. Woyczyński
Case Western Reserve University, Cleveland, OH, USA

Universitext is a series of textbooks that presents material from a wide variety of mathematical disciplines at master's level and beyond. The books, often well class-tested by their author, may have an informal, personal even experimental approach to their subject matter. Some of the most successful and established books in the series have evolved through several editions, always following the evolution of teaching curricula, into very polished texts.

Thus as research topics trickle down into graduate-level teaching, first textbooks written for new, cutting-edge courses may make their way into *Universitext*.

More information about this series at http://www.springer.com/series/223

David Harari

Galois Cohomology and Class Field Theory

David Harari
Laboratoire de Mathématiques d'Orsay
Université Paris-Saclay
Orsay, France

Translated by
Andrei Yafaev
Department of Mathematics
University College London
London, UK

ISSN 0172-5939   ISSN 2191-6675   (electronic)
Universitext
ISBN 978-3-030-43900-2   ISBN 978-3-030-43901-9   (eBook)
https://doi.org/10.1007/978-3-030-43901-9

Mathematics Subject Classification (2020): 11R34, 11R29, 11R37, 12G05

Translation from the French language edition: *Cohomologie galoisienne et théorie du corps de classes* by
David Harari, © 2017 EDP Sciences, CNRS Editions, France. http://www.edpsciences.org/ http://www.
cnrseditions.fr/. All rights reserved.
© Springer Nature Switzerland AG 2020
This work is subject to copyright. All rights are reserved by the Publisher, whether the whole or part
of the material is concerned, specifically the rights of translation, reprinting, reuse of illustrations,
recitation, broadcasting, reproduction on microfilms or in any other physical way, and transmission
or information storage and retrieval, electronic adaptation, computer software, or by similar or dissimilar
methodology now known or hereafter developed.
The use of general descriptive names, registered names, trademarks, service marks, etc. in this
publication does not imply, even in the absence of a specific statement, that such names are exempt from
the relevant protective laws and regulations and therefore free for general use.
The publisher, the authors and the editors are safe to assume that the advice and information in this
book are believed to be true and accurate at the date of publication. Neither the publisher nor the
authors or the editors give a warranty, expressed or implied, with respect to the material contained
herein or for any errors or omissions that may have been made. The publisher remains neutral with regard
to jurisdictional claims in published maps and institutional affiliations.

This Springer imprint is published by the registered company Springer Nature Switzerland AG
The registered company address is: Gewerbestrasse 11, 6330 Cham, Switzerland

Preface

The tools from Galois cohomology play a fundamental role in modern number theory. They contribute to a better understanding of the Galois group of a local or a number field, which are central objects in most of the important arithmetic problems of our times. Cohomological class field theory, that was developed after the Second World War, can be seen as the first step since it mainly concerns itself with the extensions whose Galois groups are abelian. In a certain sense it is the dimension 1 case of the Langlands program, which is today one of the most prominent areas of arithmetic.

The reason for writing this book was a shortfall in the existing literature. Books such as [39] and [42], that provide proofs of the Poitou–Tate theorems, generally assume a certain amount of results from class field theory. This makes student's understanding of the logical sequence of ideas difficult. Conversely, books on class field theory such as [9], [40] or [26] do not treat arithmetic duality theorems of Poitou–Tate type.

The aim of this monograph is therefore to provide an exposition with complete proofs of the following topics: basics of cohomology (Part I: Chaps. 1–6), local class field theory (Part II: Chaps. 7–11), global class field theory (Part III: Chaps. 12–15) and the difficult (but extremely useful) Poitou–Tate theorems (Part IV: Chaps. 16–18).

The prerequisites are limited to generalities on algebra and arithmetic as taught at the first year master's level at French universities (Galois theory, basics of arithmetic and the theory of finite groups) thus making the book accessible to the second year master's students.

We have included a review chapter on local fields (Chap. 7) and another on global fields (Chap. 12) as well as an appendix summarising results from homological algebra used in the book. We believe indeed that some knowledge of homological algebra (abelian categories, derived functors, spectral sequences) can be independently useful to a student wishing to pursue their study of algebra, arithmetic or algebraic geometry.

Furthermore, systematic usage of homological algebra provides a really elegant presentation of some results (e.g. those on class formations in Chap. 16) that otherwise would appear rather artificial. It also justifies the commutativity of diagrams, which we attempted to present as rigorously as possible throughout the book.

On the other hand, we did not consider necessary to use derived categories. Even though they may bring a new viewpoint, their understanding requires considerable effort and they are not really necessary for the topics covered in this monograph. However, they become indispensable if one is interested in more general duality theorems than those of Poitou–Tate, such as those involving fields of cohomological dimension at least 3 [19] or complexes of Galois modules [17, 20].

We have also chosen (mainly to keep the length of the main text reasonable) to relegate to an appendix some analytic topics such as the proof of the "first inequality" using the Dirichlet series (that we prove using algebraic techniques in Chap. 13), or the Čebotarev theorem (which allows to refine some results of Chap. 18). Naturally analytic techniques provide a different and important insight into global class field theory. We refer to [26] or [33] for a detailed presentation of these methods and to [29] for an overview of the subject. We have also left untreated more advanced topics such as Iwasawa theory, embedding problems or anabelian geometry of which one finds an excellent exposition in [42].

The book is designed to be particularly suitable for second year master's students in the European system or graduate students in the American system. It is a fusion of two second year master's level courses given at Orsay University over several years. The book should also be useful to researchers and Ph.D. students. Each course consisted of 44 h of lectures and 4 h of lectures per week. Both courses included most of the Parts I and II. One of the courses (entitled *Galois cohomology and number theory*) treated mainly the contents of Part IV (assuming most of the results of Part III). The other course (entitled *Class field theory*) gave a detailed exposition of Part III (without the Poitou–Tate duality). At the end of each chapter we have included a certain number of exercises (more than 110 in total). These exercises include those given during the lectures as well as those set at the exams. Some of these exercises complement the main text, however their results are not used in the proofs thus allowing the reader to have complete proofs readily available.

Many of the results presented in the book have vast generalisations, some of which are still subjects of active research. There exists for example a higher class field theory developed by Kato and Saito in the eighties (see for example [43]) to which there is now a direct approach, not based on K-theory (cf. [53]). The reader interested in studying duality theorems further may for example consult the book [39] as well as articles [10, 17, 20] for duality theorems in étale and flat cohomology (including those for group schemes such as tori and abelian varieties).

These duality theorems are now used extensively in the study of arithmetic questions about rational points on varieties, notably in the study of the local-to-global principle for linear algebraic groups. One finds an overview in [52] as well as in the articles [3, 19, 45].

Naturally the content of this book is very classical and hence largely inspired by the existing literature. In particular the influence of books [39, 40, 42, 47] and [49] is evident. I would like to extend my sincere thanks to the authors for their remarkable work which gave me a full appreciation of the beauty of the topics treated in the present monograph.

My thanks extend also to all the students who have taken my courses, pointed out errors in the online notes and suggested improvements, namely M. Chen, Z. Gao, C. Gomez, A. Jaspers, Y. Liang, S. Liu and J. Xun.

The first version of the manuscript has also benefited from valuable comments by J.-B. Bost, C. Demarche, D. Izquierdo, G. Lucchini–Arteche, A. Pirutka, J. Riou, A. Smeets, T. Szamuely and O. Wittenberg; I thank them warmly.

For the English edition, I was fortunate to receive very pertinent comments from J. Fresán and several anonymous referees.

I extend my particular thanks to C. Sabbah for bringing to my attention the possibility of writing this book and for his extremely efficient editorial work. *Last but not least*, I am especially grateful to A. Yafaev for his thorough English translation.

Orsay, France   David Harari
February 2020

Notation and Conventions

The symbol $:=$ means that the equality in question is a definition. We will some-times denote the cardinality of a finite set E by $\#E$. A commutative diagram whose vertical and horizontal arrows are exact sequences will be called an *exact com-mutative diagram*.

If A and B are finite abelian groups (which may be viewed as $\mathbf{Z}$-modules), $A \otimes B$ will denote the tensor product $A \otimes_{\mathbf{Z}} B$. Recall that if R is a ring and if M and N are right (resp. left) R-modules, then we can define the tensor product $M \otimes_R N$, which is an abelian group (and an R-module if R is commutative), [4], § 3.1.

For any $n > 0$, the *n-torsion subgroup* of an abelian group A (whose group law is written additively) is the subgroup $A[n] := \{x \in A, nx = 0\}$, and the *torsion subgroup* A_{tors} of A is the union of the $A[n]$ for $n > 0$. We call an abelian group A torsion if $A = A_{\mathrm{tors}}$. We will often write $\mathbf{Z}/n$ for the cyclic group $\mathbf{Z}/n\mathbf{Z}$.

Let p be a prime number. An abelian group A is called *p-primary* if every element of A has a p-power order. Any torsion abelian group A can be written as $\bigoplus_p A\{p\}$ where, for each prime p, we write $A\{p\}$ for the *p-primary component* of A, i.e., the subgroup of A consisting of elements of order a power of p.

An abelian group A is called *p-divisible* if multiplication by p in A is surjective, *divisible* if this property holds for all primes p (or any integer $n > 0$ instead of p). It is called *uniquely divisible* if for any integer $n > 0$, the multiplication by n in A is an isomorphism. We say that a finite group G is a *p-group* if its cardinality is a power of p. Such a group is nilpotent, in particular, it has a non-trivial centre as soon as the group itself is not trivial.

By convention, all inductive systems $(A_i)_{i \in I}$ of abelian groups, commutative rings, modules over a ring R, etc., that we will consider will be associated with non-empty *filtered* ordered sets I, such that if $i, j \in I$, there exists $k \in I$ with $i \leqslant k$ and $j \leqslant k$. This way, the inductive limit $\varinjlim_{i \in I} A_i$ will always be a well defined abelian group (resp. commutative ring, R-module...) as opposed to just being defined as a set.

If G is a group (resp. topological group) and H is a subgroup of G, we denote by G/H the set (resp. the quotient topological space) of left H-cosets, i.e., the set of aH for $a \in G$. If H is normal in G then the set G/H is canonically endowed with a structure of a quotient group (resp. topological quotient group). A *character* of a finite group (resp. of a topological group) G is a morphism (resp. a continuous morphism) from G to $\mathbf{Q}/\mathbf{Z}$ (viewed as a discrete group). We will denote by $G^* = \mathrm{Hom}_c(G, \mathbf{Q}/\mathbf{Z})$ the *dual* of a topological group G, which is also the dual of its abelianisation (it coincides with the classical Pontryagin dual $\mathrm{Hom}_c(G, \mathbf{R}/\mathbf{Z})$ if G is abelian and torsion, but not in general). A continuous morphism $f : A \to B$ of topological groups is *strict* if it induces a homeomorphism of the topological quotient space $A/\mathrm{Ker}\, f$ onto the subspace $\mathrm{Im}\, f$ of B. This is equivalent to requiring that the image of any open subset of A is open in $f(A)$ (where $f(A)$ is endowed with the induced topology by that of B). For a surjective morphism, this is equivalent to saying that it is an open map.

By definition a compact topological space will for us always be Hausdorff and countable at infinity. Same for a locally compact space. A topological space is *totally disconnected* if the only connected components are singletons. With these conventions, a continuous surjective morphism between locally compact groups is automatically strict ([22], Th. 5.29).

All fields are assumed to be commutative. A field k of strictly positive characteristic p is *perfect* if the morphism $x \mapsto x^p$ from k to k is an isomorphism (e.g. finite k), *imperfect* otherwise. For a q-power of a prime p, we will denote by $\mathbf{F}_q$ the finite field with q elements, which has characteristic p. An *extension* of a field K is a field L equipped with a field morphism (necessarily injective) $K \to L$. Such an extension is called *finite* if the K-vector space L is finite dimensional. In this case, we denote by $[L : K]$ this dimension. A Galois extension of fields will be called *abelian* if its Galois group is abelian. If k is a field, we denote by $\bar{k}$ a fixed separable closure of k (which is also an algebraic closure if k is of characteristic zero or perfect of characteristic p).

If I, J are two-sided ideals in a ring A, we denote $I.J$ or IJ the two-sided ideal consisting of finite sums of elements of the form ij with $i \in I$ and $j \in J$. In particular we write I^2 for $I.I$. A ring A is said to be an *integral domain* if it is commutative, nonzero and if the equality $ab = 0$, with $a, b \in A$, implies $a = 0$ or $b = 0$. A *principal ideal domain* A is an integral domain such that every ideal I in A is of the form aA with $a \in A$. A ring A is called *integrally closed* if it is an integral domain and every element of its field of fractions which is integral over A (i.e., a zero of a monic polynomial with coefficients in A) is in A. It is true for a unique factorisation domain (and a fortiori for a principal ideal domain). An ideal I in a commutative ring A is *prime* if the quotient A/I is an integral domain, *maximal* if A/I is a field.

Contents

Part I
Group Cohomology and Galois Cohomology: Generalities

Part I is devoted to the general results on the cohomology of finite and profinite groups, which can also be useful in areas of mathematics other than number theory. The first two chapters present the basics of the theory for finite groups. The third is slightly more specialised: it culminates in the Tate–Nakayama theorem, which will be an essential tool in our approach to the local and global class field theory. Then we move on to generalising the notions seen in the first two chapters to profinite groups and developing the very important concept of *cohomological dimension*. Finally, this part ends with one of the main applications of the general theory, the *Galois cohomology*, which is omnipresent throughout the book, and whose first properties will be seen in Chap. 6.

Chapter 1
Cohomology of Finite Groups: Basic Properties

The purpose of this chapter is to define cohomology groups associated with a finite group G and with what is called a *G-module*. This notion will be essential in all arithmetic applications that we will be dealing with in this book, especially for the class field theory.

Many definitions and properties in this chapter extend to groups G which are not necessarily finite (see Chap. VII of [47] or Chap. 6 of [54] for the case of an arbitrary group). We chose to limit ourselves to the case where G is finite which, with its slight generalisation to profinite groups (Chap. 4), will be the only one we will need in the sequel. This will make our presentation slightly simpler and will also avoid confusion between the cohomology of profinite groups as defined in Chap. 4 and cohomology of any group as defined in the references above. The two notions do not coincide in general for an arbitrary profinite group.

Throughout the chapter, G denotes a finite group (whose law is denoted multiplicatively and the neutral element is denoted by 1). Recall that *the algebra of the group G* is the set $\mathbf{Z}[G]$ of formal sums

$$\sum_{g \in G} n_g g, \quad n_g \in \mathbf{Z}$$

endowed with the obvious addition and the convolution product

$$\left(\sum_{g \in G} n_g g \right) \left(\sum_{g \in G} m_g g \right) = \sum_{(g,g') \in G \times G} n_g m_{g'} gg'.$$

In particular, $\mathbf{Z}[G]$ is a non-commutative ring if G is not abelian.

© Springer Nature Switzerland AG 2020
D. Harari, *Galois Cohomology and Class Field Theory*, Universitext,
https://doi.org/10.1007/978-3-030-43901-9_1

1.1   The Notion of G-Module

Definition 1.1 A *G-module* is an abelian group $(A, +)$ endowed with a left action $(g, x) \mapsto g \cdot x$ of G on A such that for all g in G the map $\varphi_g : x \mapsto g \cdot x$ from A to A is a morphism of abelian groups.

In other words, defining a structure of a G-module on an abelian group $(A, +)$ is equivalent to defining a map

$$G \times A \longrightarrow A, \quad (g, x) \longmapsto g \cdot x,$$

such that

$$g \cdot (g' \cdot x) = (gg') \cdot x, \quad 1 \cdot x = x, \quad g \cdot (x + y) = g \cdot x + g \cdot y$$

for all g, g' in G and all x, y in A.

Note also that with the notations of Definition 1.1, φ_g is an automorphism of A, whose inverse is $\varphi_{g^{-1}}$. If A is an abelian group, whose automorphism group we denote by $\mathrm{Aut}(A)$, to give a G-module structure on A amounts to giving a group morphism from G to $(\mathrm{Aut}(A), \circ)$. Equivalently, a G-module is a (left) module over the ring $R := \mathbf{Z}[G]$: indeed, if A is a module over the ring R, we define a structure of G-module on A by $g \cdot x = gx$ for all $(g, x) \in G \times A$. Conversely, if A is a G-module, we equip it with an R-module structure by setting $(\sum_{g \in G} n_g g) x = \sum_{g \in G} n_g (g \cdot x)$. As G is finite, for A to be of finite type as an abelian group is equivalent to be of finite type as an R-module. Therefore, in this case, we will simply say that the G-module A is of *finite type* .

Definition 1.2 A *morphism of G-modules* (or *G-morphism*, or also *G-homomorphism*) $f : A \to A'$ is a morphism of abelian groups commuting with the action of G, i.e., such that $f(g \cdot x) = g \cdot f(x)$ for any x in A and any g in G. This is equivalent to f being a morphism of R-modules.

It is straightforward to define the notions of isomorphism of G-modules, of a sub-G-module, of an exact sequence of G-modules, etc. If A and A' are G-modules, we denote $\mathrm{Hom}_G(A, A')$ the set of G-morphisms from A to A'. It is an abelian group for addition which is a subgroup of $\mathrm{Hom}_{\mathbf{Z}}(A, A')$, the latter consisting of morphisms of abelian groups from A to A' (not necessarily compatible with the action of G).

Example 1.3 (a) For any abelian group A, the trivial action of G on A (defined by $g \cdot x = x$ for any $g \in G$ and any $x \in A$) turns A into a G-module.

(b) Let $G = \{\pm 1\}$ and $M = \mathbf{Z}$. The action of G on M defined by $g \cdot x = gx$ turns M into a G-module.

(c) Let L be a finite Galois extension of a field K with group G. Then the additive group L and the multiplicative group $L^* := L - \{0\}$ are both G-modules for the action of $G = \mathrm{Aut}(L/K)$.

(d) Let A and B be G-modules. The group $\mathrm{Hom}(A, B) := \mathrm{Hom}_{\mathbf{Z}}(A, B)$ has a structure of G-module, defined by $(g \cdot f)(x) := g \cdot f(g^{-1} \cdot x)$ for any $g \in G$, $f \in \mathrm{Hom}(A, B)$, $x \in A$.

(e) The group algebra $\mathbf{Z}[G]$ is a G-module for the left action of G. More generally, if H is a subgroup of G, then the abelian group $\mathbf{Z}[G/H] := \bigoplus_{\overline{g} \in G/H} \mathbf{Z} \cdot \overline{g}$ is a G-module, where G/H is the set of left H-cosets.

We will now see how we can, if H is a subgroup of G, construct a G-module from an H-module.

Definition 1.4 Let G be a finite group and H a subgroup of G. Let A be an H-module. We define the abelian group $I_G^H(A)$ as consisting of maps $f : G \to A$ satisfying $f(hg) = h \cdot f(g)$ for any $g \in G$, $h \in H$. This abelian group is endowed with a structure of a G-module via $(g \cdot f)(g') = f(g'g)$ for all g, g' in G. We say that $I_G^H(A)$ is the *induced module* from H to G of the H-module A.

An important special case is the following.

Definition 1.5 Let G be a finite group. Consider the trivial subgroup $\{1\}$ of G. Let A be an abelian group, which we can view as a $\{1\}$-module. We denote by $I_G(A)$ the induced module $I_G^{\{1\}}(A)$ (cf. Definition 1.4 in the case $H = \{1\}$) and we say that $I_G(A)$ is the *induced G-module* of the abelian group A.

We say that the G-module M is *induced* if there exists an abelian group A such that M is isomorphic to $I_G(A)$.

Thus $I_G(A)$ is the abelian group consisting of maps from G to A, with the action of G given by $(g \cdot f)(g') = f(g'g)$.

Remark 1.6 (a) Note that if A is already endowed with a certain structure of G-module, then $I_G(A)$ is isomorphic to the G-module $\mathcal{F}(G, A)$ defined (cf. [42], p. 31) as the set of maps from G to A with the action given by $(g \cdot f)(g') := g \cdot f(g^{-1}g')$. An isomorphism of $\mathcal{F}(G, A)$ with $I_G(A)$ is given by $f \mapsto (g \mapsto g \cdot f(g^{-1}))$ (I thank Alexander Schmidt for this remark).

(b) Let A be an abelian group. We can also view the induced module $I_G(A)$ as the G-module $\mathrm{Hom}_{\mathbf{Z}}(\mathbf{Z}[G], A)$, the (left) action of G being given by the natural right action of G on the first factor. Since we have assumed G to be finite, it is easy to see that this G-module is (non-canonically) isomorphic to $\mathbf{Z}[G] \otimes A$, the action of G being this time on the left on the first factor. We sometimes call the modules of the form $\mathbf{Z}[G] \otimes A$ *co-induced*. [1] They are also modules M such that there exists a subgroup X of M such that M is the direct sum of the $g \cdot X$ for $g \in G$. The assumption that G is finite implies that there is no difference between the notions of induced and co-induced G-module.

---

[1] When G is not assumed finite, the reverse convention is often adopted (cf. for example in Chap. VII of [47]) that is to call co-induced the modules of the form $I_G(A)$ and induced the modules of the form $\mathbf{Z}[G] \otimes A$. The terminology that we use is compatible with the traditional terminology used for profinite groups that we will encounter in Chap. 4.

(c) More generally, we can view the G-module $I_G^H(A)$ from Definition 1.4 as the G-module $\mathrm{Hom}_H(\mathbf{Z}[G], A)$ of H-morphisms from $\mathbf{Z}[G]$ to A (or of morphisms of $\mathbf{Z}[H]$-modules from $\mathbf{Z}[G]$ to A), which is a sub-G-module of $\mathrm{Hom}_{\mathbf{Z}}(\mathbf{Z}[G], A)$. Here again, the assumption that G is finite implies that this G-module is isomorphic to $\mathbf{Z}[G] \otimes_{\mathbf{Z}[H]} A$ (to define the tensor product, we view $\mathbf{Z}[G]$ as a right $\mathbf{Z}[H]$- and A as a left $\mathbf{Z}[H]$-module). The action of G is on the left on the first factor. The correspondence between $I_G^H(A)$ and $\mathbf{Z}[G] \otimes_{\mathbf{Z}[H]} A$ is as follows: we associate to any function $f \in I_G^H(A)$ the element

$$\Phi(f) := \sum_{\overline{g} \in G/H} g \otimes_{\mathbf{Z}[H]} f(g^{-1})$$

of $\mathbf{Z}[G] \otimes_{\mathbf{Z}[H]} A$. Note that $g \otimes_{\mathbf{Z}[H]} f(g^{-1})$ does not depend on the representative $g \in G$ of $\overline{g}$, and for $g' \in G$, we have

$$\Phi(g' \cdot f) = \sum_{\overline{g} \in G/H} g \otimes_{\mathbf{Z}[H]} (g' \cdot f)(g^{-1}) = \sum_{\overline{g} \in G/H} g \otimes_{\mathbf{Z}[H]} f(g^{-1}g')$$
$$= \sum_{\overline{g} \in G/H} (g'g) \otimes_{\mathbf{Z}[H]} f(g^{-1}) = g' \cdot \Phi(f),$$

which shows that Φ is a morphism of G-modules. Besides, it is easy to see that Φ is bijective by choosing a set of representatives S for G/H that forms a basis of the right $\mathbf{Z}[H]$-module $\mathbf{Z}[G]$ in view of the fact that any element g of G can be uniquely written as $g = sh$ with $s \in S$ and $h \in H$.

Furthermore, if A is already endowed with a structure of G-module, then the G-module $\mathbf{Z}[G] \otimes_{\mathbf{Z}[H]} A$ is also isomorphic to $\mathbf{Z}[G/H] \otimes_{\mathbf{Z}} A$, the isomorphism being given by

$$g \otimes_{\mathbf{Z}[H]} m \longmapsto \overline{g} \otimes_{\mathbf{Z}} (g \cdot m),$$

where $m \in A$ and $g \in G$.

1.2   The Category of G-Modules

Let $f : A \to A'$ be a morphism of G-modules. It follows immediately from definitions that the kernel, the image and the cokernel of f (seen as a morphism of abelian groups) are G-modules for the evident action of G. Therefore G-modules (the group G being fixed) form an *abelian category* (cf. Appendix, Sect. A.3). We denote this category $\mathcal{M}od_G$.

Like in any abelian category (cf. Appendix, Definition A.23), for any two G-modules A and A', the covariant functor $\mathrm{Hom}_G(A, .)$ and the contravariant functor $\mathrm{Hom}_G(., A')$ are left exact.

Definition 1.7 A G-module A is called *projective* if $\mathrm{Hom}_G(A, .)$ is exact. A G-module A' is *injective* if $\mathrm{Hom}_G(., A')$ is exact.

Naturally, these definitions correspond to those of projective and injective objects in the category $\mathcal{M}od_R$ of left R-modules (cf. Appendix, Definition A.33), where $R = \mathbf{Z}[G]$.

Example 1.8 (a) Any free G-module (that is, admitting a basis) as a module over $\mathbf{Z}[G]$ is projective (more generally, a G-module is projective if and only if it is a factor of a free $\mathbf{Z}[G]$-module).

(b) Let G be the trivial group (so a G-module is simply an abelian group). Then any *divisible* abelian group A is injective (cf. Appendix, Example A.35a). The converse also holds.

(c) We immediately see that a (possibly infinite) direct sum of projective G-modules is a projective G-module. Similarly, a (possibly infinite) direct product of injective G-modules is injective. In particular, a direct sum of injective G-modules is also an injective G-module (since it identifies with a direct product of the G-modules in question).

Proposition 1.9 *Let A be a G-module. Let I be an injective sub-G-module of A. Then I is a direct factor of A (that is, there exists a sub G-module B of A such that $A = I \oplus B$).*

Proof By definition of an injective G-module, the identity $I \to I$ extends to a morphism of G-modules $r : A \to I$. It is then enough to take $B = \mathrm{Ker}\, r$. $\qquad\square$

Proposition 1.10 *For any G-module A, there exists an induced G-module I endowed with an injective G-morphism $A \to I$. In other words, any G-module embeds into an induced G-module. Moreover, we can require $I = I_G(A)$ and that A is a direct factor of I as a $\mathbf{Z}$-module.*

Proof Let $I = I_G(A) = \mathrm{Hom}_{\mathbf{Z}}(\mathbf{Z}[G], A)$ be the induced module of A. We embed A into $I_G(A)$ by associating with $a \in A$ the map $g \mapsto g \cdot a$ from G to A. It follows immediately from the definition of $I_G(A)$ that this defines a morphism of G-modules, which is injective (if $g \cdot a = 0$ for all g of G, it follows that $a = 0$ by taking $g = 1$). Finally, the map $f \overset{u}{\longmapsto} f(1)$ provides a retraction of the morphism of $\mathbf{Z}$-modules $A \to I_G(A)$, which shows that A is a direct factor of I as $\mathbf{Z}$-module (the supplement of A in I is $\mathrm{Ker}\, u$). $\qquad\square$

Remark 1.11 We can also write any G-module A as a quotient of an induced one

$$0 \longrightarrow A_{-1} \longrightarrow I_G(A) \overset{p}{\longrightarrow} A \longrightarrow 0,$$

so that the above short exact sequence of abelian groups splits.

Indeed, it suffices to define p by $f \mapsto \sum_{g \in G} g \cdot f(g^{-1})$ from $I_G(A)$ to A. This procedure will be used in the inductive arguments involving the Tate modified groups (cf. Corollary 2.7)

Note also the following property of the modules from Definition 1.4.

Proposition 1.12 *Let G be a finite group. Let H be a subgroup of G. Consider an H-module A and the induced module $I_G^H(A)$ from H to G. Then for any G-module B, the group $\mathrm{Hom}_H(B, A)$ can be identified to $\mathrm{Hom}_G(B, I_G^H(A))$.*

Proof Let $\psi \in \mathrm{Hom}_H(B, A)$. For every b in B, we define an element $\varphi(b)$ of $I_G^H(A)$ by the formula $\varphi(b)(g) = \psi(gb)$ for any g in G. The property $\varphi(b) \in I_G^H(A)$ is satisfied because of the assumption that ψ is an H-morphism.

For any g, g' in G and any b in B, one readily verifies the equality

$$\varphi(g \cdot b)(g') = \varphi(b)(g'g) = (g \cdot \varphi(b))(g'),$$

which shows that φ is a G-morphism from B to $I_G(A)$. Let $\varphi = u(\psi)$, then $u : \mathrm{Hom}_H(B, A) \to \mathrm{Hom}_G(B, I_G^H(A))$ is clearly a morphism of abelian groups. Its kernel is trivial as if $\varphi(b) = 0$ for all b of B, then in particular $\varphi(b)(1) = \psi(b)$ is zero for all b in B.

Let us finally show that u is surjective. Let θ be a G-morphism from B to $I_G^H(A)$. We have $\theta(gb)(g') = (g \cdot \theta(b))(g') = \theta(b)(g'g)$ for all g, g' in G and all b in B, by definition of the action of G on $I_G^H(A)$. Setting $g' = 1$, we obtain

$$\theta(b)(g) = \theta(gb)(1).$$

Let us define $\psi : B \to A$ by $\psi(b) = \theta(b)(1)$ for any $b \in B$. For any $h \in H$, we then have

$$\psi(hb) = \theta(hb)(1) = \theta(b)(h) = h \cdot (\theta(b)(1)),$$

the last equality resulting from the fact that $\theta(b) \in I_G^H(A)$. Thus $\psi(hb) = h \cdot \psi(b)$, showing that $\psi \in \mathrm{Hom}_H(B, A)$. Set $\varphi = u(\psi)$, we obtain from the definition of u that for all $g \in G$ and $b \in B$:

$$\varphi(b)(g) = \psi(gb) = \theta(gb)(1) = \theta(b)(g),$$

hence $u(\psi) = \varphi = \theta$. Finally u is surjective and hence an isomorphism between $\mathrm{Hom}_H(B, A)$ and $\mathrm{Hom}_G(B, I_G^H(A))$. $\qquad\square$

Corollary 1.13 *Let A be an abelian group and let $I_G(A)$ be the induced G-module. Then we have*

$$\mathrm{Hom}_{\mathbf{Z}}(B, A) = \mathrm{Hom}_G(B, I_G(A)).$$

Proof This is a special case of the above with $H = \{1\}$. $\qquad\square$

Remark 1.14 The Proposition 1.12 says exactly that the left adjoint of the functor F (from $\mathcal{M}od_H$ to $\mathcal{M}od_G$) defined by $A \mapsto I_G^H(A)$ is the forgetful functor from $\mathcal{M}od_G$ to $\mathcal{M}od_H$.

As this latter one is exact, we readily conclude that F preserves the injectives (cf. Appendix, Proposition A.34).

The fact that $I_G^H(A)$ is isomorphic to $\mathbf{Z}[G] \otimes_{\mathbf{Z}[H]} A$ (via the isomorphism made explicit in Remark 1.6, c) also implies, in conjunction with the adjunction property of the tensor product ([4], Sect. 4.1), the following proposition:

Proposition 1.15 *Let G be a finite group. Let H be a subgroup of G and A an H-module. Then for any G-module B, the group $\mathrm{Hom}_H(A, B)$ identifies with $\mathrm{Hom}_G(I_G^H(A), B)$. An isomorphism is obtained by sending any $\varphi \in \mathrm{Hom}_H(A, B)$ to the element*

$$f \longmapsto \sum_{\bar{g} \in G/H} g \cdot \varphi(f(g^{-1}))$$

of $\mathrm{Hom}_G(I_G^H(A), B)$.

Thus the forgetful functor from $\mathcal{M}od_G$ to $\mathcal{M}od_H$ is also the right adjoint for $A \mapsto I_G^H(A)$, since this last functor is isomorphic to $A \mapsto \mathbf{Z}[G] \otimes_{\mathbf{Z}[H]} A$.

However, as pointed out in the electronic version of [42] (note at the bottom of the page 61), the functoriality with respect to G and H goes in two different directions depending on whether we consider $I_G^H(A)$ or $\mathbf{Z}[G] \otimes_{\mathbf{Z}[H]} A$.

Definition 1.16 We call a G-module *relatively injective* (or weakly injective) if it is a direct factor of the induced G-module $I_G(A)$ (where A is an abelian group).

Note that by Propositions 1.10 and 1.9, every injective G-module is relatively injective.

The category $\mathcal{M}od_G$ has *enough injectives* (i.e., every G-module is isomorphic to a submodule of an injective G-module): this is a general property of categories of modules over a ring (cf. Appendix, Example A.35b).

It follows that for any additive, covariant and left exact functor $F : \mathcal{M}od_G \to \mathcal{B}$ (where $\mathcal{B}$ is an abelian category, for example $\mathcal{A}b$, the category of abelian groups), we can define right *derived functors* $R^i F$ for $i \geqslant 0$ (Appendix, Definition A.44). Recall in particular that $R^0 F = F$ and if

$$0 \longrightarrow A' \longrightarrow A \longrightarrow A'' \longrightarrow 0$$

is an exact sequence in $\mathcal{M}od_G$, we have natural morphisms (i.e., functorial with respect to morphisms of exact sequences) $\delta^i : R^i F(A'') \to R^{i+1} F(A')$ which induce a long exact sequence

$$\cdots \longrightarrow R^i F(A') \longrightarrow R^i F(A) \longrightarrow R^i F(A'') \xrightarrow{\delta^i} R^{i+1} F(A') \longrightarrow \cdots ,$$

meaning that the family of the $(R^i F)_{i \geqslant 0}$ forms a *cohomological functor* or *δ-functor* in the sense of Theorem A.46 from the appendix (where general properties of functors are recalled).

Note also that any G-module is a quotient of a projective G-module (e.g. of a free module over the ring $R := \mathbf{Z}[G]$). In other words the category of G-modules has *enough projectives* . We will see that cohomology groups $H^i(G, A)$ for a G-module A, although defined as right derived functors (thus in theory calculated using injective resolutions of A), are actually more easily calculated using a projective resolution of the G-module $\mathbf{Z}$ (equipped with the trivial action of G).

1.3   The Cohomology Groups $H^i(G, A)$

For any G-module A, we denote by A^G the submodule of A consisting of elements x satisfying $g \cdot x = x$ for all g of G. The functor $F : A \mapsto A^G$ from $\mathcal{M}od_G$ to $\mathcal{A}b$ is covariant and left exact. We can thus define right derived functors $R^i F$, and we set

$$H^i(G, A) := R^i F(A)$$

for any G-module A. These groups are covariantly "functorial in A", meaning that the morphism of G-modules $\varphi : A \to B$ induces for all $i \geqslant 0$ a homomorphism of abelian groups $\varphi_* : H^i(G, A) \to H^i(G, B)$. If G is a trivial group, we of course have $H^i(G, A) = 0$ for all $i > 0$ given that the functor $A \mapsto A^G$ is evidently exact in this case.

An *injective resolution* of A is an exact sequence

$$0 \to A \longrightarrow I_0 \longrightarrow I_1 \longrightarrow I_2 \longrightarrow \cdots,$$

where all the I_i are injective G-modules. The existence of such a resolution follows from the fact that the category of G-modules has enough injectives (Appendix, Proposition A.43). Let us recall how we obtain the groups $H^i(G, A)$ from an injective resolution as above: the groups $H^i(G, A)$ are cohomology groups of the complex

$$0 \longrightarrow I_0^G \longrightarrow I_1^G \longrightarrow I_2^G \longrightarrow \cdots$$

(e.g. $H^1(G, A)$ is obtained as the quotient of $\mathrm{Ker}[I_1^G \to I_2^G]$ by $\mathrm{Im}[I_0^G \to I_1^G]$).

More generally, one can compute the $H^i(G, A)$ using any resolution (I_j) such that all the I_j are *acyclic*, i.e., satisfy $H^i(G, I_j) = 0$ for all $i > 0$; cf. Appendix, Theorem A.46c).

General properties of derived functors (which follow from computations using injective resolutions as above) then yield the following:

Theorem 1.17
  (a) We have $H^0(G, A) = A^G$.
  (b) We have $H^i(G, A) = 0$ for any injective G-module A and any $i > 0$.
  (c) For any short exact sequence

$$0 \longrightarrow A \longrightarrow B \longrightarrow C \longrightarrow 0$$

of G-modules, we have a long exact sequence

$$0 \longrightarrow A^G \longrightarrow B^G \longrightarrow C^G \xrightarrow{\delta^0}$$

$$H^1(G, A) \longrightarrow H^1(G, B) \longrightarrow H^1(G, C) \xrightarrow{\delta^1} H^2(G, A) \longrightarrow \cdots$$

and the coboundaries δ^i depend functorially on the exact sequence under consideration.

We immediately obtain

Proposition 1.18 *For all G-modules A and B, we have $H^i(G, A \oplus B) = H^i(G, A) \oplus H^i(G, B)$.*

Note that if $\varphi : A \to A$ is the multiplication by $m > 0$, then $\varphi_* : H^i(G, A) \to H^i(G, A)$ is also the multiplication by m (observe that φ is the diagonal map $A \to A \oplus A \oplus \cdots \oplus A$ composed with the summation map $A \oplus A \oplus \cdots \oplus A \to A$). It follows that if A is m-torsion, then $H^i(G, A)$ is m-torsion.

Remark 1.19 The Proposition 1.18 readily extends to a direct sum of a finite family of G-modules or to a direct product of any family of G-modules (using the fact that a direct product of injective G-modules is injective).

Later (Proposition 1.25), we will see that the assumption that G is finite allows us to extend the proposition to a direct sum of an infinite family of G-modules.

As it is easier to construct projective G-modules (e.g. free) than injective, it is often better to use another procedure for calculating the $H^i(G, A)$. Let us begin with a lemma.

Lemma 1.20 *Let*

$$\cdots \longrightarrow P_i \longrightarrow P_{i-1} \longrightarrow \cdots \longrightarrow P_1 \longrightarrow P_0 \longrightarrow \mathbf{Z} \longrightarrow 0$$

be a projective resolution of $\mathbf{Z}$ (e.g. by free G-modules). Let $A = I_G(X)$ be an induced G-module. Then the complex $(\mathrm{Hom}_G(P_i, A))_{i \geqslant 0}$ is exact.

Proof As the P_i are projective $\mathbf{Z}[G]$-modules, they are free as $\mathbf{Z}$-modules (they are direct factors of free $\mathbf{Z}$-modules). In particular, the kernels and cokernels of the maps $P_{i+1} \to P_i$ are free $\mathbf{Z}$-modules for $i \geqslant 0$. On the other hand, Proposition 1.12 gives $\mathrm{Hom}_G(P_i, A) = \mathrm{Hom}_{\mathbf{Z}}(P_i, X)$. The result follows from the fact that if

$$0 \longrightarrow M_1 \longrightarrow M_2 \longrightarrow M_3 \longrightarrow 0$$

is an exact sequence of free $\mathbf{Z}$-modules, then it admits a section, which implies that the sequence

$$0 \longrightarrow \mathrm{Hom}_{\mathbf{Z}}(M_1, X) \longrightarrow \mathrm{Hom}_{\mathbf{Z}}(M_2, X) \longrightarrow \mathrm{Hom}_{\mathbf{Z}}(M_3, X) \longrightarrow 0$$

is still exact.   □

We deduce the following[2]:

Theorem 1.21 *Let*

$$\cdots \longrightarrow P_i \longrightarrow P_{i-1} \longrightarrow \cdots \longrightarrow P_1 \longrightarrow P_0 \longrightarrow \mathbf{Z} \longrightarrow 0$$

be a projective resolution of $\mathbf{Z}$*. Let A be a G-module. Then the* $H^i(G, A)$ *are cohomology groups of the complex*

$$0 \longrightarrow \mathrm{Hom}_G(P_0, A) \longrightarrow \mathrm{Hom}_G(P_1, A) \longrightarrow \mathrm{Hom}_G(P_2, A) \cdots$$

For example the group $H^1(G, A)$ is the quotient of $\mathrm{Ker}[\mathrm{Hom}_G(P_1, A) \to \mathrm{Hom}_G(P_2, A)]$ by $\mathrm{Im}[\mathrm{Hom}_G(P_0, A) \to \mathrm{Hom}_G(P_1, A)]$.

Proof Let $P_\bullet$ be a complex obtained by appending 0 to the right of the complex $(P_i)_{i \geqslant 0}$. As the P_i are projectives G-modules, the functor that associates to every G-module A the complex $\mathrm{Hom}_G(P_\bullet, A)$ is exact, which implies (Appendix, Theorem A.41) that the family of $(H^i(\mathrm{Hom}_G(P_\bullet, A)))_{i \geqslant 0}$ forms a δ-functor (S^i), which is immediately seen to be isomorphic to $A \mapsto H^0(G, A)$ for $i = 0$.

By universality of the derived functors (Appendix, Theorem A.46), we have a morphism of δ-functors (f^i) from $(H^i(G, .))$ to (S^i) which induces an isomorphism for $i = 0$. On the other hand, for any G-module A, we have an exact sequence

$$0 \to A \longrightarrow I \longrightarrow B \longrightarrow 0$$

with I injective in $\mathcal{M}od_G$ as the category $\mathcal{M}od_G$ has enough injectives. To show that f^i is an isomorphism for all $i \geqslant 0$, it is now sufficient (reasoning by induction on i using the long exact sequence, given that the cohomology of I in strictly positive degree is trivial) to show that the complex $\mathrm{Hom}_G(P_\bullet, I))$ is exact in > 0 degree. As I is the direct factor of the induced module $I_G(X)$ (Propositions 1.9 and 1.10), the result follows from Lemma 1.20.   □

Remark 1.22 Here is a slightly different argument to prove 1.21. Observe that the functor $A \to A^G$ from $\mathcal{M}od_G$ to $\mathcal{A}b$ identifies with the functor $A \to \mathrm{Hom}_G(\mathbf{Z}, A)$ (the action of G on $\mathbf{Z}$ being trivial). It follows that $H^i(G, A) = \mathrm{Ext}^i_G(\mathbf{Z}, A)$, the $\mathrm{Ext}^i_G(\mathbf{Z}, .)$ being by definition the derived functors of $\mathrm{Hom}_G(\mathbf{Z}, .)$. A general property of derived functors (Appendix, Theorem A.52) shows that the $\mathrm{Ext}^i_G(\mathbf{Z}, A)$ are also obtained as derived functors (applied to $\mathbf{Z}$) of the contravariant functor $\mathrm{Hom}_G(., A)$. We can thus compute them using projective resolution of $\mathbf{Z}$ as in Theorem 1.21.

We can now generalise Theorem 1.17, b).

---

[2]I thank Joël Riou who suggested this method to me which avoids invoking Theorem A.52 of the Appendix, cf. Remark 1.22 below.

Proposition 1.23 *Let A be a relatively injective G-module. Then for all $i > 0$ we have $H^i(G, A) = 0$.*

In other words, relatively injective G-modules are acyclic.

Proof As A is a factor of an induced G-module, we immediately reduce ourselves (by Proposition 1.18), to the case where A is itself induced, i.e., of the form $A = I_G(X)$ for some abelian group X. The proposition now follows from Theorem 1.21 together with Lemma 1.20.   □

Corollary 1.24 *Let*

$$0 \longrightarrow A \longrightarrow I \longrightarrow B \longrightarrow 0$$

be an exact sequence of G-modules with I relatively injective (e.g. induced). Then for all $i > 0$, we have $H^i(G, B) = H^{i+1}(G, A)$ and the map "coboundary" $H^0(G, B) \to H^1(G, A)$ is surjective.

Proof This follows from the long exact cohomology sequence and the above proposition.   □

This corollary is useful because it often allows to prove some properties of the $H^i(G, A)$ by "dimension shifting " reasoning by induction on i, as we have done in the Proof of Theorem 1.21 using an embedding of A into an injective G-module. Here is an example of this principle in action.

Proposition 1.25 *Let G be a finite group. Let$(A_j)_{j \in J}$ an inductive system of G-modules and $A := \varinjlim_j A_j$ the G-module inductive limit of the A_j. Then for all $i \geqslant 0$, we have an isomorphism*

$$\varinjlim_j H^i(G, A_j) \simeq H^i(G, A).$$

In particular we see that "the cohomology of finite groups commutes with direct sums".[3]

Proof For $i = 0$, the result is trivial. We proceed by induction on i. We immediately check that, as G is finite, the canonical map $\varinjlim_j I_G(A_j) \to I_G(A)$ is an isomorphism of G-modules. Embedding A_j in $I_G(A_j)$ (cf. Proposition 1.10), we obtain an exact sequence

$$0 \longrightarrow A_j \longrightarrow I_G(A_j) \longrightarrow B_j \longrightarrow 0.$$

Let $B = \varinjlim_j B_j$. As $\varinjlim_j$ is an exact functor in the category of abelian groups (cf. Appendix, Proposition A.30, a), we also have an exact sequence

---

[3] As J. Riou pointed out to me, the assumption that G is finite is important in this statement. For example, there is no analogue of this statement (already at the level of the H^1) for $G = \mathbf{Z}^{(\mathbf{N})}$ acting trivially on an infinite family of abelian groups.

$$0 \longrightarrow A \longrightarrow I_G(A) \longrightarrow B \longrightarrow 0$$

which, since $I_G(A)$ is induced, induces an exact sequence

$$H^0(G, I_G(A)) \longrightarrow H^0(G, B) \longrightarrow H^1(G, A) \longrightarrow 0$$

and isomorphisms $H^{i-1}(G, B) \simeq H^i(G, A)$ for $i \geqslant 2$, as well as analogous results by replacing A and B by A_j and B_j.

The case $i = 0$ then provides isomorphisms

$$\varinjlim_j H^0(G, I_G(A_j)) \simeq H^0(G, I_G(A)); \quad \varinjlim_j H^0(G, B_j) \simeq H^0(G, B),$$

hence the result for $i = 1$ thanks to the exact commutative diagram

$$
\begin{array}{ccccccc}
\varinjlim_j H^0(G, I_G(A_j)) & \longrightarrow & \varinjlim_j H^0(G, B_j) & \longrightarrow & \varinjlim_j H^1(G, A_j) & \longrightarrow & 0 \\
\downarrow \wr & & \downarrow \wr & & \downarrow & & \\
H^0(G, I_G(A)) & \longrightarrow & H^0(G, B) & \longrightarrow & H^1(G, A) & \longrightarrow & 0.
\end{array}
$$

The case $i \geqslant 2$ follows by induction thanks to an analogous commutative diagram along with the isomorphisms

$$H^{i-1}(G, B) \simeq H^i(G, A); \quad \varinjlim_j H^{i-1}(G, B_j) \simeq \varinjlim_j H^i(G, A_j). \qquad \square$$

1.4 Computation of Cohomology Using the Cochains

In small degrees ($i = 1$, $i = 2$), and also for some explicit calculations with cup-products that we will carry out in Sect. 2.5, it is useful to have an explicit description of the groups $H^i(G, A)$. To achieve this, we will construct an explicit resolution of the G-module $\mathbf{Z}$ (equipped with the trivial action of G) by free $\mathbf{Z}[G]$-modules.

For all $i \geqslant 0$, let E_i be the set of $(i + 1)$-tuples $(g_0, \ldots, g_i)$ of elements of G. Let L_i be the free $\mathbf{Z}$-module with basis E_i. The action G on E_i by translation

$$s \cdot (g_0, \ldots, g_i) := (sg_0, \ldots, sg_i) \quad s \in G, \quad (g_0, \ldots, g_i) \in L_i$$

defines a structure of G-module on L_i. As G acts without fixed points on E_i (i.e., $g \cdot x = x$ with $g \in G$ and $x \in E_i$ implies $g = 1$), the $\mathbf{Z}[G]$-module L_i is free (a basis can be obtained by choosing an element in each orbit for the action of G on E_i). Let then $d_i : L_i \to L_{i-1}$ be the morphism of G-modules defined by the formula (where the symbol $\hat{g}_j$ as usual indicates that the index j is omitted):

$$d_i(g_0, \ldots, g_i) = \sum_{j=0}^{i} (-1)^j (g_0, \ldots, \hat{g}_j, \ldots, g_i)$$

if $i > 0$ and $d_0 : L_0 \to \mathbf{Z}$ the morphism of G-modules sending the whole of (g_0) to 1.

Lemma 1.26 *The sequence*

$$\cdots \to L_2 \xrightarrow{d_2} L_1 \xrightarrow{d_1} L_0 \xrightarrow{d_0} \mathbf{Z} \to 0$$

is exact (in particular it is a resolution of $\mathbf{Z}$ by free (hence projective) $\mathbf{Z}[G]$-modules).

Proof Let us show first that $d_i \circ d_{i+1} = 0$ for all $i \geqslant 0$. It is trivial for $i = 0$. Assume $i \geqslant 1$. Then for any $(g_0, \ldots, g_{i+1}) \in L_{i+1}$, we have

$$(d_i \circ d_{i+1})((g_0, \ldots, g_{i+1})) = \sum_{k=0}^{i+1} (-1)^k d_i(g_0, \ldots, \hat{g}_k, \ldots, g_{i+1})$$

$$= \sum_{k=0}^{i+1} (-1)^k \left[\sum_{j=0}^{k-1} (-1)^j (g_0, \ldots, \hat{g}_j, \ldots, \hat{g}_k, \ldots, g_{i+1}) \right.$$

$$\left. + \sum_{j=k+1}^{i+1} (-1)^{j-1} (g_0, \ldots, \hat{g}_k, \ldots, \hat{g}_j, \ldots, g_{i+1}) \right].$$

For any pair (r, s) with $0 \leqslant r < s \leqslant i + 1$, the term

$$(g_0, \ldots, \hat{g}_r, \ldots, \hat{g}_s, \ldots, g_{i+1})$$

appears twice in the sum: once with the sign $(-1)^{r+s}$ for $j = r, k = s$, and once with the sign $(-1)^{r+s-1}$ for $k = r, j = s$. It follows that the sum is zero.

Let us define morphisms of abelian groups (they are not G-morphisms in general) $u_i : L_i \to L_{i+1}$ by $u_0(1) = (1)$, and $u_i(g_0, \ldots, g_i) = (1, g_0, \ldots, g_i)$ if $i \geqslant 1$. We then have $u_{i-1} \circ d_i + d_{i+1} \circ u_i = \mathrm{Id}_{L_i}$ for all $i \geqslant 0$. Indeed

$$(u_{i-1} \circ d_i + d_{i+1} \circ u_i)(g_0, \ldots, g_i) = \sum_{j=0}^{i} (-1)^j (1, g_0, \ldots, \hat{g}_j, \ldots, g_i)$$

$$+ (g_0, \ldots, g_i) + \sum_{j=1}^{i+1} (-1)^j (1, g_0, \ldots, \hat{g}_{j-1}, \ldots, g_i)$$

which is equal to $(g_0, \ldots, g_i)$ (the terms cancel each other pairwise). Let then x be in Ker d_i, we obtain $x = d_{i+1}(u_i(x))$ hence $x \in \operatorname{Im} d_{i+1}$. As $d_i \circ d_{i+1} = 0$, we also have $\operatorname{Im} d_{i+1} \subset \operatorname{Ker} d_i$ and therefore finally $\operatorname{Im} d_{i+1} = \operatorname{Ker} d_i$. The desired exactness follows.   □

Let A be a G-module. An element of $K^i := \operatorname{Hom}_G(L_i, A)$ identifies itself with a function $f : G^{i+1} \to A$ satisfying

$$f(s \cdot g_0, \ldots, s \cdot g_i) = s \cdot f(g_0, \ldots, g_i)$$

("homogeneous cochain"). The coboundary applications $K^i \to K^{i+1}$ are obtained via a formula analogous to the previous one. Such a function is uniquely determined by the value it takes at the elements of G^{i+1} of the form $(1, g_1, g_1 g_2, \ldots, g_1 \cdots g_i)$. Finally, we can view the elements of K^i as *non-homogeneous cochains*, namely

Theorem 1.27 *The groups $H^i(G, A)$ for $i \geqslant 1$ are obtained as cohomology groups of the complex*

$$K^0 \longrightarrow K^1 \longrightarrow K^2 \longrightarrow \cdots ,$$

where $K^0 = A$ (seen as the set of functions from $G^0 := \{1\}$ to A) and for $i \geqslant 1$, K^i is the abelian group consisting of functions $f : G^i \to A$, the coboundary $d^i : K^i \to K^{i+1}$ being defined by the formula

$$df(g_1, \ldots, g_{i+1}) = g_1 \cdot f(g_2, \ldots, g_{i+1})$$
$$+ \sum_{j=1}^{i} (-1)^j f(g_1, \ldots, g_j g_{j+1}, \ldots, g_{i+1}) + (-1)^{i+1} f(g_1, \ldots, g_i).$$

The (non-homogeneous) cochains of Ker d^i are called *i-cocycles*, or simply *cocycles* when it is clear what i is. The i-cocycles of the image d^{i-1} are called *i-coboundaries*, or simply *coboundaries*. In particular, the set of 1-cochains K^1 consists of functions from G to A. The set of 1-cocycles $Z^1(G, A) \subset K^1$ is the subgroup of functions f that further satisfy $f(g_1 g_2) = f(g_1) + g_1 f(g_2)$ for all g_1, g_2 in G. The set of 1-coboundaries $B^1(G, A)$ is the set of functions of the form $g \mapsto g \cdot a - a$ with $a \in A$. Thus we have $H^1(G, A) = Z^1(G, A)/B^1(G, A)$.

Corollary 1.28 *Let G be a finite group. Let A be a finite G-module. Then all the groups $H^i(G, A)$ are finite.*

Remark 1.29 This last corollary can be obtained by shifting, using the Corollary 1.24, noticing that A embeds into the induced module $I_G(A)$ which is also finite. We will see below that the conclusion still holds for $i \geqslant 1$ if A is only assumed to be of finite type.

Example 1.30 (a) An element of $Z^1(G, A)$ is called a *crossed homomorphism*. If the action of G on A is trivial, a crossed homomorphism is simply a homomorphism and $B^1(G, A) = 0$. It follows that $H^1(G, A)$ is the set of group morphisms from G to A.

(b) We deduce from (a) that for any finite group G acting trivially on a torsion-free abelian group A, we have $H^1(G, A) = 0$. Similarly, if $G = \mathbf{Z}/p$ acts trivially on the abelian group A, we have $H^1(G, A) \simeq A[p]$, where $A[p]$ is the p-torsion subgroup of A.

(c) A 2-cocycle is a map f from $G \times G$ to A satisfying

$$g_1 f(g_2, g_3) - f(g_1 g_2, g_3) + f(g_1, g_2 g_3) - f(g_1, g_2) = 0.$$

We call it a *system of factors*. One can show ([54], Theorem 6.6.3) that $H^2(G, A)$ classifies group extensions E of G by A such that the action (by conjugation in E) of G on A corresponding to E is the action given by the G-module structure of A. The case where this action is trivial corresponds to central extensions.

We now have a statement for projective limits analogous to Proposition 1.25 (subject to some supplementary finiteness conditions).

Proposition 1.31 *Let G be a finite group. Let $(A_n)_{n\in\mathbf{N}}$ be a projective system of finite G-modules, indexed by the integers. Let $A := \varprojlim_n A_n$ be the G-module projective limit of the A_n. Then for all $i \geqslant 0$, we have an isomorphism*

$$H^i(G, A) \simeq \varprojlim_n H^i(G, A_n).$$

Proof The statement is trivial for $i = 0$. We argue by induction on $i > 0$.

We embed each A_n in the induced G-module $I_G(A_n)$ and we set $B_n = I_G(A_n)/A_n$. It follows that $B = \varprojlim_n B_n$. It is straightforward that (the group G being finite), the canonical map $\varprojlim_n I_G(A_n) \to I_G(A)$ is an isomorphism. Moreover, we have $H^i(G, I_G(A_n)) = 0$ for $i > 0$ since $I_G(A_n)$ is induced. The proof is now identical to that of Proposition 1.25, noting that the sequence

$$0 \longrightarrow A \longrightarrow I_G(A) \longrightarrow B \longrightarrow 0$$

remains exact by the finiteness assumptions. Same holds true for the sequences:

$$\varprojlim_n H^i(G, I_G(A_n)) \longrightarrow \varprojlim_n H^i(G, B_n) \longrightarrow \varprojlim_n H^{i+1}(G, A_n)$$
$$\longrightarrow \varprojlim_n H^{i+1}(G, I_G(A_n))$$

for all $i \geqslant 0$. This follows from Proposition A.32 of the appendix. The Mittag-Leffler (ML) condition is here automatically satisfied as all modules $A_n, I_G(A_n), I_G(B_n)$ (as well as cohomology groups of the finite group G with coefficients in these modules) are finite. $\qquad\square$

Remark 1.32 Proposition 1.31 is in general false without the finiteness assumption on the A_n; see Exercise 1.8. Furthermore, there is no analogue of 1.31 with G profinite instead of finite; see Exercise 4.1.

1.5 Change of Group: Restriction, Corestriction, the Hochschild–Serre Spectral Sequence

In this paragraph, we are no longer considering one group G, but instead consider what happens when we change the acting group. Let A be a G-module and let G' be a group endowed with a morphism $f : G' \to G$. We can then endow A with a structure of a G'-module by setting

$$g' \cdot a := f(g') \cdot a, \quad g' \in G', \quad a \in A.$$

Denote by f^*A (or simply A when no confusion arises) this G'-module.

As A^G is a subgroup of $(f^*A)^{G'}$, we obtain a morphism of functors from $H^0(G, .)$ to $H^0(G', f^* \cdot)$. The universal property of derived functors (Appendix, Theorem A.46, b) shows that for any integer $i \geqslant 0$, there is a unique family of morphisms of functors

$$f_i^* : H^i(G, .) \longrightarrow H^i(G', .)$$

compatible with the coboundaries (i.e., the maps δ^i) of the long exact cohomology sequences (in an evident way). We have thus obtained a *morphism of cohomological functors* (or *morphism of δ-functors*).

Let now A' be a G'-module and $u : A \to A'$ a morphism of abelian groups. If furthermore u is f-*compatible*,[4] i.e., satisfy

$$u(f(g') \cdot a) = g' \cdot u(a) \quad g' \in G', \quad a \in A,$$

then u is a G'-homomorphism from f^*A to A' and induces an homomorphism $u_* : H^i(G', f^*A) \to H^i(G', A')$. Composing this morphism with f_i^*, we obtain a homomorphism

$$H^i(G, A) \longrightarrow H^i(G', A')$$

associated to the group homomorphism $f : G' \to G$ and to the f-compatible homomorphism $u : A \to A'$. The newly defined homomorphism has an obvious expression via the explicit definition of H^i using the cochains.

Furthermore, if we have a f-compatible morphism of short exact sequences from the sequence $0 \to A \to B \to C \to 0$ to $0 \to A' \to B' \to C' \to 0$, the corresponding morphisms of the H^i remain compatible with the maps δ^i of long exact sequences associated to these short exact sequences (we thus have a morphism of cohomological functors).

Definition 1.33

(a) Let A be a G-module and H a subgroup of G. Taking for f the canonical injection of H into G, we obtain for $i \geqslant 0$ a homomorphism Res : $H^i(G, A) \to H^i(H, A)$ that we call the *restriction* homomorphism.

---

[4]One can also say "u and f are compatible", as in [47], Chap. VII, Sect. 5.

(b) Let A be a G-module. For a normal subgroup H of G, the quotient G/H acts on A^H and the inclusion $A^H \to A$ is compatible with the canonical surjection $G \to G/H$. We deduce for all $i \geq 0$ a homomorphism $\mathrm{Inf} : H^i(G/H, A^H) \to H^i(G, A)$ that we call the *inflation* homomorphism .

Remark 1.34 The restriction and inflation homomorphisms have a very simple expression in terms of the cocycles: in the case of the restriction one simply restricts the cocycle $G^i \to A$ to the subgroup H^i; in the case of inflation, one composes the canonical surjection $G^i \to (G/H)^i$ with any cocycle $(G/H)^i \to A^H \subset A$.

Remark 1.35 Let H be a subgroup of G. Then $\mathbf{Z}[G]$ is a free left $\mathbf{Z}[H]$-module (in particular $\mathbf{Z}[G]$ is a projective object in $\mathcal{M}od_H$): indeed, a basis consists of a system of representatives $(g_j)_{j \in J}$ of right classes of G modulo H as any element g of G can be uniquely written as $g = hg_j$ with $j \in J$. We immediately deduce that an induced G-module is also an induced H-module by Remark 1.6, b) that identifies the notions of induced and co-induced G-modules for a finite group G.

Let now H be a subgroup of a finite group G and let A be an H-module. Consider the G-module $I_G^H(A)$. Associating to any $u \in I_G^H(A)$ its value at 1, we obtain a morphism of abelian groups $I_G^H(A) \to A$ compatible with the injection $H \to G$. We deduce from the above, homomorphisms (compatible with long exact sequences)

$$H^i(G, I_G^H(A)) \longrightarrow H^i(H, A).$$

Theorem 1.36 (Shapiro lemma) *The homomorphisms*

$$H^i(G, I_G^H(A)) \longrightarrow H^i(H, A)$$

defined above are isomorphisms.

Proof By Remark 1.14 (which follows form Proposition 1.12), the functor F (from $\mathcal{M}od_H$ to $\mathcal{M}od_G$) defined by $A \mapsto I_G^H(A)$ preserves the injectives. On the other hand we have the equality $(I_G^H(A))^G = A^H$ for any H-module A (indeed, if $f : G \to A$ is in $(I_G^H(A))^G$, we have $f(sg) = f(s)$ for all g, s in G hence f is constant; but this constant has to be in A^H since $f(hg) = hf(g)$ for all h in H). To obtain the result (using an injective resolutions of the H-module A, to which we apply the functor $I_G^H(.)$), it is enough to check that the functor F is exact. But we have

$$I_G^H(A) = \mathrm{Hom}_H(\mathbf{Z}[G], A)$$

(Remark 1.6, c) and the result follows from Remark 1.35 (it is also easy to check directly that F is exact). $\qquad\qquad\qquad\qquad\qquad\qquad\qquad\qquad\qquad\qquad\qquad\qquad$ $\square$

Remark 1.37 Recall (cf. Appendix, Sect. A.6) that if A and B are G-modules, the groups $\mathrm{Ext}_G^i(A, B)$ are by definition derived functors (applied to B) of the right exact functor $\mathrm{Hom}_G(A, .)$ from $\mathcal{M}od_G$ to $\mathcal{A}b$. Furthermore, for a fixed B, the family of $(\mathrm{Ext}^i(., B))$ constitutes a cohomological functor, as if

$$0 \longrightarrow B \longrightarrow B^0 \longrightarrow B^1 \longrightarrow \cdots$$

is an injective resolution of B and $0 \to A \to A' \to A'' \to 0$ is a short exact sequence of G-modules, then

$$0 \longrightarrow (\mathrm{Hom}_G(A'', B^j))_{j \geqslant 0} \longrightarrow (\mathrm{Hom}_G(A', B^j))_{j \geqslant 0}$$
$$\longrightarrow (\mathrm{Hom}_G(A, B^j))_{j \geqslant 0} \longrightarrow 0$$

is a short exact sequence of complexes (by injectivity of the B_j), which gives rise (Appendix, Theorem A.41), by definition of the groups Ext^i, to a long exact cohomology sequence

$$0 \longrightarrow \mathrm{Hom}_G(A'', B) \longrightarrow \mathrm{Hom}_G(A', B) \longrightarrow \mathrm{Hom}_G(A, B)$$
$$\longrightarrow \mathrm{Ext}^1_G(A'', B) \longrightarrow \cdots$$

Besides, we can also calculate the $\mathrm{Ext}^i_G(A, B)$ as left derived functors of $\mathrm{Hom}_G(., B)$ (Appendix, Theorem A.52), but unlike the previous construction, this does not generalise to profinite groups G (cf. Chap. 4), for the lack of enough projectives in the category of discrete G-modules.

We will also need the following lemma which will be generalised later (Proposition 4.25) to profinite groups:

Lemma 1.38 *Let G be a finite group. Let H be a subgroup of G. Then every injective of $\mathcal{M}od_G$ is also injective in $\mathcal{M}od_H$.*

Proof We already know (see Appendix, Example A.35b) that every G-module A embeds into an injective G-module of the form $I_G(I) = \mathrm{Hom}_{\mathbf{Z}}(\mathbf{Z}[G], I)$, where I is a divisible abelian group. If in addition A is an injective G-module, it is a direct factor of $I_G(I)$ and therefore it is enough to show that $I_G(I)$ is injective as an H-module. As $\mathbf{Z}[G]$ is a free $\mathbf{Z}[H]$-module of finite type, we see immediately that $I_G(I)$ is isomorphic to a direct product of a finite number of copies of $I_H(I)$, hence is injective as an H-module (Remark 1.14, which relies on Proposition A.34 of the appendix). $\qquad\square$

The following proposition is slightly similar to Shapiro's lemma. However it uses Proposition 1.15 (where the functor $I^H_G(.)$ is viewed as the left adjoint of the forgetful functor) instead of Proposition 1.12 (where we view it as right adjoint of the forgetful functor).

Proposition 1.39 *Let G be a finite group. Let H be a subgroup of G. Let A be an H-module. Then, for any G-module B we have canonical isomorphisms*

$$\mathrm{Ext}^i_H(A, B) \simeq \mathrm{Ext}^i_G(I^H_G(A), B),$$

defined for all $i \geqslant 0$. These isomorphisms are compatible with the coboundaries of the long exact exact cohomology sequences associated to the short exact sequences

$$0 \longrightarrow B \longrightarrow B' \longrightarrow B'' \longrightarrow 0,$$

as well as with the boundaries of the long homology sequences associated with the short exact sequences

$$0 \longrightarrow A \longrightarrow A' \longrightarrow A'' \longrightarrow 0.$$

Proof For a fixed A, Proposition 1.15 gives an isomorphism of functors (from $\mathcal{M}od_G$ to $\mathcal{A}b$)

$$Hom_H(A, .) \simeq \operatorname{Hom}_G(I_G^H(A), .).$$

Since any injective $\mathcal{M}od_G$ remains injective in $\mathcal{M}od_H$ (by Lemma 1.38), we obtain isomorphisms

$$\operatorname{Ext}_H^i(A, B) \simeq \operatorname{Ext}_G^i(I_G^H(A), B)$$

by taking an injective resolution of B in the category $\mathcal{M}od_G$. The compatibility with the coboundaries of the long cohomology exact sequences associated with the short exact sequences "in B" follows for the properties of derived functors. The compatibility with the boundaries of the long homology exact sequences "in A" is obtained as in the Remark 1.37, observing that an exact sequence of H-modules

$$0 \longrightarrow A \longrightarrow A' \longrightarrow A'' \longrightarrow 0$$

induces an exact sequence of G-modules

$$0 \longrightarrow I_G^H(A) \longrightarrow I_G^H(A') \longrightarrow I_G^H(A'') \longrightarrow 0,$$

and for any injective resolution $B^{\bullet}$ of the G-module B (which is also an injective resolution of B viewed as an H-module), two isomorphic exact sequences of complexes

$$0 \longrightarrow \operatorname{Hom}_H(A'', B^{\bullet}) \longrightarrow \operatorname{Hom}_H(A', B^{\bullet}) \longrightarrow \operatorname{Hom}_H(A, B^{\bullet}) \longrightarrow 0,$$
$$0 \rightarrow \operatorname{Hom}_G(I_G^H(A''), B^{\bullet}) \rightarrow \operatorname{Hom}_G(I_G^H(A'), B^{\bullet}) \rightarrow \operatorname{Hom}_G(I_G^H(A), B^{\bullet}) \rightarrow 0.$$

But the groups $\operatorname{Ext}_H^i(A, B)$ and $\operatorname{Ext}_G^i(I_G^H(A), B)$ (as well as analogous groups with A' and A'' instead of A) are the ones which appear when one takes the long exact sequences associated with these two exact sequences of complexes. $\qquad\square$

We now present another construction that will be used later. Let G be a group and A a G-module. Let $t \in G$. Set $G' = G$, $A' = A$, and denote by $f : g \mapsto t^{-1}gt$ the inner automorphism associated with t^{-1}. The homomorphism of abelian groups $u : a \mapsto t \cdot a$ from A to A is then f-compatible, and hence induces for all $i \geqslant 0$ a homomorphism $\sigma_t : H^i(G, A) \mapsto H^i(G, A)$.

Proposition 1.40 *The map σ_t is the identity.*

Proof We argue by induction on i. The case $i = 0$ is trivial. In the general case, we embed A in an induced module I and set $B = I/A$. We then have a commutative diagram whose rows are exact:

$$
\begin{array}{ccc}
H^i(G, B) & \xrightarrow{\ \delta\ } & H^{i+1}(G, A) \longrightarrow 0 \ . \\
\ \downarrow{\sigma_t} & & \ \downarrow{\sigma_t} \\
H^i(G, B) & \xrightarrow{\ \delta\ } & H^{i+1}(G, A) \longrightarrow 0
\end{array}
$$

The result follows by induction on i.   □

Remark 1.41 Let A be a G-module. If H is a normal subgroup of G, we can make G act on $H^i(H, A)$ via the conjugation action of G on H. Proposition 1.40 then implies that H acts trivially on $H^i(H, A)$. In other words, G/H acts on $H^i(H, A)$. On the other hand, an injective resolution

$$0 \longrightarrow A \longrightarrow I_0 \longrightarrow I_1 \longrightarrow I_2 \longrightarrow \cdots$$

of A by induced G-modules also yields a resolution of the H-module A by induced H-modules (by Remark 1.35).

We deduce that the groups $H^i(H, A)$ are obtained using the cohomology of the complex

$$0 \longrightarrow I_0^H \longrightarrow I_1^H \longrightarrow I_2^H \longrightarrow \cdots ,$$

they are hence naturally G/H-modules via the action of G/H on the I_j^H for $j \geqslant 0$. The corresponding action of G/H on the $H^i(H, A)$ coincides with that defined above (via the action by conjugation of G on H). Indeed, for $i = 0$, the two actions are given by $(t, a) \mapsto t \cdot a$ for $t \in G/H$ and $x \in A^H$; we deduce the result for any i by shifting.

Theorem 1.42 (restriction–inflation) *Let H be a normal subgroup of G and let A be a G-module. Then the sequence*

$$0 \longrightarrow H^1(G/H, A^H) \xrightarrow{\ \mathrm{Inf}\ } H^1(G, A) \xrightarrow{\ \mathrm{Res}\ } H^1(H, A)$$

is exact.

Proof It is straightforward (via, for example, the description using cochains) that the sequence is a complex. Let us first show the injectivity of Inf. Let $f : G/H \to A^H$ be a 1-cocycle cohomologous to 0 in $H^1(G, A)$. We can also view f as a map from G to A constant on each class mod H. Then there exists an $a \in A$ such that $f(s) = s \cdot a - a$ for all s in G, and we obtain $sa - a = sta - a$ for all t in H, hence $t \cdot a = a$ by setting $s = 1$. Lastly $a \in A^H$ and f is indeed zero in $H^1(G/H, A^H)$.

Let us now show that an element of Ker Res is in Im Inf. Let $f : G \to A$ be a 1-cocycle. If $\mathrm{Res}(f) = 0$, then there exists an $a \in A$ such that $f(t) = t \cdot a - a$ for all t in H. By if necessary replacing f by a cohomologous cocycle $t \mapsto f(t) - (t \cdot a - a)$, we may assume $f(t) = 0$ for all t in H. As $f(st) = f(s) + s \cdot f(t)$, we obtain that f factors as a map $\overline{f} : G/H \to A$. For s in H, the classes st and t in G/H are the same (recall that H is normal in G). The formula then gives $f(t) = f(st) = sf(t)$, which shows that $\overline{f}$ has coefficients in A^H. It is a cocycle that induces an element of $H^1(G/H, A^H)$ whose image by Inf is f.   □

Remark 1.43 This result can also be deduced from the Hochschild–Serre spectral sequence defined below, but it is instructive to verify it directly especially since that method generalises to "non-abelian H^1", [47], Appendix to Chap. VII.

As we have seen in Remark 1.41, the group G/H acts on the cohomology groups $H^i(H, A)$. We can then view the restriction–inflation sequence as the exact sequence of low degree terms of the spectral sequence given by the following theorem:

Theorem 1.44 (Hochschild–Serre) *Let G be a group. Let H be a normal subgroup of G and A a G-module. Then we have a spectral sequence*

$$E_2^{pq} = H^p(G/H, H^q(H, A)) \implies H^{p+q}(G, A).$$

Proof It is a special case of the composed functors spectral sequence of Grothendieck (cf. Appendix, Theorem A.67). The functor $A \mapsto A^G$ from $\mathcal{M}od_G$ to $\mathcal{A}b$ is the composed of the functor $F_1 : A \mapsto A^H$ from $\mathcal{M}od_G$ to $\mathcal{M}od_{G/H}$ and the functor $F_2 : B \mapsto B^{G/H}$ from $\mathcal{M}od_{G/H}$ to $\mathcal{A}b$. The derived functors of F_1 are the $H^i(H, .)$ by Lemma 1.38, which says [5] that an injective resolution of A in $\mathcal{M}od_G$ also provides an injective resolution of $\mathcal{M}od_H$. The derived functors of F_2 are by definition the $H^i(G/H, .)$. It is then enough to prove that F_1 preserves the injectives, which is easy by Proposition A.34 of the appendix, as it is the adjoint of the (exact) forgetful functor from $\mathcal{M}od_{G/H}$ to $\mathcal{M}od_G$. Explicitly we have for any G/H-module B (which can also be seen as a G-module with a trivial action of H) and for all G-module A:

$$\mathrm{Hom}_G(B, A) = \mathrm{Hom}_{G/H}(B, A^H).$$ □

In particular, the theorem implies (cf. Appendix, Remark A.61) that for each $n > 0$, we have a filtration of $H^n(G, A)$ by a decreasing sequence $F^0 = H^n(G, A) \supset \cdots \supset F^{n+1} = 0$, where F^p/F^{p+1} (for $p = 0, \ldots, n$) is isomorphic to a subquotient $E_\infty^{p,n-p}$ of $H^p(G/H, H^{n-p}(H, A))$. The first terms of the exact sequence of low degree terms of this spectral sequence (Appendix, Proposition A.66) are written as

$$0 \to H^1(G/H, A^H) \xrightarrow{\ \mathrm{Inf}\ } H^1(G, A) \xrightarrow{\ \mathrm{Res}\ } H^1(H, A)^{G/H}$$

$$\to H^2(G/H, A^H) \xrightarrow{\ \mathrm{Inf}\ } H^2(G, A)$$

---

[5] We can also use Theorem 1.21 along with the fact that a resolution of $\mathbf{Z}$ by the projective $\mathbf{Z}[G]$-modules provides it with a resolution by the projective $\mathbf{Z}[H]$-modules.

(the map that was not given a name is called *transgression*). We also obtain the following result, which generalises the restriction–inflation exact sequence:

Corollary 1.45 *Let G be a group. Let H be a normal subgroup of G and let A be a G-module. Let $n \geqslant 2$ be an integer, we make an additional assumption that $H^i(H, A) = 0$ for $1 \leqslant i \leqslant n - 1$. Then the sequence*

$$0 \longrightarrow H^n(G/H, A^H) \xrightarrow{\ \text{Inf}\ } H^n(G, A) \xrightarrow{\ \text{Res}\ } H^n(H, A)^{G/H}$$

is exact.

Proof The assumption tells us that the terms E_2^{pq} of the initial page of the spectral sequence are zero for $0 < q < n$.

It follows that the only terms E_∞^{pq}, with $p + q = n$, that may be nonzero are the $E_\infty^{n,0}$ and $E_\infty^{0,n}$. The assumption immediately implies that $E_\infty^{n,0} = E_2^{n,0}$ and furthermore, $E_\infty^{0,n}$ is always a subgroup of $E_2^{0,n}$.

We conclude by writing the filtration

$$0 \subset F^n \subset H^n(G, A),$$

where $F^n = E_\infty^{n,0}$ and $H^n(G, A)/F^n = E_\infty^{0,n}$. $\square$

We can also prove this corollary directly by induction on i starting by embedding A into the induced G-module $I_G(A)$, which (Remark 1.35) is also an induced H-module, cf. [47], Chap. VII, Prop. 5.

1.6   Corestriction; Applications

Let H be a subgroup of a finite group G. Let A be a G-module. We will define homomorphisms $H^i(H, A) \to H^i(G, A)$ "in the opposite direction" of the restriction. We start with the case $i = 0$. The corestriction is then defined by the *norm*:

$$N_{G/H} : a \longmapsto \sum_{s \in G/H} s \cdot a$$

from A^H to A^G, where G/H is the (finite, by assumption) set of left classes modulo H. It is immediate that $s \cdot a$ depends only on the class of s in G/H (since $a \in A^H$) and that $N_{G/H}(a) \in A^G$.

We can then extend the corestriction in degree 0 to a unique morphism from the cohomological functor $\{H^i(H, f^*.), \delta\}$ to the cohomological functor $\{H^i(G, .), \delta\}$, where f is the inclusion $H \to G$. This is possible since the former functor is *effaceable* in degree $\geqslant 1$ (and hence universal; cf. Appendix, Remark A.47). Indeed, if I is induced for G, it is induced for H (Remark 1.35) and hence we have $H^i(H, I) = 0$

for all $i > 0$. Recall now that every G-module can be embedded into an induced G-module I.

In this way we obtain *corestriction* morphisms

$$\text{Cores} : H^i(H, A) \longrightarrow H^i(G, A)$$

compatible with the usual morphisms of short exact sequences.

Example 1.46 Let K be a field. Let L be a finite Galois extension of K with Galois group $G = \text{Gal}(L/K)$. Let E be a finite extension of K with $E \subset L$. Set $H = \text{Gal}(L/E)$. Then the corestriction $H^0(H, L^*) \to H^0(G, L^*)$ is the norm $N_{E/K} : E^* \to K^*$ in the usual sense. By this we mean that for any $x \in E^*$ the element $N_{E/K}(x)$ is the product of the conjugates of x. It is also the determinant of the multiplication map by x on the K-vector space E.

Remark 1.47 Let K be a field. Let E be an étale K-*algebra*, that is, a K-algebra isomorphic to a product of fields $\prod_{i=1}^r K_i$, where the K_i are separable extensions of K. For any $x \in E$, we define $N_{E/K}(x)$ as the determinant of the multiplication by x map in the K-vector space E. This norm is the product of norms $N_{K_i/K} : K_i \to K$ associated with each extension K_i.

Theorem 1.48 *Let* $m = [G : H]$ *be the index of* H *in* G. *Then the composite* Cores $\circ$ Res *is the multiplication by* m *in* $H^i(G, A)$.

Proof The statement is clear for $i = 0$. The general case is obtained by shifting (embedding A into an induced G-module I) using Corollary 1.24. $\square$

Corollary 1.49 *Let* G *be a finite group of cardinality* m. *Let* A *be a* G-*module. Then for* $i > 0$, *the groups* $H^i(G, A)$ *are* m-*torsion.*

In particular, if A is n-torsion with n prime to m, we obtain $H^i(G, A) = 0$ for $i > 0$.

Proof Apply Theorem 1.48 in the case $H = \{1\}$. $\square$

Corollary 1.50 *Let* G *be a finite group. Then for any* G-*module of finite type* A, $H^i(G, A)$ *is finite for all* $i > 0$.

Proof The cochain description shows that the groups $H^i(G, A)$ are of finite type. As by Corollary 1.49, these groups are torsion for $i > 0$, they are finite. $\square$

Corollary 1.51 *Let* G *be a finite group. Let* A *be a uniquely divisible abelian group endowed with an action of* G *(e.g.* $A = \mathbb{Q}$ *with the trivial action by* G). *Then* $H^i(G, A) = 0$ *for all* $i > 0$.

Proof Indeed, Corollary 1.49 says that for $i > 0$, the group $H^i(G, A)$ is torsion. On the other hand, the multiplication by n in A is, by assumption, an isomorphism for all $n > 0$ and hence multiplication by n in $H^i(G, A)$ is also an isomorphism. $\square$

Example 1.52 Consider $A = \mathbf{Q}/\mathbf{Z}$ with the trivial action of G. By the above corollary and the long exact cohomology sequence associated to the exact sequence

$$0 \longrightarrow \mathbf{Z} \longrightarrow \mathbf{Q} \longrightarrow \mathbf{Q}/\mathbf{Z} \longrightarrow 0,$$

we have $H^i(G, \mathbf{Q}/\mathbf{Z}) \simeq H^{i+1}(G, \mathbf{Z})$ for all $i > 0$. In particular, $H^2(G, \mathbf{Z}) \simeq H^1(G, \mathbf{Q}/\mathbf{Z})$ is the *character group* of G. We also have $H^1(G, \mathbf{Z}) = 0$ given that $\mathbf{Z}$ has no non-trivial finite subgroup.

Remark 1.53 The disadvantage of the definition of corestriction given above is that it is not very explicit. However, it turns out to be convenient in the proof of properties of the homomorphisms Cores. In Exercise 1.1 below, a more concrete definition is given.

1.7 Exercises

In all the exercises below, G denotes a finite group.

Exercise 1.1 Let H be a subgroup of G. Let A be a G-module.
  (a) Show that one defines a surjective morphism $\pi : I_G^H(A) \to A$ of G-modules by the formula:

$$\pi(f) = \sum_{g \in G/H} g \cdot f(g^{-1}), \quad f \in I_G^H(A).$$

  (b) Let $i \geqslant 0$. Let $\pi_* : H^i(H, A) = H^i(G, I_G^H(A)) \to H^i(G, A)$ be the homomorphism induced by π on the cohomology. Show that π_* is the corestriction.

Exercise 1.2 (*suggested by J. Riou*) Let A be an injective abelian group (i.e., such that $\mathrm{Hom}_{\mathbf{Z}}(., A)$ is an exact functor). Show that $I_G(A)$ is injective in $\mathcal{M}od_G$. Deduce from this fact that if A is any abelian group, the G-module $I_G(A)$ is acyclic.

Exercise 1.3 Let I be an injective G-module. Show that I is a divisible abelian group.

Exercise 1.4 Let I be an injective G-module. Show that the G-module $I_G(I)$ is injective. Show that a direct limit of injective G-modules is an injective G-module. This is a special case of a general result on the category $\mathcal{M}od_R$ with R a left noetherian ring, see Example A.35f) of the appendix).

Exercise 1.5 Let H be a subgroup of G. Let A be a G-module. Show that for any H-module B and all $i \geqslant 0$ we have canonical isomorphisms

$$\mathrm{Ext}_G^i(A, I_G^H(B)) \simeq \mathrm{Ext}_H^i(A, B).$$

What does this result imply for $A = \mathbf{Z}$?

Exercise 1.6 Give a proof of Proposition 1.39 using a projective resolution of A instead of an injective resolution of B.

Exercise 1.7 We say that a G-module A is a *permutation* G-module if it is a direct sum of a finite number $A_1, \ldots, A_r$ of G-modules such that for each A_i, there exists a subgroup H_i of G such that A_i is isomorphic to $I_G^{H_i}(\mathbf{Z})$ (where $\mathbf{Z}$ is endowed with the trivial action).

(a) Compare $I_G^H(\mathbf{Z})$ to the G-module $\mathbf{Z}[G/H]$ consisting of formal sums $\sum_{s \in G/H} n_s s$, $s \in G/H$, the action of G on $\mathbf{Z}[G/H]$ being given by the left action of G on the set of left cosets G/H.

(b) Show that if A is a permutation G-module, then $H^1(G, A) = 0$.

(c) Is it true that $H^2(G, A) = 0$ for all permutation modules A?

(d) Show that a G-module A is permutation if and only if the $\mathbf{Z}$-module A admits a finite basis $(e_1, \ldots, e_n)$ such that there is a permutation σ of the set $\{1, \ldots, n\}$ with $g \cdot e_i = e_{\sigma(i)}$ for all i of $\{1, \ldots, n\}$. Under what condition on the integer r is this permutation transitive?

Exercise 1.8 Let $G = \mathbf{Z}/p$ (where p is a prime) and let (for $n \in \mathbf{N}$) A_n be the G-module $\mathbf{Z}$ with the trivial action of G. Let ℓ be a prime different from p. Consider the projective system (A_n), transition maps being the multiplication by ℓ. Compare $\varprojlim_n H^2(G, A_n)$ and $H^2(G, \varprojlim_n A_n)$.

Exercise 1.9 For any G-module A, we define A^* as the G-module $\operatorname{Hom}_{\mathbf{Z}}(A, \mathbf{Q}/\mathbf{Z})$, where the action of G is given by $(g \cdot f)(x) = f(g^{-1} \cdot x)$ for all $g \in G$ and all $x \in A$. In particular, if B is just an abelian group, the abelian group B^* is defined.

(a) Show that for any abelian group B, the canonical homomorphism $B \to (B^*)^*$ (that sends x to $f \mapsto f(x)$ for $x \in B$ and $f \in B^*$) is injective.

(b) Give an example of an abelian group B such that $(B^*)^*$ is not isomorphic to B. What does occur if B is finite?

(c) Show that if A is a relatively injective G-module, then A^* is also relatively injective.

Exercise 1.10 Let p be a prime. For any finite torsion p-primary abelian group A, we denote by $\operatorname{rg}(A)$ the *rank* of A, that is, the minimal number of its generators.

(a) Show that $\operatorname{rg}(A)$ is the dimension of the p-torsion subgroup $A[p]$ of A over $\mathbf{F}_p = \mathbf{Z}/p\mathbf{Z}$, and that it is also the dimension of A/pA over $\mathbf{F}_p$.

(b) Let G be a finite p-group. Set $n(G) = \dim_{\mathbf{F}_p} H^1(G, \mathbf{Z}/p\mathbf{Z})$ and $r(G) = \dim_{\mathbf{F}_p} H^2(G, \mathbf{Z}/p\mathbf{Z})$. Show that $r(G) - n(G) = \operatorname{rg}(H^3(G, \mathbf{Z}))$.

Chapter 2
Groups Modified *à la* Tate, Cohomology of Cyclic Groups

Throughout this chapter, G denotes a finite group. We continue our study from the previous chapter by introducing the Tate groups as well as some supplementary ingredients (cohomology of finite groups, Herbrand quotient, cup products). We also define a new map between cohomology groups called *transfer*, for which we prove a subtle theorem of Furtwängler (Theorem 2.13).

2.1 Tate Modified Cohomology Groups.

It appears (notably for the arithmetic duality theorems when the number fields under consideration have real places or when one wishes to develop the class formation formalism) that it is often convenient, in the case of a finite group G, to introduce the groups $\widehat{H}^i(G, A)$ for all $i \in \mathbf{Z}$. These groups coincide with the $H^i(G, A)$ for $i \geqslant 1$ but provide a little bit more information for $i \leqslant 0$. It is mostly the case $i = 0$ that will be useful except when we consider class formations where we shall also use the case $i = -2$.

Definition 2.1 The *norm* of the group algebra $\mathbf{Z}[G]$ is the element $\sum_{g \in G} g$. The *augmentation ideal* $\mathcal{I}_G$ of the group algebra $\mathbf{Z}[G]$ is the kernel of the *augmentation homomorphism* $\mathbf{Z}[G] \rightarrow \mathbf{Z}$ defined by

$$\sum_{g \in G} a_g g \longmapsto \sum_{g \in G} a_g.$$

This is the same as saying that $\mathcal{I}_G$ is the set of linear combinations of the $(g - 1)$, $g \in G$. If A is a G-module, the norm defines an endomorphism $N : A \rightarrow A$ by the formula $N(x) = \sum_{g \in G} g \cdot x$. Note that $\mathcal{I}_G A \subset \operatorname{Ker} N$ and $\operatorname{Im} N \subset H^0(G, A)$.

© Springer Nature Switzerland AG 2020
D. Harari, *Galois Cohomology and Class Field Theory*, Universitext,
https://doi.org/10.1007/978-3-030-43901-9_2

Definition 2.2 Let A be a G-module. The G-module of *co-invariants* is the G-module $A_G = H_0(G, A) := A/\mathcal{I}_G A$.

Hence A_G is the largest G-module quotient of A on which G acts trivially. Passing to the quotient, we have a homomorphism (that we can also denote by N_A^* when there is an ambiguity)

$$N^* : H_0(G, A) \longrightarrow H^0(G, A).$$

Definition 2.3 We set $\widehat{H}_0(G, A) = \operatorname{Ker} N^*$ and $\widehat{H}^0(G, A) = \operatorname{Coker} N^*$.

In other words $\widehat{H}_0(G, A) = {}_N A/\mathcal{I}_G A$ and $\widehat{H}^0(G, A) = A^G/NA$, where ${}_N A$ is the kernel of the norm in A.

Note that these groups are zero if G is the trivial group (which is not the case of $H_0(G, A)$ and $H^0(G, A)$).

The functor $A \mapsto H_0(G, A)$ is covariant and right exact. We can also define the *homology groups* $H_i(G, A)$ as left derived functors (they make sense for an arbitrary G, but like in the case of cohomology we will only need them in the case where G is finite). We obtain a homology functor (cf. Appendix, the comment after Theorem A.46), i.e., that if $0 \to A \to B \to C \to 0$ is an exact sequence of G-modules, we have a functorial long exact sequence:

$$\cdots \longrightarrow H_1(G, A) \longrightarrow H_1(G, B) \longrightarrow H_1(G, C)$$
$$\longrightarrow H_0(G, A) \longrightarrow H_0(G, B) \longrightarrow H_0(G, C) \longrightarrow 0.$$

Furthermore, $H_i(G, A) = 0$ for $i > 0$ if A is projective, or even relatively injective (in that last case, we use that A is a direct factor of a co-induced G-module $\mathbf{Z}[G] \otimes X$, because G is assumed finite). The proofs are exactly of the same type as for cohomology.

Here is an example of homology group which will be useful later on (when we will be applying the Tate–Nakayama theorem to p-adic fields in Sect. 9.1):

Proposition 2.4 *The homology group $H_1(G, \mathbf{Z})$ is the abelianisation $G^{\mathrm{ab}} = G/G'$ of G, where G' is the derived subgroup of G.*

Proof Let $\Lambda = \mathbf{Z}[G]$ be the group algebra of G. Consider the exact sequence of G-modules

$$0 \longrightarrow \mathcal{I}_G \longrightarrow \Lambda \xrightarrow{\pi} \mathbf{Z} \longrightarrow 0,$$

where $\pi : \sum_g n_g g \mapsto \sum_g n_g$ is the augmentation homomorphism. We then have $H_0(G, \mathcal{I}_G) = \mathcal{I}_G/\mathcal{I}_G^2$. It follows that the image of $H_0(G, \mathcal{I}_G)$ in $H_0(G, \Lambda) = \Lambda/\mathcal{I}_G \Lambda$ is zero. On the other hand, $H_1(G, \Lambda) = 0$ in view of the fact that Λ is free over $\mathbf{Z}[G]$. We obtain an isomorphism (via the exact homology sequence)

$$d : H_1(G, \mathbf{Z}) \longrightarrow H_0(G, \mathcal{I}_G) = \mathcal{I}_G/\mathcal{I}_G^2.$$

Let us define $f : G \longrightarrow \mathcal{I}_G/\mathcal{I}_G^2$ by $f(g) = g - 1$. The formula

$$gh - 1 = (g - 1) + (h - 1) + (g - 1)(h - 1)$$

(valid for g, h in G) shows that this is a morphism. As $\mathcal{I}_G/\mathcal{I}_G^2$ is abelian, f induces a morphism (that we again denote f) from G/G' to $\mathcal{I}_G/\mathcal{I}_G^2$, which is clearly surjective. Lastly, the group morphism $u : \mathcal{I}_G \to G/G'$ defined by $u(g - 1) = \bar{g}$ factors through the quotient by $\mathcal{I}_G^2$. Indeed, if $x = (g - 1)(h - 1)$ is in $\mathcal{I}_G^2$, then

$$u(x) = u((gh - 1) - (g - 1) - (h - 1)) = \overline{ghg^{-1}h^{-1}}$$

is the neutral element of G/G' (because $ghg^{-1}h^{-1}$ is a commutator). Thus we obtain a morphism $\bar{u} : \mathcal{I}_G/\mathcal{I}_G^2 \to G/G'$ such that $\bar{u} \circ f$ is the identity; hence f is bijective. $\qquad\square$

We will now "link together" the long exact sequences of homology and cohomology associated to an exact sequence of G-modules using the *modified cohomology groups* of Tate. This is the object of the following definition:

Definition 2.5 Let G be a finite group. Let A be a G-module. We define the groups $\widehat{H}^n(G, A)$ for $n \in \mathbf{Z}$ by the formula:

$$\widehat{H}^n(G, A) = H^n(G, A) \quad n \geqslant 1,$$
$$\widehat{H}^0(G, A) = A^G/NA,$$
$$\widehat{H}^{-1}(G, A) = \widehat{H}_0(G, A) = {}_NA/\mathcal{I}_G A,$$
$$\widehat{H}^{-n}(G, A) = H_{n-1}(G, A) \quad n \geqslant 2.$$

The reason for this definition is the following theorem.

Theorem 2.6 *Let $0 \to A \to B \to C \to 0$ be an exact sequence of G-modules. We then have the following "functorial" exact sequence:*

$$\cdots \widehat{H}^{-2}(G, C) \longrightarrow \widehat{H}^{-1}(G, A) \longrightarrow \widehat{H}^{-1}(G, B) \longrightarrow \widehat{H}^{-1}(G, C)$$
$$\xrightarrow{\delta} \widehat{H}^0(G, A) \longrightarrow \widehat{H}^0(G, B) \longrightarrow \widehat{H}^0(G, C) \longrightarrow H^1(G, A) \longrightarrow \cdots$$

Moreover, if A is relatively injective (=relatively projective), we have $\widehat{H}^n(G, A) = 0$ for all $n \in \mathbf{Z}$.

Proof The homology and cohomology exact sequences yield the following commutative diagrams with exact rows:

$$
\begin{array}{ccccccccc}
H_1(G, C) & \longrightarrow & H_0(G, A) & \longrightarrow & H_0(G, B) & \longrightarrow & H_0(G, C) & \longrightarrow & 0 \\
\downarrow & & N_A^* \downarrow & & N_B^* \downarrow & & N_C^* \downarrow & & \downarrow \\
0 & \longrightarrow & H^0(G, A) & \longrightarrow & H^0(G, B) & \longrightarrow & H^0(G, C) & \longrightarrow & H^1(G, A)
\end{array}
$$

Such a diagram defines canonically a homomorphism:

$$\delta : \operatorname{Ker} N_C^* \longrightarrow \operatorname{Coker} N_A^*$$

(if $c \in \operatorname{Ker} N_C^*$, we lift it to $b \in H_0(G, B)$, then we lift $N_B^*(b)$ to $a \in H^0(G, A)$. We then verify that the class $\bar{a} := \delta(c)$ of a in $\operatorname{Coker} N_A^*$ is independent of the choice of b). We have thus defined $\delta : \widehat{H}_0(G, C) \to \widehat{H}^0(G, A)$. The snake lemma (Appendix, Lemma A.29), applied to the homology and cohomology long exact sequences, then gives the desired exact sequence in view of the fact that, for any G-module M, $\widehat{H}_0(G, M)$ is a subgroup of $H_0(G, M)$ and $\widehat{H}^0(G, M)$ is a quotient of $H^0(G, M)$ (which allows to make the "links").

Let us now show that if A is relatively injective, then all the $\widehat{H}^n(G, A)$ are zero. For $n \geqslant 1$, it is Proposition 1.23. The proof for $n \leqslant -2$ is exactly similar (replacing the induced by co-induced, the injectives by projectives, etc). Let us verify it for $n = 0$ (the case $n = -1$ is similar). It is enough to consider the case where $A = \mathbf{Z}[G] \otimes_{\mathbf{Z}} X$ is co-induced (= induced as G is finite). Then X is identified with a subgroup of A, and A is the direct sum of gX for $g \in G$. Every element x of A can be uniquely written as $x = \sum_{g \in G} g x_g$ with $x_g \in X$, and such an element is in A^G if and only is all the x_g are equal, i.e., if and only if $a = Nx$ with $x \in X$. Hence, $\widehat{H}^0(G, A) = 0$. $\square$

Thus, the $\widehat{H}^n$ form a cohomological functor, satisfying $\widehat{H}^n(G, A) = 0$ for any induced (or co-induced) G-module A and any $n \in \mathbf{Z}$. In particular, we have:

Corollary 2.7 *Let*

$$0 \longrightarrow A \longrightarrow I \longrightarrow B \longrightarrow 0$$

be an exact sequence of G-modules with I relatively injective (e.g. induced). Then for all $n \in \mathbf{Z}$, we have $\widehat{H}^n(G, B) = \widehat{H}^{n+1}(G, A)$.

This allows us to prove the properties by shifting, through writing an arbitrary G-module as a sub-G-module or a quotient G-module of an induced one. For example, the analogue of the Proposition 1.25 remains true for the modified cohomology groups.

Homology groups can be computed in a way analogous to the cohomology groups using explicit complexes.

In particular, one easily obtains formulas for the Tate modified groups in the following way: for any $n \geqslant 0$, denote by $X_n = X_{-1-n} = \mathbf{Z}[G^{n+1}]$ the algebra of the finite group G^{n+1} (endowed with its evident structure of a G-module). We then obtain an exact complex $X_{\bullet}$ ("complete standard resolution of $\mathbf{Z}$"):

$$\cdots \longrightarrow X_2 \longrightarrow X_1 \longrightarrow X_0 \longrightarrow X_{-1} \longrightarrow X_{-2} \longrightarrow \cdots, \qquad (2.1)$$

the differentials $\partial_n : X_n \to X_{n-1}$ for $n > 0$ being defined by

$$\partial_n(g_0, \ldots, g_n) = \sum_{i=0}^{n} (-1)^i (g_0, \ldots, \widehat{g_i}, \ldots, g_n)$$

and $\partial_{-n} : X_{-n} \longrightarrow X_{-n-1}$ for $n > 0$ by

$$\partial_{-n}(g_0, \ldots, g_{n-1}) = \sum_{g \in G} \sum_{i=0}^{n} (-1)^i (g_0, \ldots, g_{i-1}, g, g_i, \ldots, g_{n-1}).$$

Lastly, we define $\partial_0 : X_0 \to X_{-1}$ by $\partial_0(g_0) = \sum_{g \in G} g$. The verification of the exactness of $X_\bullet$ is similar to the Lemma 1.26, by constructing a family of homomorphisms of abelian groups $u_i : X_i \to X_{i+1}$ satisfying $u_{i-1} \circ \partial_i + \partial_{i+1} \circ u_i = 0$ for all $i \in \mathbf{Z}$: for $i \geq 0$, we take the same u_i as in Lemma 1.26. For $i < -1$, we set $u_i(g_0, \ldots, g_{-i-1}) = 0$ if $g_0 \neq 1$ and $u_i(1, g_1, \ldots, g_{-i-1}) = (g_1, \ldots, g_{-i-1})$. Lastly, we take $u_{-1}(g) = 0$ if $g \neq 1$ and $u_{-1}(1) = 1$.

Let A be a G-module. We obtain the Tate modified group $\widehat{H}^n(G, A)$ for all $n \in \mathbf{Z}$ as

$$\widehat{H}^n(G, A) = H^n((\mathrm{Hom}(X_n, A))^G), \tag{2.2}$$

that is, as the nth cohomology group of the complex given by $((\mathrm{Hom}(X_n, A))^G)_{n \in \mathbf{Z}}$ (the differentials being the transposed of those of the X_n *without adding a sign*[1]).

This last complex is obtained by applying the functor $.^G$ to the exact complex $(\mathrm{Hom}(X_n, A))_{n \in \mathbf{Z}}$, which for $n \geq 0$ coincides with the complex of maps from G^{n+1} to A.

2.2 Change of Group. Transfer

Let A be a G-module. Let H be a subgroup of G. Denote by $N_G : A \to A$ and $N_H : A \to A$ the norm maps associated to the G-module and the H-module A, defined, respectively, by $N_G(x) = \sum_{g \in G} g \cdot x$ and $N_H(x) = \sum_{h \in H} h \cdot x$.

We have $N_G A \subset N_H A$ (to see this, group the elements of G in classes modulo H). By passing to the quotient, this induces a restriction homomorphism $\mathrm{Res} : \widehat{H}^0(G, A) \to \widehat{H}^0(H, A)$.

On the other hand we can also define restriction and corestriction homomorphisms for homology.

The corestriction $H_i(H, A) \to H_i(G, A)$ $(i \geq 0)$ is simply the morphism of homological functors which for $i = 0$, corresponds to the canonical surjection $A/\mathcal{I}_H A \to A/\mathcal{I}_G A$. The restriction $H_i(G, A) \to H_i(H, A)$ is the morphism of homological functors which, for $i = 0$, is given by the homomorphism $N'_{G/H} : A/\mathcal{I}_G A \to A/\mathcal{I}_H A$ defined for any x in A (with class $\overline{x}$ in $A/\mathcal{I}_G A$), by

$$N'_{G/H}(\overline{x}) = \sum_{s \in G/H} s^{-1} \cdot x$$

---

[1] As J. Riou pointed out to me, this convention is different from the one used, for example, in [12], which would be indispensable if we have replaced A by a complex of G-modules.

This is a valid definition since on the one hand if s and t are in the same left class modulo H, then $s^{-1} \cdot x$ and $t^{-1} \cdot x$ coincide modulo $\mathcal{I}_H A$ and on the other hand the right hand term is zero if $x \in \mathcal{I}_G A$.

Note that we also have

$$N'_{G/H}(\overline{x}) = \sum_{s \in H \backslash G} s \cdot x, \tag{2.3}$$

where $H \backslash G$ is this time the set of *right* classes modulo H.

One checks that the restriction homomorphism $H_0(G, A) \to H_0(H, A)$ induces a homomorphism Res : $\widehat{H}_0(G, A) \to \widehat{H}_0(H, A)$.

Finally, we have defined for all $n \in \mathbf{Z}$, homomorphisms Res : $\widehat{H}^n(G, A) \to \widehat{H}^n(H, A)$. Using the following lemma, we obtain a morphism of cohomological functors.

Lemma 2.8 *The family of homomorphisms* Res : $\widehat{H}^n(G, .) \to \widehat{H}^n(H, .)$ *is compatible with exact sequences.*

Proof Let $0 \to A \to B \to C \to 0$ be an exact sequence of G-modules. We need to check that the following diagram is commutative:

$$
\begin{array}{ccc}
\widehat{H}^n(G, C) & \xrightarrow{\delta} & \widehat{H}^{n+1}(G, A) \\
{\scriptstyle\text{Res}}\downarrow & & {\scriptstyle\text{Res}}\downarrow \\
\widehat{H}^n(H, C) & \xrightarrow{\delta} & \widehat{H}^{n+1}(H, A)
\end{array}
$$

We have seen this already for $n \geqslant 0$ and for $n \leqslant -2$. There remains the case of $n = -1$. Then $\widehat{H}^n(G, C) = {_{N_G}}C / \mathcal{I}_G C$ and $\widehat{H}^{n+1}(G, A) = A^G / N_G A$ (and the same for H). Let then $c \in {_N}C$ whose class we denote by $\overline{c}$ in $\widehat{H}^{-1}(G, C)$. The class of $\delta(\overline{c})$ is obtained by lifting c to some $b \in B$ and taking the class $N_G(b)$ in $A^G / N_G A$. The class of $N_G(b)$ in $A^G / N_H A$ is therefore Res$(\delta(\overline{c}))$.

On the other hand, Res$(\overline{c})$ is the class of $\sum_{i \in I} s_i c$, where $(s_i)_{i \in I}$ is a system of representatives of right classes $H \backslash G$. Then $\delta(\text{Res}(\overline{c}))$ is the class (in $A^H / N_H A$) of $N_H(\sum_{i \in I} s_i b)$, which is equal to that of $N_G(b)$. $\qquad\square$

Similarly, we have corestriction morphisms (giving homomorphisms of cohomological functors) Cores : $\widehat{H}^n(H, A) \to \widehat{H}^n(G, A)$. For $n = 0$, the corestriction is induced by $x \mapsto \sum_{g \in G/H} g \cdot x$ from A^H to A^G and for $n = -1$, it is induced by the canonical surjection $H_0(H, A) \to H_0(G, A)$ by passing to subgroups $\widehat{H}_0(H, A)$ and $\widehat{H}_0(G, A)$.

We then obtain a generalisation of the results of Chap. 1:

Theorem 2.9 *Let G be a finite group. Let H be a subgroup of G of index m. Let $n \in \mathbf{Z}$. Then:*
   (a) The composite Cores $\circ$ Res *is multiplication by m in $\widehat{H}^n(G, A)$.*
   (b) The group $\widehat{H}^n(G, A)$ is killed by the order of G.
   (c) If A is of finite type, all the groups $\widehat{H}^n(G, A)$ are finite.

Note in particular that unlike what is happening for $H^0(G, A)$ (non-modified), the results presented here are valid for $n = 0$.

Proof This is analogous to Theorem 1.48 and its corollaries, once one has (directly) checked that Cores ∘ Res is multiplication my m in $\widehat{H}^0(G, A)$.   □

Remark 2.10 As J. Riou has pointed out, it is possible to rather formally construct the restriction and corestriction for the modified groups $\widehat{H}^n$ ($n \in \mathbf{Z}$). This is done using the following general result from homological algebra (that is, verified by shifting): let $(F^n)_{n \in \mathbf{Z}}$ and $(G^n)_{n \in \mathbf{Z}}$ be two cohomological functors from an abelian category $\mathcal{C}$ (with enough projectives and injectives) to an abelian category $\mathcal{D}$. Assume that (F^n) vanishes on the injectives and (G^n) vanishes on the projectives. Then any natural transformation $F^0 \to G^0$ extends uniquely to a morphism of cohomological functors $(F^n)_{n \in \mathbf{Z}} \to (G^n)_{n \in \mathbf{Z}}$. We apply this result to $F^n = \widehat{H}^n(G, .)$, $G^n = \widehat{H}^n(H, .)$ for the restriction (and vice versa for the corestriction), where H is a subgroup of a finite group G.

To finish this paragraph, we will study the particularly important case of the restriction morphism $\widehat{H}^{-2}(G, \mathbf{Z}) \to \widehat{H}^{-2}(H, \mathbf{Z})$, namely that of the restriction morphism $H_1(G, \mathbf{Z}) \to H_1(H, \mathbf{Z})$ on homology. Recall (Proposition 2.4) that the group $H_1(G, \mathbf{Z})$ is identified with the abelianisation of G, i.e., the quotient of G by its derived subgroup G' and the same for H.

Definition 2.11 Let G be a finite group. Let H be a subgroup of G. The homomorphism $V : G/G' \to H/H'$ induced by the restriction $H_1(G, \mathbf{Z}) \to H_1(H, \mathbf{Z})$ is called the *transfer* (*Verlagerung* in German).

Let $\mathcal{I}_G$ be the augmentation ideal of $\mathbf{Z}[G]$ (recall that it is the two-sided ideal of the ring $\mathbf{Z}[G]$ defined as the kernel of the augmentation morphism $\mathbf{Z}[G] \to \mathbf{Z}$).

As we have already seen in the proof of the Proposition 2.4, the boundary d of the homology exact sequence induces an isomorphism

$$H_1(G, \mathbf{Z}) \simeq H_0(G, \mathcal{I}_G) = \mathcal{I}_G / \mathcal{I}_G^2,$$

and the same when one replaces G by H. The same argument applied to the induced H-module $\mathbf{Z}[G]$ and its sub-H-module $\mathcal{I}_G$ yields an injection $d_H : H_1(H, \mathbf{Z}) \hookrightarrow H_0(H, \mathcal{I}_G) = \mathcal{I}_G / \mathcal{I}_H \mathcal{I}_G$. The exact sequence

$$0 \longrightarrow \mathcal{I}_G \longrightarrow \mathbf{Z}[G] \longrightarrow \mathbf{Z} \longrightarrow 0$$

and the compatibility of the restriction with the boundary maps in homology yield a commutative diagram (whose top horizontal arrow is induced by $g \mapsto (g - 1)$):

$$
\begin{array}{ccc}
H_1(G, \mathbf{Z}) = G/G' & \xrightarrow{\;\;\underset{\sim}{d}\;\;} & \mathcal{I}_G / \mathcal{I}_G^2 \\
{\scriptstyle V}\big\downarrow & & \big\downarrow{\scriptstyle N'_{G/H}} \\
H_1(H, \mathbf{Z}) = H/H' & \xrightarrow{\;\;d_H\;\;} & \mathcal{I}_G / \mathcal{I}_H \mathcal{I}_G .
\end{array}
\qquad (2.4)
$$

Let us recall (formula (2.3) applied to $A = \mathcal{I}_G$) that the homomorphism $N'_{G/H}$: $\mathcal{I}_G/\mathcal{I}_G^2 \to \mathcal{I}_G/\mathcal{I}_H\mathcal{I}_G$ is defined by

$$N'_{G/H}(x \bmod \mathcal{I}_G^2) = \sum_{s \in H \backslash G} s \cdot x \quad \bmod \mathcal{I}_H\mathcal{I}_G, \quad \forall x \in \mathcal{I}_G. \tag{2.5}$$

We also have the following obvious commutative diagram:

$$
\begin{array}{ccccccccc}
0 & \longrightarrow & \mathcal{I}_H & \longrightarrow & \mathbf{Z}[H] & \longrightarrow & \mathbf{Z} & \longrightarrow & 0 \\
& & \downarrow & & \downarrow & & \downarrow & & \\
0 & \longrightarrow & \mathcal{I}_G & \longrightarrow & \mathbf{Z}[G] & \longrightarrow & \mathbf{Z} & \longrightarrow & 0.
\end{array}
$$

As d_H is injective, so is the natural homomorphism $\mathcal{I}_H/\mathcal{I}_H^2 \to \mathcal{I}_G/\mathcal{I}_H\mathcal{I}_G$. In other words $\mathcal{I}_H \cap \mathcal{I}_H\mathcal{I}_G = \mathcal{I}_H^2$. One deduces an isomorphism

$$H/H' \simeq \mathcal{I}_H/\mathcal{I}_H^2 = (\mathcal{I}_H + \mathcal{I}_H\mathcal{I}_G)/\mathcal{I}_H\mathcal{I}_G$$

induced by $h \mapsto (h-1)$, which can also be directly verified (cf. Exercise 2.5).

The next proposition computes V explicitly. Item (a) is the classical definition of transfer in group theory.

Proposition 2.12 *With the above notations, let R (with $1 \in R$) be a system of representatives of right classes of G modulo H. For all $\sigma \in G$ and $\rho \in R$, we write $\rho\sigma = \sigma_\rho \rho'$ with $\rho' \in R$ and $\sigma_\rho \in H$ (this decomposition is unique by definition of R). Then:*
(a) The transfer $V : G/G' \to H/H'$ is given by the formula

$$V(\sigma \bmod G') = \left(\prod_{\rho \in R} \sigma_\rho \right) \quad \bmod H'.$$

(b) We have a commutative diagram:

$$
\begin{array}{ccc}
G/G' & \xrightarrow{\quad V \quad} & H/H' \\
\delta_G \downarrow & & \downarrow \delta_H \\
\mathcal{I}_G/\mathcal{I}_G^2 & \xrightarrow{\quad S \quad} & (\mathcal{I}_H + \mathcal{I}_H\mathcal{I}_G)/\mathcal{I}_H\mathcal{I}_G,
\end{array}
$$

where the isomorphisms δ_G, δ_H are induced by $\sigma \mapsto \sigma - 1$ and the homomorphism S is defined by:

$$S(x \bmod \mathcal{I}_G^2) = \left(\sum_{\rho \in R} \rho \right) x \quad \bmod \mathcal{I}_H\mathcal{I}_G.$$

We recover the fact that the expression for the transfer given in (a) does not depend on the choice of R and the choice of the representative $\sigma \in G$ in a given class of G/G'. Note also that in the statement (b), the isomorphism δ_G is the same as the isomorphism d of the first row of the diagram (2.4), while the isomorphism δ_H is induced by the injection d_H of the second line of the same diagram.

Proof (a) Let $\sigma \in G$, whose image is $(\sigma - 1) \in \mathcal{I}_G/\mathcal{I}_G^2$ by d. We still denote by σ the class of σ in G/G'. By definition of $N'_{G/H}$, we have

$$N'_{G/H}(d(\sigma)) = \sum_{\rho \in R} \rho(\sigma - 1) \in \mathcal{I}_G/\mathcal{I}_H \mathcal{I}_G.$$

Using diagram (2.4), we deduce the following equalities in $\mathcal{I}_G/\mathcal{I}_H \mathcal{I}_G$

$$d_H(V(\sigma)) = \sum_{\rho \in R} \sigma_\rho \rho' - \sum_{\rho \in R} \rho = \sum_{\rho \in R} \sigma_\rho \rho' - \sum_{\rho \in R} \rho'$$

as (for a fixed σ) the map $\rho \mapsto \rho'$ is a bijection from R to R. Moreover, in $\mathcal{I}_G/\mathcal{I}_H \mathcal{I}_G$, we have

$$0 = \sum_{\rho \in R} (\sigma_\rho - 1)(\rho' - 1) = \left(\sum_{\rho \in R} \sigma_\rho \rho' - \sum_{\rho \in R} \rho' \right) - \left(\sum_{\rho \in R} (\sigma_\rho - 1) \right).$$

Finally, we obtain

$$d_H\big(V(\sigma)\big) = \sum_{\rho \in R} (\sigma_\rho - 1)$$

in $\mathcal{I}_G/\mathcal{I}_H \mathcal{I}_G$. But $\sum_{\rho \in R}(\sigma_\rho - 1)$ is the element of $\mathcal{I}_H/\mathcal{I}_H^2$ corresponding to $\prod_{\rho \in R} \sigma_\rho$ via the isomorphism $H/H' \simeq \mathcal{I}_H/\mathcal{I}_H^2$ seen above.

(b) The top horizontal arrow d of the diagram (2.4) is an isomorphism. We deduce that all the elements of the image of the morphism $N'_{G/H} : \mathcal{I}_G/\mathcal{I}_G^2 \to \mathcal{I}_G/\mathcal{I}_H \mathcal{I}_G$ are represented by elements of $\mathcal{I}_H \subset \mathcal{I}_G$. The desired commutative diagram follows from the diagram (2.4), along with the explicit expression (2.5) of $N'_{G/H}$. $\qquad \square$

The next theorem (whose first proof was given by Furtwängler) about transfer will be used later in the proof of the *principal ideal theorem* (Sect. 15.4).

The proof we give here (following Neukirch [40], Theorem 7.6 with a bit more details) is due to Witt.

Theorem 2.13 *Let G be finite group whose derived subgroup is denoted by G'. Let G'' be the derived subgroup of G'. Then the transfer $V : G/G' \to G'/G''$ is the trivial homomorphism.*

Proof After, if necessary, replacing G by G/G'' (whose derived subgroup is G'/G''), we can assume that $G'' = \{1\}$, i.e., that G' is abelian. For any $x \in G$, let $\delta x = \delta(x) := (x - 1) \in \mathcal{I}_G$. We will need two lemmas:

Lemma 2.14

(a) *Let* $x_1, \ldots, x_r$ *be elements of* G. *Then there exist elements* $j_2, \ldots, j_r$ *of* $\mathcal{I}_G$ *such that*

$$\delta(x_1 \cdots x_r) = (1 + j_r)\delta x_r + (1 + j_{r-1})\delta x_{r-1} + \cdots + (1 + j_2)\delta x_2 + \delta x_1.$$

(b) *If* $\tau = [x, y] = x^{-1}y^{-1}xy$ $(x, y \in G)$ *is a commutator in* G, *then there exist* $i, j \in \mathcal{I}_G$ *such that* $\delta(\tau) = i\delta(x) + j\delta(y)$.

To prove Lemma 2.14, we first observe that if $x, y \in G$, then $\delta(xy) = (1 + \delta x)\delta y + \delta x$ with $\delta x \in \mathcal{I}_G$. This proves (a) by induction on r. Item (b) follows from (a) and the fact that $\delta(x^{-1}) = -(1 + \delta(x^{-1}))\delta x$ with $\delta(x^{-1}) \in \mathcal{I}_G$.

We will now show the second lemma:

Lemma 2.15 *There is an element* $\mu \in \mathbf{Z}[G]$ *satisfying the two following properties:*
(a) *For all* $\sigma \in G$, *we have* $\mu(\delta\sigma) = 0 \bmod \mathcal{I}_{G'}\mathbf{Z}[G]\mathcal{I}_G = \mathcal{I}_{G'}\mathcal{I}_G$.
(b) $\mu = [G : G'] \bmod \mathcal{I}_G$, *where* $[G : G'] \in \mathbf{Z}$ *is the index of* G' *in* G.

Proof of Lemma 2.15 Let $G = \{\sigma_1, \ldots, \sigma_n\}$. Let $\pi : \mathbf{Z}^n \to G/G'$ be the morphism of $\mathbf{Z}$-modules sending the ith vector ε_i of the canonical basis of $\mathbf{Z}^n$ to σ_i. The kernel $\mathrm{Ker}\,\pi$ is of finite index in $\mathbf{Z}^n$, hence isomorphic to $\mathbf{Z}^n$. We thus obtain an exact sequence of $\mathbf{Z}$-modules

$$0 \longrightarrow \mathbf{Z}^n \overset{f}{\longrightarrow} \mathbf{Z}^n \overset{\pi}{\longrightarrow} G/G' \longrightarrow 0.$$

Let $f(\varepsilon_k) = \sum_{i=1}^n m_{ki}\varepsilon_i$, and denote by M the matrix (m_{ki}) (it is the transpose of the matrix of f in the canonical basis). We can assume that the determinant of M is positive and hence equals $[G : G']$. For each $k \in \{1, \ldots, n\}$, we have $\pi(f(\varepsilon_k)) = 0$ in G/G'. Therefore, there exist elements τ_k of G' such that

$$\prod_{i=1}^n \sigma_i^{m_{ki}} = \tau_k. \tag{2.6}$$

Noticing that each τ_k is a product of commutators $[\sigma_i, \sigma_j]$, the equality (2.6) and Lemma 2.14 show the existence of elements $\mu_{ki} \in \mathbf{Z}[G]$, satisfying $\mu_{ki} = m_{ki} \bmod \mathcal{I}_G$, and such that

$$\sum_{i=1}^n \mu_{ki}(\delta\sigma_i) = 0, \quad k = 1, \ldots, n. \tag{2.7}$$

We work in the commutative ring $A = \mathbf{Z}[G/G'] = \mathbf{Z}[G]/\mathcal{I}_{G'}\mathbf{Z}[G]$. Let U be the class of the matrix (μ_{ki}) in $M_n(A)$ and let $\mu \in \mathbf{Z}[G]$ whose class $\overline{\mu}$ in A is $\det U$. In particular, we already have the property (b) of the lemma, namely $\mu = [G : G']$ $\bmod \mathcal{I}_G$. The comatrix $L \in M_n(A)$ of U then satisfies the following relation: $LU =$

$UL = \overline{\mu}I_n$ in $M_n(A)$. By lifting L to a matrix (λ_{jk}) with coefficients in $\mathbf{Z}[G]$, we obtain that the product matrix $[\lambda_{jk}] \cdot [\mu_{ki}]$ coincides with μI_n mod $\mathcal{I}_{G'}\mathbf{Z}[G]$. By multiplying on the right by the column vector $(\delta\sigma_1, \ldots, \delta\sigma_n)$ (whose entries are in $\mathcal{I}_G$), the equality (2.7) gives

$$\mu(\delta\sigma_j) = \sum_{k=1}^{n} \lambda_{jk} \sum_{i=1}^{n} \mu_{ki}(\delta\sigma_i) = 0 \quad \text{mod } \mathcal{I}_{G'}\mathbf{Z}[G]\mathcal{I}_G,$$

hence

$$\mu(\delta\sigma) = 0 \quad \text{mod } \mathcal{I}_{G'}\mathbf{Z}[G]\mathcal{I}_G \tag{2.8}$$

for any $\sigma \in G$. The property (a) of Lemma 2.15 follows. □

End of the Proof of Theorem 2.13 We fix a system R of representatives for the classes (left or right, this does not matter since G' is a normal subgroup of G) of G/G', with $1 \in R$. Let $\mu \in \mathbf{Z}[G]$ be as in Lemma 2.15. Let us show that $\mu = \sum_{\rho\in R} \rho$ mod $\mathcal{I}_{G'}\mathbf{Z}[G]$. The element $\overline{\mu}$ of $A = \mathbf{Z}[G/G']$ can be written

$$\overline{\mu} = \sum_{\rho\in R} n_\rho \overline{\rho}$$

with $n_\rho \in \mathbf{Z}$, where $\overline{\rho}$ is the class of ρ in A. Property (a) of Lemma 2.15 yields $\overline{\mu}\overline{\sigma} = \overline{\mu}$ for any $\overline{\sigma} \in G/G'$, which already implies that all the n_ρ equal a fixed integer m. Hence $\mu = m\sum_{\rho\in R} \rho$ mod $\mathcal{I}_{G'}\mathbf{Z}[G]$. Now, property (b) of Lemma 2.15 implies $m = 1$, since $\sum_{\rho\in R} \rho = [G:G']$ mod $\mathcal{I}_G$ in view of the fact that R is of cardinality $[G:G']$.

Let $\sigma \in G$ (we still denote by σ its class in G/G'). By Proposition 2.12 (b), the image of an element $\delta_G(\sigma)$ by the homomorphism $S \in \mathcal{I}_G/\mathcal{I}_G^2$ is then given by

$$S(\delta_G(\sigma)) = \left(\sum_{\rho\in R} \rho\right) \cdot \delta\sigma \in \mathbf{Z}[G]/\mathcal{I}_{G'}\mathcal{I}_G.$$

On the other hand, as $\delta\sigma \in \mathcal{I}_G$ and $\mu = \sum_{\rho\in R} \rho$ mod $\mathcal{I}_{G'}\mathbf{Z}[G]$, we have:

$$\left(\sum_{\rho\in R} \rho\right) \cdot \delta\sigma = \mu \cdot (\delta\sigma) = 0 \quad \text{mod } \mathcal{I}_{G'}\mathcal{I}_G,$$

which shows that the transfer $V : G/G' \to G'$ is the zero map. □

2.3   Cohomology of a Cyclic Group

Let G be a cyclic group of cardinality n. One of the advantages of the Tate modified cohomology is that in this case, the sequence of groups $\widehat{H}^q(G, A)$ (for $q \in \mathbf{Z}$) is 2-periodic, which allows us to easily calculate them by reducing the computation to $\widehat{H}^0(G, A)$ and $\widehat{H}^{-1}(G, A)$, which admit explicit descriptions. We have the following theorem:

Theorem 2.16 *Let G be a finite cyclic group of order n. Let A be a G-module. Then for all $q \in \mathbf{Z}$, the groups $\widehat{H}^q(G, A)$ and $\widehat{H}^{q+2}(G, A)$ are isomorphic.*

Proof The idea of the proof is to identify the $\widehat{H}^q(G, A)$ with the cohomology of a certain complex which, by construction, will be 2-periodic.

Fix a generator s of G and let $D = s - 1$ in $\mathbf{Z}[G]$. On the other hand

$$N = \sum_{t \in G} t = \sum_{i=0}^{n-1} s^i.$$

Let us define a complex $K(A)$ by $K^i(A) = A$ for all i, the coboundaries $d^i :$ $K^i(A) \to K^{i+1}(A)$ being defined as: d^i is the multiplication by D if i is even and d^i is the multiplication by N if i is odd (note that this is a complex since $ND = DN = 0$).

For any exact sequence $0 \to A \to B \to C \to 0$ of G-modules, we have a corresponding sequence of complexes

$$0 \longrightarrow K(A) \longrightarrow K(B) \longrightarrow K(C) \longrightarrow 0$$

from which we obtain an associated long exact sequence with coboundary operators δ^i. We thus obtain a cohomology functor $(H^q(K(.)), \delta)$ which is clearly 2-periodic with respect to q. In order to conclude, it would be enough to show that it is isomorphic to the functor $(\widehat{H}^q(G, .), \delta)$.

As G is generated by s, we have $A^G = \operatorname{Ker} D$ and $\mathcal{I}_G A = \operatorname{Im} D$. It follows that for $q = 0$ and $q = -1$, we have $\widehat{H}^q(G, A) = H^q(K(A))$ and the coboundary operator between degrees 0 and 1 are the same. In particular, if A is relatively injective, we have $H^q(K(A)) = 0$ for $q = -1, 0$, hence, as the complex $K(A)$ is 2-periodic, for every $q \in \mathbf{Z}$. We conclude that for every G-module A and every $q \in \mathbf{Z}$, the groups $\widehat{H}^q(G, A)$ and $H^q(K(A))$ are isomorphic. This is done, for example, by arguing by shifting using Corollary 2.7 (cf. also Remark 2.10), after having written A as a submodule (resp. as a quotient) of an induced G-module I (resp I').   □

Remark 2.17 The isomorphism given by the Theorem 2.16 depends on the choice of a generator of G. As we will see later (Remark 2.29), such a choice also makes explicit the isomorphism of Theorem 2.16 using the *cup-product*.

Example 2.18 In the case of a cyclic group $G = \langle \sigma \rangle$ of order n and of a subgroup $H = \langle \sigma^d \rangle$ of index d, we have simple formulas for restriction and corestriction. Let A be a G-module. As we know that Res : $\widehat{H}^0(G, A) \to \widehat{H}^0(H, A)$

is induced by passing to the quotient from the canonical injection $A^G \to A^H$, and Cores : $\widehat{H}^{-1}(H, A) \to \widehat{H}^{-1}(G, A)$ is induced by restricting to subgroups from the canonical surjection $A/\mathcal{I}_H A \to A/\mathcal{I}_G A$, Theorem 2.16 (cf. also Remark 2.29) gives an expression for Res in even degree and for Cores in odd degree. We obtain opposite parities by noting (using formulas from Sect. 2.2) that Res : $\widehat{H}^{-1}(G, A) \to \widehat{H}^{-1}(H, A)$ is induced by the map $x \mapsto \sum_{k=0}^{d-1} \sigma^k \cdot x$ from $A/\mathcal{I}_G A$ to $A/\mathcal{I}_H A$, and Cores : $\widehat{H}^0(H, A) \to \widehat{H}^0(G, A)$ is induced by $x \mapsto \sum_{k=0}^{d-1} \sigma^k \cdot x$ from A^H to A^G.

2.4 Herbrand Quotient

Let G be a finite *cyclic* group. Let A be a G-module. When $\widehat{H}^0(G, A)$ and $\widehat{H}^1(G, A)$ are finite groups, we denote by $h^0(A)$ and $h^1(A)$ their respective cardinalities.

Definition 2.19 Assume that $\widehat{H}^0(G, A)$ and $\widehat{H}^1(G, A)$ are finite. We define the *Herbrand quotient* $h(A)$ of a G-module A to be the rational number

$$h(A) := \frac{h^0(A)}{h^1(A)}.$$

The properties of the Herbrand quotient are summarised in the following proposition:

Theorem 2.20 *Let G be a cyclic group.*
(a) Let

$$0 \longrightarrow A \longrightarrow B \longrightarrow C \longrightarrow 0$$

be an exact sequence of G-modules. If two Herbrand quotients among $h(A)$, $h(B)$, $h(C)$ are defined, then so is the third one and we have $h(B) = h(A)h(C)$.
(b) If A is a finite G-module, then $h(A) = 1$.
(c) Let $f : A \to B$ be a homomorphism of G-modules with finite kernel and cokernel. If one of the Herbrand quotients $h(A)$, $h(B)$ is defined, then so is the other and $h(A) = h(B)$.

Proof (a) The modified long exact cohomology sequence is written as (taking into account the periodicity Theorem 2.16):

$$\cdots \longrightarrow \widehat{H}^1(G, C) \longrightarrow \widehat{H}^0(G, A) \longrightarrow \widehat{H}^0(G, B) \longrightarrow \widehat{H}^0(G, C)$$
$$\longrightarrow \widehat{H}^1(G, A) \longrightarrow \widehat{H}^1(G, B) \longrightarrow \widehat{H}^1(G, C) \longrightarrow \cdots$$

This long exact sequence is hence 6-periodic ("the exact hexagon "). We deduce an exact sequence:

$$0 \longrightarrow I_1 \longrightarrow \widehat{H}^0(G, A) \longrightarrow \widehat{H}^0(G, B) \longrightarrow \widehat{H}^0(G, C)$$
$$\longrightarrow \widehat{H}^1(G, A) \longrightarrow \widehat{H}^1(G, B) \longrightarrow \widehat{H}^1(G, C) \longrightarrow I_1 \longrightarrow 0,$$

where

$$I_1 = \text{Im}[\widehat{H}^1(G, C) \longrightarrow \widehat{H}^0(G, A)] = \text{Ker}[\widehat{H}^0(G, A) \longrightarrow \widehat{H}^0(G, B)].$$

It is then immediate that if two of the quotients $h(A), h(B), h(C)$ are defined, then so is the third. If this is the case, the alternating product of the cardinalities in the eight term exact sequence below is 1, which gives, (denoting by s the cardinality of I_1):

$$h^0(A)h^0(C)h^1(B)s = h^0(B)h^1(A)h^1(C)s,$$

hence

$$h(B) = h(A)h(C).$$

(b) Let s be a generator of the cyclic group G. Let D be the multiplication by $s - 1$ in A, hence we have an exact sequence

$$0 \longrightarrow H^0(G, A) \longrightarrow A \xrightarrow{\ D\ } A \longrightarrow H_0(G, A) \longrightarrow 0$$

as an element x of A is in $H^0(G, A)$ if and only if $s \cdot x - x = 0$, and the image of D is precisely $I_G(A)$. Similarly we have an exact sequence

$$0 \longrightarrow \widehat{H}^{-1}(G, A) \longrightarrow H_0(G, A) \xrightarrow{\ N\ } H^0(G, A) \longrightarrow \widehat{H}^0(G, A) \longrightarrow 0.$$

If the cardinality m of A is finite, we obtain that $H^0(G, A)$ and $H_0(G, A)$ have the same cardinality by the first sequence. Then we obtain that $\widehat{H}^{-1}(G, A) \simeq H^1(G, A)$ and $\widehat{H}^0(G, A)$ have the same cardinality thanks to the second sequence. It follows that $h(A) = 1$.

(c) Let I be the image of f. We have exact sequences

$$0 \longrightarrow \text{Ker}\, f \longrightarrow A \longrightarrow I \longrightarrow 0.$$
$$0 \longrightarrow I \longrightarrow B \longrightarrow \text{Coker}\, f \longrightarrow 0.$$

From (b), we deduce that $h(\text{Ker}\, f) = h(\text{Coker}\, f) = 1$. We then obtain that, according to (a), if $h(A)$ (resp. $h(B)$) is defined, then $h(I)$ is also defined and therefore $h(B)$ (resp. $h(A)$) is defined, with $h(A) = h(I) = h(B)$. □

2.5 Cup-Products

Let G be a finite group. Let A and B be two G-modules, then we can view $A \otimes B :=$ $A \otimes_{\mathbf{Z}} B$ as a G-module via the formula $g \cdot (a \otimes b) = (g \cdot a) \otimes (g \cdot b)$ for all $g \in G$

and all $(a, b) \in A \times B$. From this we deduce a bilinear map on homogeneous cochain groups (defined in the Sect. 1.4):

$$K^p(G, A) \times K^q(G, B) \xrightarrow{\cup} K^{p+q}(G, A \otimes B)$$

defined (for all integers p, q) by the formula:

$$(a \cup b)(g_0, \ldots, g_{p+q}) = a(g_0, \ldots, g_p) \otimes b(g_p, \ldots, g_{p+q}).$$

We then have (with the convention that throughout this Sect. 2.5, a "cocycle" and a "cochain" denote a homogeneous cocycle and a homogeneous cochain respectively):

Theorem 2.21 *If a and b are cocycles, then $a \cup b$ is also a cocycle.*

If one of the two cochains a or b is a coboundary and the other is a cocycle, then $a \cup b$ is a coboundary.

The map $\cup$ induces for all $p, q \geqslant 0$ a bilinear map

$$H^p(G, A) \times H^q(G, B) \longrightarrow H^{p+q}(G, A \otimes B)$$

again denoted by $\cup$ and called the cup-product.

Proof It is enough to verify the formula:

$$d(a \cup b) = (da) \cup b + (-1)^p (a \cup db), \tag{2.9}$$

where d is the coboundary between the groups of cochains. On the other hand we have

$$d(a \cup b)(g_0, \ldots, g_{p+q+1})$$

$$= \sum_{i=0}^{p} (-1)^i a(g_0, \ldots, \widehat{g_i}, \ldots, g_{p+1}) \otimes b(g_{p+1}, \ldots, g_{p+q+1})$$

$$+ \sum_{i=p+1}^{p+q+1} (-1)^i a(g_0, \ldots, g_p) \otimes b(g_p, \ldots, \widehat{g_i}, \ldots, g_{p+q+1})$$

and

$$(da \cup b)(g_0, \ldots, g_{p+q+1})$$

$$= \sum_{i=0}^{p+1} (-1)^i a(g_0, \ldots, \widehat{g_i}, \ldots, g_{p+1}) \otimes b(g_{p+1}, \ldots, g_{p+q+1})$$

and also

$(a \cup db)(g_0, \ldots, g_{p+q+1})$

$$= a(g_0, \ldots, g_p) \otimes \sum_{i=0}^{q+1} (-1)^i b(g_p, \ldots, \widehat{g}_{p+i}, \ldots, g_{p+q+1})$$

$$= a(g_0, \ldots, g_p) \otimes \sum_{i=p}^{p+q+1} (-1)^{i-p} b(g_p, \ldots, \widehat{g}_i, \ldots, g_{p+q+1}).$$

We thus obtain the formula modulo the following term

$$(-1)^{p+1} a(g_0, \ldots, \widehat{g}_{p+1}, \ldots, g_{p+1}) \otimes b(g_{p+1}, \ldots, g_{p+q+1})$$
$$+ (-1)^p a(g_0, \ldots, g_p) \otimes b(g_p, \ldots, \widehat{g}_p, \ldots, g_{p+q+1})$$

which cancels out, hence the result.   □

In [16], Chap. 3, Sect. 4, one finds a definition of the cup product which does not use explicit computations with cocycles.

Remark 2.22 In the special case $p = 0$, the cup-product of an element $a \in H^0(G, A)$ by an element $b \in H^q(G, B)$ is simply given by $f_*(b)$, where $f_* : H^q(G, B) \to H^q(G, A \otimes B)$ is the homomorphism induced by the morphism of G-modules $x \mapsto a \otimes x$ from B to $A \otimes B$. This follows immediately from the definition of the cup-product and from the description of f_* in terms of the cocycles.

In particular, for $p = q = 0$, the cup-product is simply the map $(a, b) \mapsto a \otimes b$ from $A^G \times B^G$ to $(A \otimes B)^G$. It is clear that it is functorial in A and B. We can then more generally associate a cup-product to a *pairing* (i.e., a bilinear map compatible with the action of G) $\varphi : A \times B \to C$ between G-modules, using the fact that such a map factors through $A \otimes B$. We will often denote by $\cup$ the cup-product thus obtained if it is clear what φ is.

In addition, we have the two following compatibility properties of the cup-product.

Proposition 2.23

(a) Let B be a G-module. Let

$$0 \longrightarrow A' \longrightarrow A \longrightarrow A'' \longrightarrow 0 \quad and \quad 0 \longrightarrow C' \longrightarrow C \longrightarrow C'' \longrightarrow 0$$

be exact sequences of G-modules. We assume that we have a pairing $A \times B \to C$ which induces pairings $A' \times B \to C'$ and $A'' \times B \to C''$. Then for all $\alpha'' \in H^p(G, A'')$ and all $\beta \in H^q(G, B)$, we have

$$(\delta \alpha'') \cup \beta = \delta(\alpha'' \cup \beta) \in H^{p+q+1}(G, C'),$$

where δ is the coboundary between the cohomology groups.

(b) Let $0 \to B' \to B \to B'' \to 0$ and $0 \to C' \to C \to C'' \to 0$ be exact sequences of G-modules. Let A be a G-module. Assume we are given a pairing

$A \times B \to C$ which induces pairings $A \times B' \to C'$ and $A \times B'' \to C''$. Then for all $\alpha \in H^p(G, A)$ and all $\beta'' \in H^q(G, B'')$, we have

$$\alpha \cup (\delta\beta'') = (-1)^p \delta(\alpha \cup \beta'') \in H^{p+q+1}(G, C').$$

Proof Let us for instance prove (b). Lift α to a cocycle $a \in Z^p(G, A)$ and β'' to a cocycle $b'' \in Z^q(G, B'')$. We can lift b'' to a homogeneous cochain $b \in K^q(G, B)$ (indeed, if M is a G-module, we have $K^q(G, M) = X^q(G, M)^G$, where $X^q(G, M)$ is the induced G-module consisting of the functions from G^{q+1} to M. It follows that the functor $K^q(G, .)$ is exact).

Let us identify B' with its image in B. Then $\delta(\beta'')$ is represented by $db \in Z^{q+1}(G, B')$ and $\delta(\alpha \cup \beta'')$ is represented by $d(a \cup b) \in Z^{p+q+1}(G, C')$. As $da = 0$, the formula (2.9) gives

$$d(a \cup b) = (-1)^p (a \cup db).$$

The result follows from this by passing to the cohomology classes.   □

Remark 2.24 Let A be a G-module. Then A embeds into the induced module $I_G(A)$ as a direct factor of it as a $\mathbf{Z}$-module (Proposition 1.10). It follows that the sequence

$$0 \longrightarrow A \longrightarrow I_G(A) \longrightarrow A/I_G(A) \longrightarrow 0$$

remains exact when tensored with any G-module B. The Proposition 2.23, along with the fact that the cup-product is "bifunctorial" and its definition for $p = q = 0$, characterises the cup-product in a unique way (by shifting). It is also possible to go in the other direction by Remark 1.11.

2.6 Cup-Products for the Modified Cohomology

We will now (using Remark 2.24), extend the definition of the cup-product to Tate modified cohomology. This definition will be in particular used in the case $p = -2, q = 2$ to apply the Tate–Nakayama theorem to the Galois cohomology of local fields (Sect. 9.1).

Theorem 2.25 *Let G be a finite group. Then there exists a unique family of bilinear maps*

$$\widehat{H}^p(G, A) \times \widehat{H}^q(G, B) \xrightarrow{\cup} \widehat{H}^{p+q}(G, C)$$

associated to all $p, q \in \mathbf{Z}$ and any pairing $A \times B \to C$ of G-modules, satisfying:
(1) For $p = q = 0$, these maps are induced by the natural map

$$A^G \times B^G \longrightarrow C^G.$$

(2) These maps are functorial in G-modules.

(3) These maps satisfy properties analogous to (a) and (b) of Proposition 2.23 with respect to the exact sequences for all $p, q \in \mathbf{Z}$.

Naturally, for $p, q \geqslant 0$, we recover the usual cup-product defined in Theorem 2.21, which in this case gives us an explicit description in terms of cocycles.

Proof We immediately reduce to the case $C = A \otimes B$. Consider the complete standard resolution $(X_{\bullet}, \partial)$ (i.e., the complex (2.1)) of $\mathbf{Z}$. For any G-module M, recall that we denote by $\operatorname{Hom}_G(X_{\bullet}, A)$ the complex whose nth term is $\operatorname{Hom}_G(X_n, A)$, with the same sign convention as for formula (2.2). We then have (using Definition A.51 of the appendix for the total tensor product of the two complexes) a morphism of complexes

$$\operatorname{Hom}_G(X_{\bullet}, A) \otimes \operatorname{Hom}_G(X_{\bullet}, B) \longrightarrow \operatorname{Hom}_G(X_{\bullet} \otimes X_{\bullet}, C). \qquad (2.10)$$

Lemma 2.26 *There exists a family of homomorphisms*

$$\varphi_{p,q} : X_{p+q} \longrightarrow X_p \otimes X_q, \quad p, q \in \mathbf{Z}$$

satisfying:

$$(a) \qquad \varphi_{p,q}\partial = (\partial \otimes 1)\varphi_{p+1,q} + (-1)^p (1 \otimes \partial)\varphi_{p,q+1}.$$
$$(b) \qquad (\pi \otimes \pi)\varphi_{0,0} = \pi,$$

where $\pi : X_0 = \mathbf{Z}[G] \to \mathbf{Z}$ is the augmentation homomorphism.

Assume for now that Lemma 2.26 is proved. The $\varphi_{p,q}$ induce [2] a homomorphism of complexes (where $X_{\bullet} \otimes X_{\bullet}$ in the left side term designates, as above, the total complex)

$$\operatorname{Hom}_G(X_{\bullet} \otimes X_{\bullet}, C) \longrightarrow \operatorname{Hom}_G(X_{\bullet}, C).$$

Composing with (2.10), we obtain a morphism

$$\operatorname{Hom}_G(X_{\bullet}, A) \otimes \operatorname{Hom}_G(X_{\bullet}, B) \longrightarrow \operatorname{Hom}_G(X_{\bullet}, C), \quad f \otimes g \longmapsto f \cdot g$$

between complexes of G-modules.

Then, the property (a) gives

$$\partial(f \cdot g) = (\partial f) \cdot g + (-1)^p f \cdot (\partial g),$$

---

[2] There is a subtlety here (pointed out by J. Riou): the $\varphi_{p,q}$ do not generally induce morphisms of complexes $X_{\bullet} \to X_{\bullet} \otimes X_{\bullet}$ because $X_{\bullet}$ is not *à priori* bounded from either side, and hence the image of X_n in $\prod_{p+q=n} X_p \otimes X_q$ is not contained in $\bigoplus_{p+q=n} X_p \otimes X_q$. Nevertheless, the passage to $\operatorname{Hom}_G(., C)$ allows to define the desired homomorphism in an evident way.

which implies that if f and g are cocycles, the same holds for $f \cdot g$ (and the cohomology class of $f \cdot g$ depends only on the classes of f and g). From this we deduce, using formula, (2.2), a family of bilinear maps

$$\widehat{H}^p(G, A) \times \widehat{H}^q(G, B) \longrightarrow \widehat{H}^{p+q}(A, C)$$

which clearly satisfy property (2) of Theorem 2.25. Property (1) follows from property (b) of Lemma 2.26. Lastly, property (3) is proved in exactly the same way as Proposition 2.23. Thus we have proved the existence assertion in Theorem 2.25, and uniqueness follows from property (1) and Remark 2.24.

It remains to prove Lemma 2.26. For this, we define

$$\varphi_{p,q}(g_0, \dots, g_{p+q}) = (g_0, \dots, g_p) \otimes (g_p, \dots, g_{p+q})$$

if $p, q \geqslant 0$. For $p, q \geqslant 1$, we set

$$\varphi_{-p,-q}(g_1, \dots, g_{p+q}) = (g_1, \dots, g_p) \otimes (g_{p+1}, \dots, g_{p+q}).$$

Lastly, the definition of the $\varphi_{p,q}$ if one of the indexes p, q is > 0 and the other is < 0 depends on the sign of $p + q$. More precisely, we set, for all $p \geqslant 0$ and all $q \geqslant 1$:

$$\varphi_{p,-p-q}(g_1, \dots, g_q) = \sum_{(h_1,\dots,h_p)\in G^p} (g_1, h_1, \dots, h_p) \otimes (h_p, \dots, h_1, g_1, \dots, g_q).$$

$$\varphi_{-p-q,p}(g_1, \dots, g_q) = \sum_{(h_1,\dots,h_p)\in G^p} (g_1, \dots, g_q, h_1, \dots, h_p) \otimes (h_p, \dots, h_1, g_q).$$

$$\varphi_{p+q,-q}(g_0, \dots, g_p) = \sum_{(h_1,\dots,h_q)\in G^q} (g_0, \dots, g_p, h_1, \dots, h_q) \otimes (h_q, \dots, h_1).$$

$$\varphi_{-q,p+q}(g_0, \dots, g_p) = \sum_{(h_1,\dots,h_q)\in G^q} (h_1, \dots, h_q) \otimes (h_q, \dots, h_1, g_0, \dots, g_p).$$

The lemma follows from a direct computation (which is not difficult but very tedious).   □

The following proposition provides a "set of formulas" describing the behaviour of the cup-product with respect to the usual cohomological operations:

Proposition 2.27

(a) If we identify $(A \otimes B) \otimes C$ with $A \otimes (B \otimes C)$, then

$$(\alpha \cup \beta) \cup \gamma = \alpha \cup (\beta \cup \gamma)$$

for all $\alpha \in \widehat{H}^p(G, A)$, $\beta \in \widehat{H}^q(G, B)$, $\gamma \in \widehat{H}^r(G, C)$ ("associativity" of the cup-product).

(b) If we identify $A \otimes B$ with $B \otimes A$, we have $(\alpha \cup \beta) = (-1)^{pq}(\beta \cup \alpha)$ for all $\alpha \in \widehat{H}^p(G, A)$ and $\beta \in \widehat{H}^q(G, B)$ ("anticommutativity" of the cup-product).

(c) If H is a subgroup of G, A and B are G-modules and Res is the restriction, then $\mathrm{Res}(\alpha \cup \beta) = \mathrm{Res}(\alpha) \cup \mathrm{Res}(\beta)$ for all $\alpha \in \widehat{H}^p(G, A)$ and $\beta \in \widehat{H}^q(G, B)$.

(d) If H is a normal subgroup of G, A and B are G/H-modules and Inf is the inflation, then

$$\mathrm{Inf}(\alpha \cup \beta) = \mathrm{Inf}(\alpha) \cup \mathrm{Inf}(\beta)$$

for all $\alpha \in \widehat{H}^p(G/H, A^H)$ and $\beta \in \widehat{H}^q(G/H, B^H)$.

(e) If H is a subgroup of G, A and B are G-modules and Cores the corestriction, then $\mathrm{Cores}(\alpha \cup \mathrm{Res}(\beta)) = \mathrm{Cores}(\alpha) \cup \beta$ for all $\alpha \in \widehat{H}^p(H, A)$ and $\beta \in \widehat{H}^q(G, A)$.

Proof All these properties are proved by shifting, using Theorem 2.25 (which defines the cup-product) and Remark 2.24. Let us prove (b), for example; we embed $A_0 := A$ in the induced module $I_G(A)$ and set $A_1 = I_G(A)/A$, then, by induction, define $A_p = I_G(A_{p-1})/A_{p-1}$ for all $p > 0$. Denoting by J_G the quotient module of $\mathbf{Z}[G]$ by $\mathbf{Z}$, we easily see that (the latter viewed as a submodule of $\mathbf{Z}[G]$ via $1 \mapsto \sum_{g \in G} g$), we have

$$A_1 \simeq A \otimes J_G. \tag{2.11}$$

Corollary 2.7 implies that the iterated coboundary $\delta^p : H^n(G, A_p) \to H^{n+p}(G, A)$ is an isomorphism for all $n \in \mathbf{Z}$. We adopt similar notations for B. Applying property (3) of Theorem 2.25 (p times in the case (a) and q times in the case (b)), we obtain, for $p, q \geqslant 0$, the following commutative diagram:

$$
\begin{array}{ccc}
\widehat{H}^0(G, A_p) \times \widehat{H}^0(G, B_q) \xrightarrow{\ \cup\ } & \widehat{H}^0(G, (A \otimes B_q)_p) \simeq \widehat{H}^0(G, A_p \otimes B_q) \\
\downarrow{\scriptstyle \delta^p} \qquad \downarrow{\scriptstyle \mathrm{Id}} & \downarrow{\scriptstyle \delta^p} \\
\widehat{H}^p(G, A) \times \widehat{H}^0(G, B_q) \xrightarrow{\ \cup\ } & \widehat{H}^p(G, (A \otimes B)_q) \simeq \widehat{H}^p(G, A \otimes B_q) \\
\downarrow{\scriptstyle \mathrm{Id}} \qquad \downarrow{\scriptstyle \delta^q} & \downarrow{\scriptstyle (-1)^{pq}\delta^q} \\
\widehat{H}^p(G, A) \times \widehat{H}^q(G, B) \xrightarrow{\ \cup\ } & \widehat{H}^{p+q}(G, A \otimes B).
\end{array}
$$

Indeed, we have isomorphisms $(A \otimes B_q)_p \simeq A_p \otimes B_q$ and $(A \otimes B)_q \simeq A \otimes B_q$ by (2.11).

As the desired formula is clear for $p = q = 0$, the result for $p, q \geqslant 0$ then follows from the fact that vertical arrows are isomorphisms. The case $p < 0$ is similar: writing the exact sequence

$$0 \longrightarrow A_{-1} \longrightarrow I_G(A) \longrightarrow A \longrightarrow 0,$$

then setting $A_{-n-1} = \mathrm{Ker}[I_G(A_{-n}) \to A_{-n}]$ for all $n \geqslant 1$.

The proof of (a), (c) and (d) is completely analogous, the formulas being clear for $p = q = 0$.

To prove (e), the same shifting argument reduces it to verification of the commutativity of the following diagram

$$
\begin{array}{ccc}
\widehat{H}^0(H, A_p) \times \widehat{H}^0(H, B_q) & \xrightarrow{\ \cup\ } & \widehat{H}^0(H, A_p \otimes B_q) \\
\text{Cores} \downarrow \quad\quad \uparrow \text{Res} & & \downarrow \text{Cores} \\
\widehat{H}^0(G, A_p) \times \widehat{H}^0(G, B_q) & \xrightarrow{\ \cup\ } & \widehat{H}^0(G, A_p \otimes B_q),
\end{array}
$$

for which it is enough to verify the commutativity of

$$
\begin{array}{ccc}
A_p^H \times B_q^H & \xrightarrow{\ \otimes\ } & (A_p \otimes B_q)^H \\
N_{G/H} \downarrow \quad\quad \uparrow \text{Id} & & \downarrow N_{G/H} \\
A_p^G \times B_q^G & \xrightarrow{\ \otimes\ } & (A_p \otimes B_q)^G.
\end{array}
$$

It follows from a computation which is valid for all $a \in A_p^H$, $b \in A_q^G$:

$$
N_{G/H}(a \otimes b) = \sum_{g \in G/H} ga \otimes gb = \sum_{g \in G/H} ga \otimes b = N_{G/H}(a) \otimes b.
$$

$\square$

Lastly, this is the final compatibility which will be used in the proofs of duality theorems:

Proposition 2.28 *Let*

$$
0 \longrightarrow A' \longrightarrow A \longrightarrow A'' \longrightarrow 0; \quad 0 \longrightarrow B' \longrightarrow B \longrightarrow B'' \longrightarrow 0
$$

be exact sequences of G-modules. Let C be a G-module and $\varphi : A \times B \longrightarrow C$ be a bilinear map (compatible with the action of G) such that $\varphi(A' \times B')=0$, so that φ induces pairings

$$
\varphi' : A' \times B'' \longrightarrow C \quad and \quad \varphi'' : A'' \times B' \longrightarrow C.
$$

Then, for $p, q \geqslant 0$, the induced cup-products

$$
H^p(G, A'') \times H^q(G, B') \longrightarrow H^{p+q}(G, C)
$$

and

$$
H^{p+1}(G, A') \times H^{q-1}(G, B'') \longrightarrow H^{p+q}(G, C)
$$

are compatibles (up to a sign) with the coboundaries δ. More precisely, we have

$$(\delta\alpha) \cup \beta + (-1)^p (\alpha \cup \delta\beta) = 0$$

for all $\alpha \in H^p(G, A'')$ and $\beta \in H^{q-1}(G, B'')$.

Proof We lift α to a cocycle $a'' \in Z^p(G, A'')$ and similarly β to $b'' \in Z^{q-1}(G, B'')$, then a'' and b'', respectively, to cochains a and b of $K^p(G, A)$ and $K^{q-1}(G, B)$. Then da and db come from $a' \in K^{p+1}(G, A')$ and $b' \in K^q(G, B')$, respectively, which represent $\delta\alpha$ and $\delta\beta$ respectively. Using formula (2.9) one then verifies that

$$a' \cup b'' + (-1)^p (a'' \cup b') = d(a \cup b)$$

is a coboundary; hence it is zero in $H^{p+q}(G, C)$.   □

Remark 2.29 Let $G = \langle s \rangle$ be a finite cyclic group of order m. Consider the exact sequence

$$0 \longrightarrow \mathbf{Z} \xrightarrow{\ i\ } \mathbf{Z}[G] \xrightarrow{\ s-1\ } \mathbf{Z}[G] \xrightarrow{\ \varepsilon\ } \mathbf{Z} \longrightarrow 0,$$

where ε is the augmentation and i is the natural injection. This sequence of free $\mathbf{Z}$-modules remains exact when tensored with A. Cutting the thus obtained sequence up into two short exact sequences, we obtain (by Corollary 2.7) isomorphisms $\theta_q(A)$: $\widehat{H}^q(G, A) \to \widehat{H}^{q+2}(G, A)$. In particular, for $q = 0$ and $A = \mathbf{Z}$, this gives us an isomorphism $\theta : \mathbf{Z}/m\mathbf{Z} = \widehat{H}^0(G, \mathbf{Z}) \to \widehat{H}^2(G, \mathbf{Z})$ (which depends on the choice of a generator s of G which we made at the beginning). Let $a = \theta(1)$. We then obtain an explicit isomorphism (cf. Theorem 2.16) between $\widehat{H}^q(G, A)$ and $\widehat{H}^{q+2}(G, A)$ by forming a cup-product with a. Indeed, the compatibility of the cup-product with the coboundaries (property (3) of Theorem 2.25) gives, for any G-module A, a commutative diagram (up to a sign):

$$
\begin{array}{ccc}
\widehat{H}^q(G, A) & =\!=\!=\!=\!= & \widehat{H}^q(G, A) \\
{\scriptstyle \cup 1}\big\downarrow & & \big\downarrow{\scriptstyle \cup a} \\
\widehat{H}^q(G, A) & \xrightarrow{\ \theta_q(A)\ } & \widehat{H}^{q+2}(G, A).
\end{array}
$$

The cup-product by $1 \in \mathbf{Z}/m\mathbf{Z} = \widehat{H}^0(G, \mathbf{Z})$ is the identity in $\widehat{H}^q(G, A)$ (it is immediate for $q = 0$ and then can be deduced by shifting). As $\theta_q(A)$ is also an isomorphism, we obtain the desired isomorphism, using the formula (valid up to a sign) $(\theta_q(A))(x) = x \cup a$.

Remark 2.30 Remark 2.22 extends easily to the modified cohomology classes $a \in \widehat{H}^0(G, A)$ and $b \in \widehat{H}^n(G, B)$ for $n \in \mathbf{Z}$: one sees this, for example, by shifting from the case $n = 0$.

We finish this chapter with two explicit computations of the cup-product in low degrees. Proposition 2.32 will be in particular used in the proof of an important compatibility formula that we will establish later on (Proposition 9.3). If A is a

G-module and $a \in A$ has norm zero (resp. is in A^G), we denote by $\bar{a}_0$ its class in $\widehat{H}^{-1}(G, A)$ (resp. $\bar{a}^0$ its class in $\widehat{H}^0(G, A)$).

Lemma 2.31 *Let G be a finite group. Let A and B be G-modules. Consider an element $a \in A$ with $N_G(a) = 0$ and a cocycle $f \in Z^1(G, B)$, whose cohomology class we denote by $\bar{f} \in H^1(G, B)$. We set*

$$c = -\sum_{t \in G}(t \cdot a) \otimes f(t) \in A \otimes B.$$

Then

$$\bar{a}_0 \cup \bar{f} = \bar{c}^0 \in \widehat{H}^0(G, A \otimes B).$$

Proof We embed B into the induced module $B' = I_G(B)$ and we set $B'' = B'/B$. Recall that as B is a direct factor of B' as an abelian group, the sequence

$$0 \longrightarrow A \otimes B \longrightarrow A \otimes B' \longrightarrow A \otimes B'' \longrightarrow 0$$

remains exact.

As $H^1(G, B') = 0$, the image of the cocycle f in $Z^1(G, B')$ is a coboundary. Thus we obtain an element $b' \in B'$ such that $f(t) = t.b' - b'$ for all $t \in G$. If $t \in G$, the image of $t.b' - b'$ in B'' is zero as $t.b' - b' \in B$. In other words, the image b'' of b' in B'' belongs to $H^0(G, B'')$. By definition of the coboundary d, we then have

$$d((\bar{b}'')^0) = \bar{f} \in H^1(G, B),$$

hence

$$\bar{a}_0 \cup \bar{f} = -d(\bar{a}_0 \cup (\bar{b}'')^0)$$

by property (3) of Theorem 2.25 (more precisely, by the modified cohomology analogue of Proposition 2.23, b)

Remark 2.30 implies $\bar{a}_0 \cup (\bar{b}'')^0 = \overline{a \otimes (\bar{b}'')^0}$. As the coboundary

$$d : \widehat{H}^{-1}(G, A \otimes B'') \to \widehat{H}^0(G, A \otimes B)$$

is induced by the norm, setting $d = N_G(a \otimes b') = \sum_{t \in G}(t \cdot a) \otimes (t \cdot b')$ we obtain:

$$\bar{a}_0 \cup \bar{f} = -\bar{d}^0.$$

Now, we observe that as $N_G(a) = 0$, we have

$$d = \sum_{t \in G}(t \cdot a) \otimes (f(t) + b') = -c,$$

whence the result. □

Recall (Proposition 2.4) that the exact sequence

$$0 \to I_G \to \mathbf{Z}[G] \to \mathbf{Z} \to 0 \tag{2.12}$$

induces an isomorphism $d : \widehat{H}^{-2}(G, \mathbf{Z}) \to \widehat{H}^{-1}(G, I_G) = I_G/I_G^2$. Hence we have an isomorphism $s \mapsto \bar{s}$ from G^{ab} to $\widehat{H}^{-2}(G, \mathbf{Z})$, obtained by sending $s \in G$ to $d^{-1}(s-1)$, i.e., to the element $\bar{s}$ such that $d(\bar{s}) = \overline{(s-1)}_0$.

Proposition 2.32 *Let B be a G-module. Let $f : G \to B$ be a cocycle in $Z^1(G, B)$, whose class is $\bar{f} \in H^1(G, B)$. Let $s \in G$. Then, with the above notations, we have*

$$\bar{s} \cup \bar{f} = \overline{f(s)}_0 \in \widehat{H}^{-1}(G, B).$$

In particular, if G is abelian and the action of G on B is trivial, the cup-product $\bar{s} \cup \bar{f}$ is obtained by simply evaluating the morphism $f = \bar{f} : G \to B$ at s. The element of B thus obtained is then in the subgroup $\widehat{H}^{-1}(G, B)$ of B consisting of elements of norm zero.

Proof The exact sequence (2.12) splits as an exact sequence of abelian groups. Thus, it induces another exact sequence

$$0 \longrightarrow I_G \otimes B \longrightarrow \mathbf{Z}[G] \otimes B \longrightarrow B \longrightarrow 0$$

from which we obtain ($\mathbf{Z}[G] \otimes B$ being an induced module) an isomorphism

$$d : \widehat{H}^{-1}(G, B) \longrightarrow \widehat{H}^0(G, I_G \otimes B).$$

Thus it suffices to show that the images of $\bar{s} \cup \bar{f}$ and $\overline{f(s)}_0$ under this isomorphism, coincide. The description of d (cf. Proof of Theorem 2.6) yields (setting $x = \sum_{t \in G} t \otimes (t \cdot f(s))$):

$$d(\overline{f(s)}_0) = \bar{x}^0.$$

On the other hand, property (3) of Theorem 2.25 and Lemma 2.31 yield the equality

$$d(\bar{s} \cup \bar{f}) = \overline{(s-1)}_0 \cup \bar{f} = \bar{y}^0,$$

where $y := -\sum_{t \in G}(t \cdot (s-1)) \otimes f(t)$. Thus

$$y = \sum_{t \in G}(t - ts) \otimes f(t) = \sum_{t \in G} t \otimes f(t) - \sum_{t \in G} ts \otimes f(t)$$

$$= \sum_{t \in G} t \otimes f(t) - \sum_{t \in G} ts \otimes f(ts) + \sum_{t \in G} ts \otimes (t \cdot f(s))$$

because f is a 1-cocycle. As $t \mapsto ts$ is a bijection from G to G, the two first terms simplify and we finally find

$$y = \sum_{t \in G} ts \otimes (t \cdot f(s)).$$

Hence

$$x - y = \sum_{t \in G} t(1 - s) \otimes (t \cdot f(s)) = N_G((1 - s) \otimes f(s)),$$

which implies that $\bar{x}^0 = \bar{y}^0$, as desired. $\square$

2.7 Exercises

In all these exercises, G is a finite group.

Exercise 2.1 Let A be a G-module. We denote by A^* the G-module $\mathrm{Hom}_{\mathbf{Z}}(A, \mathbf{Q}/\mathbf{Z})$ (cf. Exercise 1.9). Show that for any G-module A and any integer $i \geqslant 0$, the group $H_i(G, A)^*$ is isomorphic to $H^i(G, A^*)$ (start with the case $i = 0$).

Exercise 2.2 Extend Shapiro's lemma to the modified cohomology $\widehat{H}^n$ for all $n \in \mathbf{Z}$.

Exercise 2.3 Let H be a subgroup of G. Let A be a G-module and B an H-submodule of A. For any $\sigma \in G$, we set $^\sigma H = \sigma H \sigma^{-1}$.
  (a) Show that the homomorphisms

$$^\sigma H \longrightarrow H, h \longmapsto \sigma^{-1} h \sigma; \quad B \longrightarrow \sigma B, b \longmapsto \sigma b$$

are compatible and induce isomorphisms

$$\sigma_* : H^n(H, B) \longrightarrow H^n(^\sigma H, \sigma B)$$

for all $n \geqslant 0$.
  (b) Show that if C is a G-module, we have for all $\alpha \in H^p(H, A), \gamma \in H^q(H, C)$, $\sigma \in G$, the formula

$$\sigma_*(\alpha \cup \gamma) = \sigma_* \alpha \cup \sigma_* \gamma.$$

Exercise 2.4 (*sequel to Exercise* 1.10) Let p be a prime number and G a finite p-group.
  (a) Show that $\widehat{H}^{-1}(G, \mathbf{Z}) = 0$.
  (b) Show that $\widehat{H}^{-2}(G, \mathbf{Z}/p\mathbf{Z}) = G^{\mathrm{ab}}/pG^{\mathrm{ab}}$.
  (c) With the notations of the Exercise 1.10, show that:

$$\mathrm{rg}(\widehat{H}^{-3}(G, \mathbf{Z})) = \mathrm{rg}(\widehat{H}^{-3}(G, \mathbf{Z}/p\mathbf{Z})) - \mathrm{rg}(\widehat{H}^{-2}(G, \mathbf{Z}/p\mathbf{Z})).$$

Exercise 2.5 Let G be a finite group. Let H be a subgroup of G. Let R be a system of representatives of right classes of G modulo H, with $1 \in R$. Set $H' = H - \{1\}$ and $R' = R - \{1\}$. Let $(n_{\rho,\tau})_{\rho \in R', \tau \in H'}$ be a family of integers.

(a) Show that if we have:

$$\sum_{\rho \in R', \tau \in H'} n_{\rho, \tau} (\tau - 1) \rho \in \mathbf{Z}[H],$$

then all the $n_{\rho, \tau}$ are zero.

(b) Deduce that if the family $(m_{\rho, \tau})_{\rho \in R, \tau \in H}$ of integers satisfies

$$\sum_{\rho \in R, \tau \in H} m_{\rho, \tau} (\tau - 1)(\rho - 1) \in I_H,$$

then this sum is in fact zero.

(c) Recover that $I_H I_G \cap I_H = I_H^2$ (observe that if $\tau, \tau' \in H$ and $\rho \in R$, then $(\tau - 1)(\tau' \rho - 1) = (\tau - 1)(\rho - 1) \bmod I_H^2$).

Exercise 2.6 (*after* [47], *Appendix to the third part, Lemma 4*) Let B be a G-module. We embed B into the induced module $B' = I_G(B)$ and we set $B'' = B'/B$. Let $u : G \times G \to B$ be a 2-cocycle with cohomology class $\bar{u} \in H^2(G, B)$. For each $b \in B^G$, we denote by $\bar{b}^0$ its image in $\widehat{H}^0(G, B)$. For each $s \in G$, we denote by $\bar{s}$ its image in $\widehat{H}^{-2}(G, \mathbf{Z})$ via the isomorphism $G^{\mathrm{ab}} \to \widehat{H}^{-2}(G, \mathbf{Z})$ (cf. Proposition 2.32).

(a) Show that there exists a 1-cochain $f' : G \to B'$ satisfying

$$u(x, y) = x \cdot f'(y) - f'(xy) + f'(x)$$

for all $x, y \in G$.

(b) Show that the composite $f'' : G \to B''$ is a 1-cocycle whose class $\overline{f}''$ satisfies $d(\overline{f}) = \bar{u}$, where $d : H^1(G, B'') \to H^2(G, B)$ is the coboundary map. Verify that $N_G(f'(s)) \in B$ for all $s \in G$.

(c) Using Proposition 2.32, deduce that for all $s \in G$, we have

$$\bar{s} \cup \bar{u} = \overline{N_G(f'(s))}^0 \in \widehat{H}^0(G, B).$$

(d) Let $s \in G$. We set $a = \sum_{t \in G} u(t, s)$. Show that

$$\bar{s} \cup \bar{u} = \bar{a}^0.$$

(You may use a) taking $(x, y) = (t, s)$, and sum the equality you obtain over $t \in G$.)

Chapter 3
p-Groups, the Tate–Nakayama Theorem

Let p be a prime number. Recall that a finite p-group is a group whose cardinality is a power of p. Such a group is nilpotent, in particular the centre and the abelianisation of a non-trivial p-group are non-trivial. A p-Sylow subgroup (abbreviated as p-Sylow) H of a finite group G is a subgroup which is a p-group whose index $[G : H]$ is prime to p. In this chapter we will see that the p-groups possess special cohomological properties which often allow us to study the cohomology of a finite group by reducing to that of its p-Sylows.

3.1 Cohomologically Trivial Modules

Lemma 3.1 *Let p be a prime number. Let G be a finite p-group and A a nonzero G-module. We assume that A is torsion and p-primary. Then $A^G \neq 0$.*

Proof We reduce ourselves to the case where A is finite by considering the G-module generated by a nonzero element of A. This module is finite as it is a torsion module of finite type over $\mathbf{Z}$.

Let $A_1, \ldots, A_r$ be the orbits other than the elements of A^G for the action of G on A. The cardinality of each A_i is $[G : G_i]$, where G_i is the stabiliser of an element of A_i. This cardinality is therefore divisible by p (recall that $\#A_i > 1$). The class equation

$$\#A = \#A^G + \sum_i \#A_i$$

then shows that A and A^G have the same cardinality modulo p and their cardinality is a power of p. It follows that A^G can not be of cardinality 1. $\qquad\square$

Lemma 3.2 *Let G be a finite group and let A be a G-module. Let p be a prime number and let H be a subgroup of G. Then if p does not divide $[G : H]$, the map*

© Springer Nature Switzerland AG 2020
D. Harari, *Galois Cohomology and Class Field Theory*, Universitext,
https://doi.org/10.1007/978-3-030-43901-9_3

Res : $H^q(G, A)\{p\} \rightarrow H^q(H, A)$ is injective for all $q > 0$. The same result holds for all $q \in \mathbf{Z}$ if we replace the groups H^q by the modified groups $\widehat{H}^q$.

Proof This follows from the formula $(\mathrm{Cores} \circ \mathrm{Res})(x) = [G : H] \cdot x$ for all x in $H^q(G, A)$ (Theorem 1.48). The argument with $\widehat{H}^q$ is identical by Theorem 2.9. $\square$

Lemma 3.3 Let G be a finite group. Let A and B be G-modules.
   (a) Assume that B is induced. Then the G-module $\mathrm{Hom}(A, B) := \mathrm{Hom}_{\mathbf{Z}}(A, B)$ is induced.
   (b) Assume that A is a projective G-module. Then $A \otimes B$ is a relatively injective G-module.

Recall (Example 1.3, d) that the action of G on $\mathrm{Hom}(A, B)$ is given by $(g \cdot f)(x) := g \cdot f(g^{-1} \cdot x)$ for all $g \in G$, $f \in \mathrm{Hom}(A, B)$, $x \in A$.

Proof (a) (See also [47], Chap. IX, Prop. 1 for a more general statement.) To say that B is induced means (Remark 1.6, b) that it is a direct sum of the $g \cdot X$ for $g \in G$, where X is a fixed subgroup of B. Then $\mathrm{Hom}(A, B)$ is a direct sum of the $\mathrm{Hom}(A, g \cdot X) = g \cdot \mathrm{Hom}(A, X)$ (indeed, $\mathrm{Hom}(A, g \cdot X)$ is simply the subgroup of $\mathrm{Hom}(A, B)$ consisting of the f whose image is contained in $g \cdot X$, which is equivalent to $\mathrm{Im}(g^{-1} \cdot f) \subset X$), hence is induced as it is the direct sum of the $g \cdot \mathrm{Hom}(A, X)$ ($\mathrm{Hom}(A, X)$ being a subgroup of $\mathrm{Hom}(A, B)$)).
(b) follows from the fact that if A is projective, it is a direct factor of a free $\mathbf{Z}[G]$-module $\bigoplus_I \mathbf{Z}[G]$, and from the fact that $\otimes$ commutes with direct sums (Appendix, Proposition A.30, b). $\square$

We will also need the following notation (cf. Proposition 1.10, Remarks 1.11 and 2.24): let G be a finite group and A be a G-module. We have (Proposition 1.10) a natural embedding $A \hookrightarrow I_G(A)$. We denote by A_1 the quotient $I_G(A)/A$. More generally, we define by induction $A_q = (A_{q-1})_1$ for all $q > 0$. Similarly, we can write A as a quotient of $I_G(A)$ via the surjective homomorphism $f \mapsto \sum_{g \in G} g \cdot f(g^{-1})$. Let us then call A_{-1} the kernel of $I_G(A) \rightarrow A$ and we set $A_q = (A_{q+1})_{-1}$ if $q < 0$. Then we have by shifting

$$\widehat{H}^q(G, A) = \widehat{H}^{q-r}(G, A_r) \tag{3.1}$$

for all $q, r \in \mathbf{Z}$ by Corollary 2.7.

Definition 3.4 Let G be a finite group. We say that a G-module A is *cohomologically trivial* if for all $n > 0$ and every subgroup H of G, we have $H^n(H, A) = 0$.

In particular, such a G-module is also a cohomologically trivial H-module for any subgroup H of G. Note that an induced G-module is cohomologically trivial: this follows from the fact that A is also induced as an H-module (Remark 1.35).

Lemma 3.5 Let G_p be a p-Sylow of G. A G-module A is cohomologically trivial if and only if A is a cohomologically trivial G_p-module for every prime number p.

Note that if G'_p is another p-Sylow of G, then A is cohomologically trivial as G_p-module if and only if it is cohomologically trivial as G'_p-module.

Indeed, G_p and G'_p are conjugated (say by $t \in G$), which gives for any subgroup H' of G'_p a bijective homomorphism from A to A (defined by $x \mapsto t^{-1} \cdot x$) compatible with the isomorphism $g \mapsto tgt^{-1}$ from H' to $H := tH't^{-1}$. We thus obtain an isomorphism from $H^n(H, A)$ to $H^n(H', A)$ (see also Exercise 2.3). We could of course limit ourselves to the primes that divide the cardinality of G since for the others, the group G_p is trivial.

Proof Let H be a subgroup of G. Let H_p be a p-Sylow of H. Then H_p is contained in a p-Sylow of G, which we can assume to be G_p by the above remark. The result then follows from the fact that for $n \geq 1$, the restriction $H^n(H, A)\{p\} \to H^n(H_p, A)$ is injective (Lemma 3.2) as the index $[H : H_p]$ is prime to p. $\square$

Theorem 3.6 *Let p be a prime number. Let G be a finite p-group and A a p-torsion G-module. Assume there exists $q \in \mathbf{Z}$ such that $\widehat{H}^q(G, A) = 0$. Then A is an induced G-module (in particular it is cohomologically trivial).*

(The proof will even show that A is a free $\mathbf{F}_p[G]$ module.)

Proof We start by finding an induced G-module V such that V^G is isomorphic to A^G. To do this, set $\Lambda = \mathbf{F}_p[G]$, and choose a basis I of the $\mathbf{F}_p$-vector space A^G. Set $V = \bigoplus_I \Lambda$, then V is an induced G-module (as a direct sum of induced G-modules). As $\Lambda^G = \mathbf{F}_p$, we obtain an isomorphism $j_G : A^G \simeq V^G$. We now try to extend this isomorphism to a G-morphism from A to V.

As the functor $\mathrm{Hom}(., V)$ is exact in the category of $\mathbf{F}_p$-vector spaces, we have an exact sequence of G-modules

$$0 \longrightarrow \mathrm{Hom}(A/A^G, V) \longrightarrow \mathrm{Hom}(A, V) \longrightarrow \mathrm{Hom}(A^G, V) \longrightarrow 0$$

and on the other hand the fact that V is induced implies, by Lemma 3.3, that the G-module $M := \mathrm{Hom}(A/A^G, V)$ is induced. Thus we have $H^1(G, M) = 0$ hence a surjection:

$$u : \mathrm{Hom}_G(A, V) \longrightarrow \mathrm{Hom}_G(A^G, V) = \mathrm{Hom}_G(A^G, V^G).$$

It follows that j_G extends to a G-homomorphism $j : A \to V$.

Observe that the G-module $\mathrm{Ker}\, j$ satisfies $(\mathrm{Ker}\, j)^G = 0$ since the restriction of j to A^G is an isomorphism from A^G to V^G. As G is a p-group, this implies $\mathrm{Ker}\, j = 0$ by Lemma 3.1. Hence j is injective. Let C be its cokernel. The long exact cohomology sequence

$$0 \longrightarrow A^G \longrightarrow V^G \longrightarrow C^G \longrightarrow H^1(G, A)$$

yields the following implications

$$H^1(G, A) = 0 \implies C^G = 0 \implies C = 0$$

since j induces an isomorphism $A^G \simeq V^G$ (the last implication comes again from Lemma 3.1). We have shown that if $H^1(G, A) = 0$, then $A \simeq V$ is an induced G-module.

Let $q \in \mathbf{Z}$ such that $\widehat{H}^q(G, A) = 0$. By shifting, we will reduce ourselves to the case $q = 1$. We have, by formula (3.1):

$$\widehat{H}^q(G, A) = H^1(G, A_{q-1}).$$

This shows that $H^1(G, A_{q-1}) = 0$. By what precedes, this implies that A_{q-1} (which is again p-torsion) is induced. But then we have

$$H^1(G, A) = \widehat{H}^{2-q}(G, A_{q-1}) = 0$$

and A is induced by the case $q = 1$.   □

Theorem 3.7 *Let G be a finite group. Let A be a G-module.*

(a) *Assume that for every prime number p, there exists an integer $q \in \mathbf{Z}$ (which may depend on p) such that*

$$\widehat{H}^q(G_p, A) = \widehat{H}^{q+1}(G_p, A) = 0,$$

where G_p is a p-Sylow of G.
Then A is cohomologically trivial. Suppose moreover that A is a free $\mathbf{Z}$-module, then A is a projective $\mathbf{Z}[G]$-module (hence relatively injective).

(b) *Assume that A is cohomologically trivial; then $\widehat{H}^q(H, A) = 0$ for any subgroup H of G and any $q \in \mathbf{Z}$. Moreover, there is an exact sequence*

$$0 \longrightarrow R \longrightarrow F \longrightarrow A \longrightarrow 0$$

of G-modules with F free over $\mathbf{Z}[G]$ and R projective over $\mathbf{Z}[G]$.

Proof We start by writing A as a quotient of a free $\mathbf{Z}[G]$-module F, hence we have an exact sequence of G-modules

$$0 \longrightarrow R \longrightarrow F \longrightarrow A \longrightarrow 0.$$

Fix a prime number p and let us first place ourselves in the situation of (a), i.e., $\widehat{H}^q(G_p, A) = \widehat{H}^{q+1}(G_p, A) = 0$. Then as F is in particular relatively injective, we obtain

$$\widehat{H}^{q+1}(G_p, R) = \widehat{H}^{q+2}(G_p, R) = 0 \tag{3.2}$$

which, using the exact sequence

$$0 \longrightarrow R \xrightarrow{\;\cdot p\;} R \longrightarrow R/pR \longrightarrow 0 \tag{3.3}$$

(note that R is torsion-free since it is a submodule of F) implies the equality $\widehat{H}^{q+1}(G_p, R/pR) = 0$. Theorem 3.6 then implies that R/pR is an induced G_p-module.

We start with the case where A is assumed to be free over $\mathbf{Z}$. We will then show that A is a direct factor of F (hence a projective $\mathbf{Z}[G]$-module). Let M be the G-module $M := \mathrm{Hom}(A, R)$.

Lemma 3.8 *We have $H^1(G, M) = 0$.*

Proof The exact sequence (3.3) and the fact that A is free over $\mathbf{Z}$ yield an isomorphism of G-modules
$$M/pM \simeq \mathrm{Hom}(A, R/pR).$$

This shows that M/pM is an induced G_p-module by Lemma 3.3 since R/pR is an induced G_p-module. It follows that $H^1(G_p, M)[p] = 0$ by Theorem 2.6 and the exact sequence of modified cohomology
$$0 = \widehat{H}^0(G_p, M/pM) \longrightarrow H^1(G_p, M) \xrightarrow{\;\cdot p\;} H^1(G_p, M)$$

associated to the exact sequence
$$0 \longrightarrow M \xrightarrow{\;\cdot p\;} M \longrightarrow M/pM \longrightarrow 0.$$

Hence we have $H^1(G, M)\{p\} = 0$ (recall that the restriction $H^1(G, M)\{p\} \to H^1(G_p, M)$ is injective by Lemma 3.2). This holds for all p, thus we obtain $H^1(G, M) = 0$. $\qquad\qquad\square$

Let us now proceed with the proof of the fact that A is a direct factor of F, keeping the assumption that A is free over $\mathbf{Z}$.

This assumption implies that we have an exact sequence of G-modules
$$0 \longrightarrow M = \mathrm{Hom}(A, R) \longrightarrow \mathrm{Hom}(A, F) \longrightarrow \mathrm{Hom}(A, A) \longrightarrow 0$$

and $H^1(G, M) = 0$ implies that $\mathrm{Hom}_G(A, F)$ surjects onto $\mathrm{Hom}_G(A, A)$, which allows us to obtain (by considering Id_A) a section of the surjective homomorphism of G-modules $F \to A$. Thus F is isomorphic (*as G-module*) to the direct sum $A \oplus R$, as desired. Thus we have proved a) in the special case where A is free over $\mathbf{Z}$.

Let us now prove (a) in general. What precedes applies to R by (3.2) as R is free over $\mathbf{Z}$ as a submodule of F. Hence we obtain that R is projective (in particular relatively injective), hence cohomologically trivial. Since this also holds for F (which is induced), the G-module $A = F/R$ is also cohomologically trivial (using the long exact sequence) and (a) follows.

Let us show (b). Let A be a cohomologically trivial G-module. A fortiori, A satisfies the assumptions of (a) (e.g. for $q = 1$). Let us choose a surjection $F \to A$ whose kernel is R with F free over $\mathbf{Z}[G]$. We have just seen that R is then projective (as a G-module, hence also as an H-module for any subgroup H of G) hence for any $q \in \mathbf{Z}$, we obtain

$$\widehat{H}^q(H, A) = \widehat{H}^{q+1}(H, R) = 0$$

for any subgroup H of G, which proves (b).   □

Remark 3.9 By Theorem 3.7 (b), we could have defined cohomological triviality of a G-module A (for G finite) by the property of vanishing (for any subgroup H of G) of all Tate cohomology groups $\widehat{H}^q(H, A)$.

3.2   The Tate–Nakayama Theorem

In this paragraph, G is a finite group and for every prime number p, we fix a p-Sylow G_p of G.

Lemma 3.10 *Let A be a cohomologically trivial G-module. Let B be a torsion-free G-module. Then the G-module $A \otimes B$ is cohomologically trivial.*

Proof By Theorem 3.7, we have an exact sequence

$$0 \longrightarrow R \longrightarrow F \longrightarrow A \longrightarrow 0$$

with F free over $\mathbf{Z}[G]$ and R a direct factor of a free $\mathbf{Z}[G]$-module. Then the G-modules $F \otimes B$ and $R \otimes B$ are both relatively injective (Lemma 3.3, b). They are therefore cohomologically trivial. As B is torsion-free, the sequence

$$0 \longrightarrow R \otimes B \longrightarrow F \otimes B \longrightarrow A \otimes B \longrightarrow 0$$

remains exact. Using the long exact sequence, we immediately deduce that the G-module $A \otimes B$ is also cohomologically trivial.   □

Remark 3.11 The same proof shows that it is sufficient to assume that $\mathrm{Tor}_{\mathbf{Z}}(A, B) = 0$, where $\mathrm{Tor}_{\mathbf{Z}}(., B)$ is the first derived functor of the left functor $. \otimes_{\mathbf{Z}} B$ in the category of modules over the ring $\mathbf{Z}$ (cf. Appendix, Proposition A.56).

Proposition 3.12 *Let A and A' be two G-modules. Let $f : A' \to A$ be a G-homomorphism. Assume that for every prime p, there exists an integer n_p such that the homomorphism*

$$f_*^i : \widehat{H}^i(G_p, A') \longrightarrow \widehat{H}^i(G_p, A)$$

is surjective for $i = n_p$, bijective for $i = n_p + 1$, and injective for $i = n_p + 2$. Let B be a torsion-free G-module. Then for any subgroup H of G, the homomorphism

$$\widehat{H}^i(H, A' \otimes B) \longrightarrow \widehat{H}^i(H, A \otimes B)$$

induced by $f \otimes 1$ is bijective for all $i \in \mathbf{Z}$. In particular, the homomorphism $\widehat{H}^i(H, A') \to \widehat{H}^i(H, A)$ induced by f is bijective for all $i \in \mathbf{Z}$.

Proof We start with the case where f is injective. Let A'' be its cokernel. The long exact sequence associated to the exact sequence

$$0 \longrightarrow A' \longrightarrow A \longrightarrow A'' \longrightarrow 0$$

and the assumption made on the f_*^i yields

$$\widehat{H}^{n_p}(G_p, A'') = \widehat{H}^{n_p+1}(G_p, A'') = 0$$

for every prime p.

By Theorem 3.7, the G-module A'' is cohomologically trivial. By Lemma 3.10, the G-module $A'' \otimes B$ is also cohomologically trivial. As B be is torsion-free, the sequence

$$0 \longrightarrow A' \otimes B \longrightarrow A \otimes B \longrightarrow A'' \otimes B \longrightarrow 0$$

is exact. This implies that $\widehat{H}^q(H, A' \otimes B) \to \widehat{H}^q(H, A \otimes B)$ is bijective for all $q \in \mathbf{Z}$ and all subgroups H of G, as desired.

The general case reduces to the special case where f is injective by the following procedure: embed A' into the induced module $\overline{A}' := I_G(A')$ and set $A^* = A \oplus \overline{A}'$. We then obtain (via f and the embedding $j : A' \to \overline{A}'$) an injection $\theta = (f, j) : A' \to A^*$. As $\overline{A}'$ and $\overline{A}' \otimes B$ are cohomologically trivial, we have $\widehat{H}^q(H, A) = \widehat{H}^q(H, A^*)$ and $\widehat{H}^q(H, A \otimes B) = \widehat{H}^q(H, A^* \otimes B)$. This allows us to reduce ourselves to the previous case by replacing f by θ. □

Note that the Proposition 3.12 remains valid provided $\mathrm{Tor}_{\mathbf{Z}}(A, B) = \mathrm{Tor}_{\mathbf{Z}}(A', B) = 0$ with a slightly different proof (the above argument goes not go through directly since we can not a priori deduce that $\mathrm{Tor}_{\mathbf{Z}}(A'', B) = 0$ from the assumptions). See Exercise 3.6.

Proposition 3.13 *Let A, B, C be three G-modules. Let $\varphi : A \times B \to C$ be a bilinear map compatible with the action of G. Let $q \in \mathbf{Z}$ and $a \in \widehat{H}^q(G, A)$. For a subgroup H of G and a G-module D, we denote by a_H the restriction of a to H and by*

$$f(n, H, D) : \widehat{H}^n(H, B \otimes D) \longrightarrow \widehat{H}^{n+q}(H, C \otimes D)$$

the homomorphism defined by the cup-product with a_H (with respect to the bilinear map induced by φ).

Assume that for all prime numbers p, there is an integer n_p such that $f(n, G_p, \mathbf{Z})$ is surjective for $n = n_p$, bijective for $n = n_p + 1$, and injective for $n = n_p + 2$. Then $f(n, H, D)$ is bijective for all $n \in \mathbf{Z}$, all subgroups H of G, and all torsion-free G-modules D (here again $\mathrm{Tor}_{\mathbf{Z}}(B, D) = \mathrm{Tor}_{\mathbf{Z}}(C, D) = 0$ would suffice).

Proof Consider first the case $q = 0$. Then $a \in \widehat{H}^0(G, A) = A^G/N_G A$ comes from an element (again denoted a) of A^G. Let $f : B \to C$ be the G-homomorphism defined by $f(b) = \varphi(a, b)$. Then $f_*^n : \widehat{H}^n(G_p, B) \to \widehat{H}^n(G_p, C)$ is simply $f(n, G_p, \mathbf{Z})$ and the Proposition 3.12 implies that $f(n, H, D)$ is bijective as $f(n, H, D)$ is then the homomorphism induced by $f \otimes \text{Id} : B \otimes D \to C \otimes D$.

The case of an arbitrary q is dealt with by shifting. Let us for example show how to pass from $q - 1$ to q by embedding A into the induced module $\overline{A} = I_G(A)$ (to go in the other direction, we write A as the quotient of the induced module $\overline{A}$) and C into $\overline{C} = I_G(C)$. Set $A_1 = \overline{A}/A$ and $C_1 = \overline{C}/C$. This gives a bilinear map $\varphi_1 : A_1 \times B \to C_1$ induced by the bilinear map

$$\overline{\varphi} : \overline{A} \times B \longrightarrow \overline{C}, \quad (f, b) \longmapsto (g \longmapsto \varphi(f(g), g \cdot b)), \quad \forall f \in \overline{A}, b \in B, g \in G.$$

We can then write $a = \delta(a_1)$ with $a_1 \in \widehat{H}^{q-1}(G, A_1)$ and a_1 defines, by cup-product, homomorphisms

$$f_1(n, H, D) : \widehat{H}^n(H, B \otimes D) \longrightarrow \widehat{H}^{n+q-1}(H, C_1 \otimes D).$$

As $\overline{C} \times D$ is again cohomologically trivial (Lemma 3.10), the coboundary $\delta : \widehat{H}^{n+q-1}(H, C_1 \otimes D) \to \widehat{H}^{n+q}(H, C \otimes D)$ is an isomorphism. But, in view of the compatibility of cup-products with the coboundaries (Proposition 2.23), we know that $f(n, H, D)$ is obtained (up to a sign) by composing f_1 with δ. If the desired result holds for a_1, it holds therefore for a, hence the result follows by induction on q.   □

Theorem 3.14 (Tate–Nakayama) *Let A be a G-module. We consider an element a of $H^2(G, A)$. Assume that for all primes p, the following assumptions hold:*

(a) We have $H^1(G_p, A) = 0$.
(b) The group $H^2(G_p, A)$ is of cardinality $m_p := \#G_p$ and is generated by the restriction a_p of a to $H^2(G_p, A)$.

Then for each torsion-free G-module D and each subgroup H of G, the cup-product by $a_H = \text{Res}_H(a) \in H^2(H, A)$ induces isomorphisms

$$\widehat{H}^n(H, D) \longrightarrow \widehat{H}^{n+2}(H, A \otimes D)$$

for all $n \in \mathbf{Z}$. In particular, the cup-product by a_H induces isomorphisms

$$\widehat{H}^n(H, \mathbf{Z}) \longrightarrow \widehat{H}^{n+2}(H, A).$$

Proof We apply the previous proposition with $B = \mathbf{Z}$, $C = A$, $q = 2$, taking for $\varphi : A \otimes \mathbf{Z} \to A$ the obvious map. We choose $n_p = -1$. The cup-product

$$\widehat{H}^n(G_p, \mathbf{Z}) \longrightarrow \widehat{H}^{n+2}(G_p, A)$$

induced by a_p is surjective for $n = -1$ (using assumption a). For $n = 0$, it is the map

$$\mathbf{Z}/m_p\mathbf{Z} \longrightarrow H^2(G_p, A)$$

that sends the canonical generator of $\mathbf{Z}/m_p\mathbf{Z}$ to a_p. The assumption b) implies that this map is bijective. For $n = 1$, the group $H^1(G_p, \mathbf{Z})$ is zero hence the cup-product is injective. □

The special case $\widehat{H}^n(H, \mathbf{Z}) \simeq \widehat{H}^{n+2}(H, A)$ is due to Tate. Later we will apply it with $n = -2$, $G = \mathrm{Gal}(L/K)$ and $A = L^*$ for a finite Galois extension L of a p-adic field K (Sect. 9.1). We will deduce an isomorphism from $H^{-2}(G, \mathbf{Z}) = G^{\mathrm{ab}}$ to $\widehat{H}^0(G, L^*) = K^*/NL^*$, once we have verified the assumptions of the theorem in this setting. We will also see (Sect. 15.1) that an analogous statement is valid in the global case, replacing L^* by *the idèle class group* of L.

3.3 Exercises

Exercise 3.1 Let p be a prime number. Let G be a finite p-group. Let A be a p-torsion G-module.

(a) Show that if $H_0(G, A) = 0$, then $A = 0$ (consider the G-module $A' := \mathrm{Hom}_{\mathbf{Z}}(A, \mathbf{Z}/p)$ and apply Exercise 2.1).
(b) Does this result still hold if we only assume that A is p-primary, i.e., that each element of A is of order a power of p? You may consider $G = \mathbf{Z}/2$ and make its non-trivial element act by multiplication by -1 on $\mathbf{Q}_2/\mathbf{Z}_2$.

Exercise 3.2 Show that the conclusion of the Lemma 3.1 can become false if one removes either the assumption that G is a p-group, or the assumption that A is a p-primary torsion group.

Exercise 3.3 Give an example of a finite group G and of a G-module A such that there exists a $q \in \mathbf{Z}$ satisfying $\widehat{H}^q(G, A) = \widehat{H}^{q+1}(G, A) = 0$, but A is not cohomologically trivial (you may take $G = \mathbf{Z}/6\mathbf{Z}$ and make it act on $\mathbf{Z}/3\mathbf{Z}$ by a well-chosen action).

Exercise 3.4 Let G be a finite group. Let

$$0 \longrightarrow X_1 \longrightarrow X_2 \longrightarrow \cdots \longrightarrow X_n \longrightarrow 0$$

be an exact sequence of G-modules. Let $j \in \{1, \ldots, n\}$. Show that if X_i is cohomologically trivial for all $i \neq j$, then so is X_j.

Exercise 3.5 Let G be a finite group. Let A be a cohomologically trivial G-module. Show that if N is a torsion-free G-module, then $\mathrm{Hom}(N, A)$ is again a cohomologically trivial G-module.

Exercise 3.6 (*suggested by J. Riou*) Extend Proposition 3.12 to the case where we only assume $\text{Tor}_{\mathbf{Z}}(A, B) = \text{Tor}_{\mathbf{Z}}(A', B) = 0$, and not that B is necessarily torsion-free. You may first assume that f is surjective and show (using Proposition A.56 of the appendix) that if we set $K = \text{Ker}\, f$, the G-modules K and $K \otimes B$ are cohomologically trivial. Then, reduce to this special case by replacing A by $A \oplus F$, where F is a free $\mathbf{Z}[G]$ module.

Chapter 4
Cohomology of Profinite Groups

When studying Galois cohomology of a field k, one often needs to work with the absolute Galois group $\Gamma_k := \mathrm{Gal}(\bar{k}/k)$ (where $\bar{k}$ is the separable closure of k) and not that of a finite extension. We will see that one can define cohomology of a profinite group by a limit procedure from that of its finite quotients (which in the case of Γ_k correspond to finite Galois extensions of k).

4.1 Basic Facts About Profinite Groups

Definition 4.1 A topological group G is called *profinite* if it is a projective limit of finite groups (each endowed with the discrete topology).

Recall that if (G_i) is a projective system of topological spaces, the topology on the projective limit is that induced by the product topology on $\prod G_i$. This product topology, by definition, admits as a basis of open sets, the sets of the form $\prod U_i$, where each U_i is open in G_i and almost all the U_i (i.e., all but a finite number) are equal to G_i.

In addition, in any topological group, any open subgroup is closed (as complement of the union of the other cosets). The same argument shows that a subgroup of finite index is closed if and only if it is open. On the other hand, a subgroup of a topological group is open if and only if it is a neighbourhood of the identity element. This implies that a subgroup containing an open subgroup is itself open.

Let G be a profinite group whose identity element we denote by 1. Then 1 admits a basis $\{G_i\}$ of neighbourhoods which are open, normal, finite index subgroups and G identifies with the projective limit of the G/G_i.

A topological group G is profinite if and only if it is compact and totally disconnected ([42], Prop. 1.1.3). Changing the projective system defining G does not affect its structure as topological group.

© Springer Nature Switzerland AG 2020
D. Harari, *Galois Cohomology and Class Field Theory*, Universitext,
https://doi.org/10.1007/978-3-030-43901-9_4

Profinite groups form a category where morphisms are the *continuous* group morphisms. On the other hand, if H is a closed subgroup of a profinite group G, it is profinite itself and the continuous map between compact topological spaces $\pi : G \to G/H$ admits a *section* (i.e., a continuous map $s : G/H \to G$ such that $\pi \circ s = \mathrm{Id}_{G/H}$), cf. [49], Part. I, Sect. 1, Prop. 1. It follows:

Proposition 4.2 *Let H be a closed subgroup of a profinite group G. Then the topological space G is homeomorphic to the product space $H \times G/H$. In particular, every continuous map on H extends to a continuous map on G.*

Proof Let $s : G/H \longrightarrow G$ be a section, then the map

$$H \times G/H \longrightarrow G, \quad (h, g_1) \longmapsto s(g_1)h$$

is a homeomorphism whose inverse is $g \mapsto (s(\pi(g))^{-1}g, \pi(g))$, where $\pi : G \to G/H$ is the canonical surjection. $\qquad\qquad\square$

Below are some examples of profinite groups.

Example 4.3 (a) All finite groups are profinite (!).

(b) The absolute Galois group $\Gamma_k = \mathrm{Gal}(\bar{k}/k)$ of a field k is profinite: it is by definition the group of k-automorphisms of the field $\bar{k}$, and identifies with the projective limit of the $\mathrm{Gal}(L/k)$ when L ranges through the finite Galois extensions of k contained in $\bar{k}$.

(c) If M is a torsion discrete abelian group, its *Pontryagin dual* $M^* := \mathrm{Hom}(M, \mathbf{Q}/\mathbf{Z})$ is a profinite group when endowed with the simple convergence topology (that is, the "open-compact" topology). Indeed, M is the inductive limit of its finite subgroups N, hence M^* is the projective limit of finite groups N^*. One can show that $M \mapsto M^*$ induces an anti-equivalence of categories between discrete torsion abelian groups and profinite abelian groups. This is a special case of Pontryagin duality for locally compact abelian groups; cf. [42], Th. 1.1.11.

Note that the dual of an abelian profinite group G is by definition the discrete torsion group $\mathrm{Hom}_c(G, \mathbf{Q}/\mathbf{Z})$ consisting of *continuous* homomorphisms from G to $\mathbf{Q}/\mathbf{Z}$. This duality does not work very well for a discrete group which is not torsion. For example for $M = \mathbf{Z}$ we have $M^* = \mathbf{Q}/\mathbf{Z}$ and $M^{**} = \widehat{\mathbf{Z}}$, the profinite completion of $\mathbf{Z}$ (it is the projective limit of the $\mathbf{Z}/n\mathbf{Z}$).

(d) Every *closed* subgroup of a profinite group is profinite. Similarly, every quotient by a closed normal subgroup is profinite (see Lemma 4.7 for a slightly more precise statement).

(e) The additive group of the ring of integers $\mathcal{O}_K$ of a complete local field K (= field complete for a discrete valuation with finite residue field) is profinite. For example $\mathbf{Z}_p$ is profinite. Same for the multiplicative group $\mathcal{O}_K^*$.

(f) Let K be a p-adic field, that is, a finite extension of $\mathbf{Q}_p$ (p-adic fields are local fields of characteristic zero, see recollection of facts in Chap. 7). Let A be an *abelian variety* (:= a connected *projective* algebraic group; the case of dimension 1 corresponds to elliptic curves) over K. The group $A(K)$ of K-points of A is profinite.

In the last three examples, the easiest way to see that the groups in question are profinite is the characterisation "compact+ totally disconnected".

Remark 4.4 Note that in a profinite group G, open subgroups are of finite index but the converse is generally false even when G is abelian. One example is the additive group $\mathcal{O}_K = \mathbf{F}_q[[t]]$ of the ring of integers of the field of Laurent series $K = \mathbf{F}_q((t))$ over a finite field. To see this, it is enough to take the kernel of a $\mathbf{F}_q$-linear form which is not continuous on $\mathcal{O}_K$. Another example is when G is the multiplicative group $\mathcal{O}_K^*$ (see Exercise 11.3) or the Galois group of the maximal abelian extension of $\mathbf{Q}$ (that will be studied in Chap. 15).

It turns out that for a profinite group, it is possible to define the index of a subgroup even when the subgroup in question is not of finite index. This is done using the following notion.

Definition 4.5 A *supernatural number* is a formal product $\prod_p p^{n_p}$, where p ranges through the set of prime numbers and $n_p \in \mathbf{N} \cup \{+\infty\}$. We define in an evident manner the gcd, the lcm and the product for an arbitrary family of supernatural numbers.

Definition 4.6 Let G be a profinite group. Let H be a closed subgroup of G. The *index* of H in G (denotes by $[G : H]$) is the supernatural number defined as the lcm of indexes $[G/U : H/(H \cap U)]$ (the latter are usual integers) when U ranges through the set of open normal subgroups of G. The *order* of a profinite group is the supernatural number $[G : \{1\}]$.

It is not entirely obvious that[1] the usual notion of "finite index" is the same as "finite index" in the sense of the previous definition. The following lemma shows that this is indeed the case.

Lemma 4.7 *Let H be a closed subgroup of a profinite group G. Let (U_i) be a family of open normal subgroups of G. Set $G_i = G/U_i$. Let $H_i = H/(H \cap U_i)$ be the image of H in G_i. Then:*

(a) The group H identifies with $\varprojlim H_i$, and the set G/H with $\varprojlim(G_i/H_i)$. Moreover, the transition arrows (associated with the inclusions $U_j \subset U_i$) are surjective in both cases.

(b) The subgroup H is of finite index (in the usual sense) if and only if there is an index j such that for all $U_i \subset U_j$, we have $G_i/H_i = G_j/H_j$. In particular, this definition is equivalent to $[G : H]$ (defined as in Definition 4.6) being a natural number. In this case $[G : H]$ is the usual index of H. Hence H is open if and only if the supernatural number $[G : H]$ is a natural number.

Proof (a) The canonical map $H \to \varprojlim H_i$ has a trivial kernel since $\bigcap U_i = \{1\}$ (the U_i form a basis of open neighbourhoods of $\{1\}$). It is also surjective as every element of $\varprojlim H_i$ comes from a $g \in G$ (by definition of a profinite group). But, if

---

[1]Thanks to Miaofen Chen for pointing out this difficulty, and to J. Riou for suggesting the simple method relying on Lemma 4.7 to deal with it. My initial argument was more complicated.

$g \notin H$, there exists an index i such that $gU_i \cap H = \emptyset$ (because H is closed), which contradicts the fact that the image of g in G/U_i is in H_i. Finally we have $H \simeq \varprojlim H_i$. The other isomorphism is obtained in exactly the same way and the surjectivity of the transition maps is in both cases evident.

(b) We observe that if $U_j \subset U_i$, then the cardinality of the finite set G_j/U_j is at least equal to that of G_i/U_i. The first assertion follows from the identification of G/H with $\varprojlim (G_i/H_i)$, combined with the surjectivity of transition maps. The other assertions are trivial.   □

Definition 4.8 We say that G is a *pro-p-group* if its order is a power of p (it comes down to saying that G is the projective limit of finite p-groups). A *p-Sylow subgroup* (or *p-Sylow* for short) of a profinite group G is a closed subgroup H of G which is a pro-p-group and such that the index $[G : H]$ is prime to p.

The proposition below follows from the properties of indexes and of Sylow subgroups. Its proof uses the following easy lemma ([6], Sect. 9.6, Prop. 8).

Lemma 4.9 *Every projective limit of a sequence $(X_n)_{n \in \mathbb{N}}$ of non-empty finite sets is itself non-empty. Same holds for the projective limit of a sequence of compact non-empty topological spaces.*

Proposition 4.10 *(a) Let $K \subset H \subset G$ be profinite groups. Then we have an equality of supernatural numbers:*

$$[G : K] = [G : H] \cdot [H : K].$$

(b) Let G be a profinite group. For any prime number p, the group G admits p-Sylows, and they are pairwise conjugate.

(c) Let G be a profinite group. Then any pro-p-subgroup of G is contained in a p-Sylow.

(d) Let G be a profinite abelian group. Then it is a direct product of its p-Sylows.

Proof (a) Let U be an open normal subgroup of G. For any closed (hence profinite) subgroup G' of G, let $G'_U = G'/(G' \cap U)$ be its image by the canonical surjection $G \to G/U$. It is a finite group. By multiplicativity of the indexes for a finite group, we have

$$[G_U : K_U] = [G_U : H_U] \cdot [H_U : K_U].$$

As any normal open subgroup of H contains a subgroup of the form $H \cap U$ with U a normal subgroup of G, we obtain the formula by taking the lcm over these Us.

(b) Let $\mathcal{D}$ be a family of open normal subgroups of G.

For all $U \in \mathcal{D}$, the set $S(U)$ of p-Sylows of a finite group G/U is finite and nonempty by the first classical Sylow theorem. By Lemma 4.9, the projective limit of the $S(U)$ for U in $\mathcal{D}$ is then non-empty, which yields a projective system $(H_U)_{U \in \mathcal{D}}$ where H_U is a p-Sylow of G/U. We see immediately that $\varprojlim_{U \in \mathcal{D}} H_U$ is a p-Sylow of G.

Let H and K be two p-Sylow subgroups of G. The set $C(U)$ of the $c \in G/U$ such that $c H_U c^{-1} = K_U$ is finite and non-empty by the second classical Sylow theorem (since H_U and K_U are p-Sylows of the finite group H_U). Hence we have $\varprojlim_{U \in \mathcal{D}} C(U) \neq \varnothing$, which yields an $x \in G$ such that $x H x^{-1} = K$.

(c) Let H be a pro-p-subgroup of G. Then, for all U in $\mathcal{D}$, the set $S'(U)$ of p-Sylows of the finite group G_U containing H_U is finite and non-empty (again by classical Sylow theorems). We obtain a projective system $(S'(U))_{U \in \mathcal{D}}$ and by Lemma 4.9, there exists an element $(H'_U)_{U \in \mathcal{D}}$ in the projective limit of this family. Thus H'_U is a p-Sylow of G_U containing H_U. We then consider $\varprojlim_{U \in \mathcal{D}} H'_U$. It is a pro-$p$-Sylow of G containing H.

(d) can be deduced easily from an analogous statement for finite abelian groups, which follows for example from the theorem on the structure of these groups. $\square$

Remark 4.11 Another consequence of Lemma 4.9 (which will be used in the proof of theorem 15.11, b) is the following. Let G be an abelian Hausdorff topological group. Let A be a compact subgroup of G and let (B_n) be a decreasing sequence of compact subgroups of G. Then

$$A + \bigcap_n B_n = \bigcap_n (A + B_n).$$

Indeed, the $\subset$ is obvious. Conversely, if $x \in \bigcap_n (A + B_n)$, then the set X_n of pairs (a, b) of $A \times B_n$ such that $x = a + b$ forms a decreasing sequence of non-empty compact sets, hence their intersection is non-empty. That means that $x \in A + \bigcap_n B_n$.

Example 4.12 (a) The group $\mathbf{Z}_p$ is a pro-p-group. It is the p-Sylow of

$$\widehat{\mathbf{Z}} := \varprojlim_{n \in \mathbf{N}^*} \mathbf{Z}/n = \prod_{p \in \mathcal{P}} \mathbf{Z}_p,$$

where $\mathcal{P}$ denotes the set of prime numbers.

(b) Let G be a discrete group. The *profinite completion* $\widehat{G}$ of G is the projective limit of the finite quotients of G. The *p-completion* $\widehat{G}_p$ of G is the projective limit of quotients of G which are finite p-groups. It is the largest quotient of $\widehat{G}$ which is a pro-p-group.

(c) Let K be a p-adic field. Let K_{nr} be its maximal unramified extension and K^{mr} be the maximal tamely ramified extension of K_{nr} (cf. Sects. 7.3 and 7.4). Let $U = \mathrm{Gal}(\bar{K}/K_{\mathrm{nr}})$ be the inertia group. The theory of ramification groups (cf. Example 7.15) tells us that $U_p := \mathrm{Gal}(\bar{K}/K^{\mathrm{mr}})$ is the unique p-Sylow of U, and the quotient U/U_p is isomorphic to $\prod_{\ell \neq p} \mathbf{Z}_\ell$.

4.2 Discrete G-Modules

In what follows, G is a profinite group. Let A be a discrete abelian group endowed with an action of G. We will say that G acts *continuously* on A if for all x in A, the map $g \mapsto g \cdot x$ from G to A is continuous.

Remark 4.13 As the G-module A is discrete, the property that G acts continuously is here equivalent to require that the stabiliser of every element of A is an open subgroup of A. Indeed, if G acts continuously, then for all x of A the stabiliser of x is the inverse image of $\{x\}$ (which is an open subset of the discrete module A) via $g \mapsto g \cdot x$. It is therefore an open subset of G. Conversely, assume that all the stabilisers are open. Let then $g_0 \in G$ and $x \in A$. As the stabiliser of x is an open subgroup U, the coset $g_0 U$ is open in G, it contains g_0, and the map $g \mapsto g \cdot x$ is constant on $g_0 U$. It is therefore continuous at g_0.

Note also that we can consider the groups A endowed with another topology than the discrete one and thus obtain "continuous cohomology", but in this book we will only consider the case where A is discrete.

Definition 4.14 A *discrete G-module* (or simply a G-module where there is no ambiguity) is an abelian group A endowed with an action of G such that G acts continuously on A.

We thus require, in addition to the usual definition of a G-module, that the stabiliser of any point is open. Naturally, for G finite, this coincides with the usual notion of G-module. We denote by C_G the category of discrete G-modules (it is a full abelian subcategory of $\mathcal{M}od_G$). It A is a discrete G-module, we have $A = \bigcup_U A^U$, where U runs through the set of all open subgroups of G.

Remark 4.15 Let A be a discrete G-module. Then A is of finite type over $\mathbf{Z}[G]$ if and only if it is of finite type over $\mathbf{Z}$. This is because the stabiliser of every element of A is a subgroup of finite index in G. Therefore, we can refer to a discrete G-module of finite type without any ambiguity.

Example 4.16 The case of particular interest to us is that of $G = \Gamma_k = \mathrm{Gal}(\bar{k}/k)$, the absolute Galois group of a field k or a quotient of this group.

The main theorem of "infinite Galois theory" implies that the map $\Gamma \mapsto L^\Gamma$ establishes a one-to-one correspondence between the *closed* subgroups Γ of Γ_k and the field extensions L of k contained in $\bar{k}$. Open subgroups ($=$ closed of finite index) are the ones corresponding to finite extensions of k (note that there may exist finite index subgroups which are not closed. These subgroups do not correspond to any extension of k).

(a) Take the trivial action $\gamma \cdot x = x$ for all $\gamma \in \Gamma_k, x \in M$. We will use it often with $M = \mathbf{Z}$ and $M = \mathbf{Z}/n\mathbf{Z}$. By convention, when talking of the "Γ_k-module $\mathbf{Z}/n\mathbf{Z}$", we will understand that the action of Γ_k is trivial.

(b) Let n be an integer coprime with the characteristic of k. We obtain discrete Γ_k-modules by taking the action of the Galois group on the multiplicative group $\bar{k}^*$ or on the nth roots of unity in $\bar{k}$ (we will denote this latter Γ_k-module by μ_n). We can also, for any integer $m > 0$, consider the discrete Γ_k-module $\mu_n^{\otimes m} := \mu_n \otimes_{\mathbf{Z}} \cdots \otimes_{\mathbf{Z}} \mu_n$ (m factors), with the convention that $\mu_n^{\otimes 0} = \mathbf{Z}/n\mathbf{Z}$.

This yields Γ_k-modules which are isomorphic to $\mathbf{Z}/n\mathbf{Z}$ as abelian groups, but not generally as Γ_k-modules. Lastly, for all $m \geqslant 0$, we have a discrete Γ_k-module $(\mathbf{Q}/\mathbf{Z})(m) := \varinjlim_n \mu_n^{\otimes m}$.

(c) More generally, if S is a *commutative algebraic group* over k, we can make Γ_k act on the set $S(\bar{k})$ of its $\bar{k}$-points. If Car $k = 0$, the case of a finite group over k (taken in the sense of schemes) corresponds to $S(\bar{k})$ being finite. It is for example the case of $\mathbf{Z}/n\mathbf{Z}$ and μ_n. The trivial action corresponds to the fact that all $\bar{k}$-points of S are defined over k.

The case of the module $\bar{k}^*$ corresponds to the *multiplicative group* $\mathbf{G}_m$. We can also take S to be an abelian variety.

(d) Assume Car $k = 0$. If M is a finite Γ_k-module (corresponding to an algebraic k-group again denoted, with a slight abuse of notation, by M), we can define its *dual* M' using the finite algebraic group $\mathrm{Hom}(M, \mathbf{G}_m)$ ("Cartier dual"). That means that the Γ_k-module M' consists of morphisms φ from M to $\bar{k}^*$ (or to μ_n if M is n-torsion). The action of Γ_k on M' is given by $(\gamma \cdot \varphi)(x) := \gamma(\varphi(\gamma^{-1} \cdot x))$ for $\gamma \in \Gamma_k$, $\varphi \in M'$, $x \in M$ (this formula may seem complicated, but is is necessary to send $M \otimes M'$ to $\bar{k}^*$ in a way which is compatible with the action of Γ_k. It is also consistent with the formula in the Example 1.3, d).

For example, $\mathbf{Z}/n\mathbf{Z}$ and μ_n are Γ_k-modules Cartier dual of each other. More generally, we can define the Cartier dual of a torsion Γ_k-module M as $\mathrm{Hom}(M, \mathbf{Q}/\mathbf{Z}(1))$.

4.3 Cohomology of a Discrete G-Module

There are several ways to define cohomology groups $H^n(G, A)$ where G is a profinite group and A is a discrete G-module. The category C_G possesses enough injectives (cf. Appendix, Example A.35, c). We can hence use the derived functors of the functor $A \mapsto A^G$ from C_G to $\mathcal{A}b$. Nevertheless, there is a small difficulty with computing it: unlike $\mathcal{M}od_G$, the category C_G does not possess enough projectives if G is infinite (note for example that for G profinite and infinite, the module $\mathbf{Z}[G]$ *is not* a discrete G-module, the stabilisers being trivial), cf. Exercise 4.2. Since our aim is to reduce ourselves to the cohomology of finite groups, it is easier to adapt the definition using cochains.

Definition 4.17 Let $A \in C_G$. For $q \geqslant 0$, we denote by $K^q(G, A)$ the set of *continuous* maps (i.e., locally constant) from G^q to A. Let $d : K^q(G, A) \to K^{q+1}(G, A)$ be the coboundary defined in the usual way (cf. Theorem 1.27). We then define cohomology groups $H^q(G, A)$ as cohomology groups of the complex $(K^q(G, A))$.

We then have:

Proposition 4.18 *Let (G_i) be a projective system of profinite groups. Let (A_i) be an inductive system of discrete G_i-modules, the transition maps being compatible with those of the $(G_i)s$. Let $G = \varprojlim G_i$ and $A = \varinjlim A_i$. Then for all $q \in \mathbf{N}$:*

$$H^q(G, A) = \varinjlim H^q(G_i, A_i).$$

Recall that by convention, all the inductive systems considered in this book are associated to ordered non-empty filtered sets.

Proof It is enough to show that the canonical homomorphisms

$$u_q : \varinjlim K^q(G_i, A_i) \longrightarrow K^q(G, A)$$

are isomorphisms. This is what we will now do.

The fact that we do not assume the transition map to be surjective complicates the proof slightly.

(a) Injectivity of u_q. Let $f_i : G_i^q \to A_i$ be locally constant. By compactness of G_i^q, the set F of values that it takes is finite. Assume that f_i induces the zero function on $K^q(G, A)$. By finiteness of F, after possibly increasing i, we can then assume that the induced function: $G^q \to A_i$ is zero. For each $j \geqslant i$, let E_j be the subset of G_j^q consisting of x whose image in A_i (by the composite g_j of the projection on G_i^q and of f_i) is nonzero. As F is finite, each E_j is a finite intersection of closed subsets of the G_j^q, hence is compact. The E_j then form a projective system of compact sets whose limit is empty. One of the sets E_j is therefore empty (Lemma 4.9). That means that the map $g_j : G_j^q \to A_i$ is zero, hence a fortiori so is $f_j : G_j^q \to A_j$. This shows in particular that the image of f_i in $\varinjlim K^q(G_i, A_i)$ is zero. This proves the injectivity.

(b) Surjectivity of u_q. We will need two lemmas.

Lemma 4.19 *Let $f : G^q \to A$ be a locally constant function. Then there exists an index j such that for any y in G^q, the image $f(y)$ depends only on the image $p_j(y)$ of x in G_j^q.*

Proof As f is locally constant and G^q is compact, there exists a partition of G^q into a finite number $V_1, \ldots, V_s$ of open sets, such that f is constant on each of these sets. By definition of the topology of $G^q = \varprojlim G_i^q$ (which is induced by the direct product $\prod G_i^q$), we can assume that each $V_1, \ldots, V_s$ is defined by a finite number of conditions of the type $p_i(x) \in U_i$, where U_i is open in G_i^q. Let I be the finite set of indexes thus obtained. Let j be an index larger than all the $i \in I$. Then, if two elements y, y' of G^q satisfy $p_j(y) = p_j(y')$, they satisfy $p_i(y) = p_i(y')$ for all $i \in I$. Hence they belong to the same open set among the $V_1, \ldots, V_s$. Therefore, $f(y) = f(y')$. This proves Lemma 4.19.

Lemma 4.20 *Let $f : G^q \to A$ be a locally constant function. Then there exists an index j and a locally constant function $f_j : G_j^q \to A$ such that $f = f_j \circ p_j$.*

Proof Let j be an index defined as in Lemma 4.19. By continuity of p_j, the image $p_j(G^q)$ is a compact of G_j^q. Let $f_j : p_j(G^q) \to A$ defined by $f_j(p_j(y)) = f(y)$ for all $y \in G^q$. This makes sense since $f(y)$ depends on $p_j(y)$ only. For all $a \in A$, the inverse image of a by f_j is $p_j(f^{-1}(a))$, which is compact (by continuity of f and p_j and compactness of G^q), hence closed in $p_j(G^q)$. As the map f_j only takes (like f) a finite number of values, it is continuous on $p_j(G^q)$. The subgroup $p_j(G^q)$ of G_j^q is compact, thus we can extend (by Proposition 4.2) f_j to a continuous function (still denoted f_j) from G_j^q to A, which clearly satisfies $f = f_j \circ p_j$. This proves Lemma 4.20.

We can now finish the proof of the surjectivity of u_q. Let $f : G^q \to A$ be a locally constant function. Then f only takes a finite number of values. We can, by definition of $A = \varinjlim A_i$, therefore find an index i so that all these values are in the image of the map $A_i \to A$. Let j be as in Lemma 4.20, that we may assume $\geqslant i$ (replace it with $\max(i, j)$ if necessary). Then, by construction, the function $f_j : G_j^q \to A$ comes from a function $g : G_j^q \to A_j$, which then satisfies $u_q(g) = f$. $\qquad\square$

We immediately deduce the following corollaries (the first of which can be taken as a definition of the groups $H^q(G, A)$).

Corollary 4.21 *Let A be a discrete G-module. Then*

$$H^q(G, A) = \varinjlim_U H^q(G/U, A^U),$$

where U runs through the set of open normal subgroups of G.

Proof We indeed have $G = \varprojlim_U (G/U)$ and $A = \varinjlim_U A^U$. $\qquad\square$

Corollary 4.22 *Let A be a discrete G-module. Then*

$$H^q(G, A) = \varinjlim H^q(G, B),$$

where B runs through the set of finite type sub-G-modules of A.

Proof This follows from the fact that A is the union (hence inductive limit) of such Bs. $\qquad\square$

Corollary 4.23 *For $q \geqslant 1$, the groups $H^q(G, A)$ are torsion.*

Proof This follows from Corollaries 4.21 and 1.49. $\qquad\square$

Remark 4.24 If I is an injective in C_G, it is immediate that the I^U are injective in $C_{G/U}$ for any open normal subgroup U of G. Thus, we see that the $H^q(G, A)$ are obtained as derived functors of $A \mapsto A^G$ from C_G to $\mathcal{A}b$. Indeed, the $H^q(G, .)$ form a cohomological functor which coincides with $A \mapsto A^G$ in degree zero, and is effaceable since if we embed a discrete G-module A in an injective I of C_G, by Corollary 4.21, we have for all $q > 0$:

$$H^q(G, I) = \varinjlim_U H^q(G/U, I^U) = 0$$

as I^U is injective in $C_{G/U}$. We conclude using the Remark A.47 from the appendix.

Proposition A.36 from the appendix gives a converse to the first statement in the above remark. That is, that a discrete G-module I is injective in C_G if and only if there exists a system of neighbourhoods of 1 in G consisting of open normal subgroups U with I^U injective in $C_{G/U}$. We deduce from it the following statement (which generalises Lemma 1.38). This statement is probably well-known but we were not able to find a reference in the existing literature.

Proposition 4.25 *Let G be a profinite group. Let H be a closed subgroup of G. Let A be an injective in C_G. Then A is also injective in C_H.*

Proof (*after J. Riou; see Exercise 4.4 for a similar result*) We have already dealt with the case where G is finite (Lemma 1.38). Let us deal with the general case. Let V be an open subgroup of H. By Proposition A.36, it is enough to show that for any open subgroup V of H, the H/V-module A^V is injective. Let U be an open normal subgroup of G; then A^U is an injective G/U-module (Remark 4.24) and by the case of a finite group, A^U is also injective as an $H' := H/(H \cap U)$-module.

Let $V' := V/(V \cap U)$, it is an open normal subgroup of H' and Remark 4.24 tells us that $(A^U)^{V'}$ is an injective H'/V'-module. In other words A^{UV} is an injective H/V-module where UV is the subgroup of G generated by U and V (it is the set of products uv with $u \in U$ and $v \in V$ since U is normal in G). We conclude by noting that A^V is the inductive limit of H/V-modules A^{UV} where U ranges through normal open subgroups of G, and therefore A^V remains an injective H/V-module (Appendix, Example A.35, (f) or Exercise 1.4), since the group H/V is finite. □

Using Corollary 4.21, properties of the cohomology of a profinite group G can be immediately deduced from that of a finite group. One just needs to be careful about properties involving a subgroup H: to remain in the category of profinite groups, H needs to be *closed* in G. For modules defined by functions from G or G^n to a G-module A, one also needs to take *continuous* functions. In particular:

(1) For any closed subgroup H of G and any discrete G-module A, we have a G-module $I_G^H(A)$ (the definition is the same provided we assume the functions from G to A to be *continuous*). In particular, any induced G-module $I_G(A)$ is acyclic for the functor $H^0(G, .)$ (this allows us to use the usual shifting arguments). Note however that there is no good notion of co-induced module in C_G since $\mathbf{Z}[G]$ is not a discrete G-module if G is infinite.

(2) For any closed subgroup H of G, the restriction and inflation homomorphisms are defined as in Sect. 1.5, and Shapiro's lemma is still valid (one directly verifies that the functor $A \mapsto I_G^H(A)$ is still exact on C_G). Proposition 1.40 and the

Hochschild–Serre spectral sequence (as well as its consequences) also continue to hold[2] when H is a normal closed subgroup of G.

(3) When H is a closed finite index subgroup (i.e., an open subgroup) of G, the corestriction $H^q(H, A) \to H^q(G, A)$ is well defined. Theorem 1.48 and Corollary 1.51 remain true in this context.

(4) For discrete G-modules A and B, the groups $\mathrm{Ext}^i_G(A, B)$ are still defined as derived functors (applied to B) of $\mathrm{Hom}_G(A, .)$. We can no longer view them as left derived functors (indeed, the category C_G does not have enough projectives when G is infinite). Nevertheless, Remark 1.37 is still valid. If H is an open subgroup of G, Propositions 1.15 and 1.39 remain valid (with the same proofs).

(5) If p and q are two integers, the cup-product

$$H^p(G, A) \times H^q(G, B) \longrightarrow H^{p+q}(G, A \otimes B)$$

is defined as in Sect. 2.5 (one just needs to take continuous cochains in the definition), and possesses the same properties.

In Chap. 6 we will see examples of computations of cohomology groups when $G = \Gamma_k$ is the absolute Galois group of a field. We can also extend the notion of cohomologically trivial G-module (cf. Exercise 4.11 for basic properties of this notion).

Definition 4.26 Let G be a profinite group. We say that a discrete G-module A is *cohomologically trivial* if for any $n > 0$ and any closed subgroup H of G, we have $H^n(H, A) = 0$.

Remark 4.27 Until now, we have been using Proposition 4.18 in the case where transition maps between the groups G_j are surjective, but there is another case of interest. It is the case where the family G_j is a filtered family of open subgroups of a profinite group K, transition maps being the inclusions. Set $G = \bigcap G_j$. It is a closed subgroup of K. We then have $\varprojlim G_j = \bigcap G_j = G$. This allows us to identify $H^q(G, A)$ with $\varinjlim H^q(G_j, A)$ for any K-module A. We will sometimes encounter this situation when computing the Brauer group of infinite algebraic extensions of a field.

We also sometimes need the group $\widehat{H}^0(G, A)$. To define it, consider two normal open subgroups of U and V of G with $V \subset U$.

We define a *deflation* map:

$$\mathrm{Def} : \widehat{H}^0(G/V, A^V) \longrightarrow \widehat{H}^0(G/U, A^U)$$

as the one induced on quotients by the identity, as we have

$$\widehat{H}^0(G/V, A^V) = A^G / N_{G/V} A^V, \quad \widehat{H}^0(G/U, A^U) = A^G / N_{G/U} A^U$$

---

[2] There is a small subtlety here: to imitate the proof for G finite, we must either use Proposition 4.25, or the slightly weaker fact that an induced G-module $I_G(A)$ is acyclic for $H^0(H, .)$, which is the object of Exercise 4.4.

and $N_{G/V} A^V \subset N_{G/U} A^U$ (group the elements of G/V in classes modulo U/V).

Definition 4.28 Let G be a profinite group. Let A be a discrete G-module. We define a modified cohomology group $\widehat{H}^0(G, A)$ to be the *projective* limit of the $\widehat{H}^0(G/U, A^U)$ for U a normal subgroup of G, the transition maps being the *deflation* maps.

Then the restriction and corestriction are still well defined (under the usual assumptions) for $\widehat{H}^0$, with the formula Cores ∘ Res $= \cdot [G : H]$ when H is an open subgroup of G (we can also define the groups $\widehat{H}^q(G, A)$ for $q < 0$, cf. [42], Déf. 1.9.3., but we will not use it. In particular, the definition of cup-products is more delicate in this setting).

Remark 4.29 There is a subtlety here: if U and V are open normal subgroups of a profinite group G with $V \subset U$, we also have an inflation map $\widehat{H}^0(G/U, A^U) \to \widehat{H}^0(G/V, A^V)$ (which allows us to take the inductive limit), obtained via the norm map. The situation is in some sense the inverse of the one which allows us to define the deflation map. In the case where the groups G/U and G/V are cyclic, the groups $H^n(G/U, A^U)$ and $H^n(G/V, A^V)$ for n even are isomorphic to $\widehat{H}^0(G/U, A^U)$ and $\widehat{H}^0(G/V, A^V)$ respectively (Theorem 2.16). The isomorphisms are compatible with the inflation maps. Thus, if G is *procyclic* (i.e., isomorphic to a projective limit of cyclic groups), we can compute the $H^n(G, A)$ for n even as the inductive limits of the $\widehat{H}^0(G/U, A^U)$ *where the transition maps are induced by the norms*.

4.4 Exercises

Exercise 4.1 Show that the analogue of the Proposition 1.31 does not hold if G is only assumed to be profinite, even if the G-module $\varprojlim A_n$ is assumed to be discrete (you may take $G = \mathbf{Z}_p$ and $A_n = \mathbf{Z}/p^n \mathbf{Z}$, endowed with the trivial action of G).

Exercise 4.2 Let G be an infinite profinite group. Let $f : P \to \mathbf{Z}$ be a surjective morphism of discrete G-modules. Let $x \in P$ be such that $f(x) = 1$. Let U be the stabiliser of x.

(a) Show that there is an open normal subgroup V of G which does not contain U.

(b) Consider the augmentation morphism $\pi : \mathbf{Z}[G/V] \to \mathbf{Z}$ and assume that there is a morphism of G-modules $\tilde{f} : P \to \mathbf{Z}[G/V]$ such that $f = \pi \circ \tilde{f}$. Let

$$\tilde{f}(x) = \sum_{g \in G/V} a_g g,$$

with $a_g \in \mathbf{Z}$. Show that $\sum_{g \in G/V} a_g = 1$.

(c) Let g_0 be a non-trivial element of $U/(V \cap U)$. Show that if $h \in G/V$ belongs to the subgroup $H \subset G/V$ generated by g_0, then $a_{h \cdot g} = a_g$ for all $g \in G/V$.

(d) Let m be the order of g_0 in G/V. Making the subgroup H act by translation on G/V, show that there exists an integer d such that $md = 1$, and derive a contradiction. Deduce that the category C_G does not have enough projectives.

Exercise 4.3 Solve Exercise 1.7 assuming this time that G is a profinite group and H is an open subgroup of G.

Exercise 4.4 Let G be a profinite group. Let H be a closed subgroup of G. Let A be an abelian group.

(a) Show that H is the projective limit of the H/V for V an open subgroup of H normal in G.

(b) For such a V, show that $I_G(A)^V = I_{G/V}(A)$ (that one can view as a H/V-module) is the inductive limit of a family of induced H/V-modules.

(c) Deduce from (b) that for all $n > 0$, we have $H^n(H, I_G(A)) = 0$.

Exercise 4.5 Using Proposition A.36 of the appendix, generalise Exercise 1.4 to the case of a profinite group G.

Exercise 4.6 Let G be a profinite group. Let A be a finite G-module.

(a) Show that for any element $f \in I_G(A)$, there exists a normal open subgroup U of G such that for any $x \in G$, the value $f(x)$ of f at x depends only on the class of x in G/U.

(b) Deduce that $I_G(A)$ is the inductive limit (or the union) of the $I_G^U(A)$, the limit being taken over open normal subgroups U of G.

Exercise 4.7 Let G be a profinite group. Let A be a finite G-module. Let $n > 0$.

(a) Assume that $H^n(G, A)$ is finite. Show that there exists a *finite* G-module B and an injective morphism i of G-modules from A to B, such that the map $H^n(G, A) \to H^n(G, B)$ induced by i is zero.

(b) Give an example where the conclusion of (a) is no longer valid if we do not assume that $H^n(G, A)$ is finite.

Exercise 4.8 Let G be a profinite abelian group. We assume that for any integer $n > 0$, the group G/nG is finite.

(a) Show that nG is an open subgroup of G.

(b) Let U be an open subgroup of G. Show that nU is an open subgroup of G (you may compare G/U and nG/nU).

(c) Deduce that if A is a discrete finite G-module, then $H^1(G, A)$ is finite.

Exercise 4.9 Let G be a profinite group.

(a) Let M be a G-module isomorphic to $\mathbf{Z}$ as an abelian group. Show that there exists an open normal subgroup H of G, acting trivially on M, and such that $[G : H] \leqslant 2$.

(b) Fix an open normal subgroup H of G with $[G : H] = 2$. How many isomorphism classes of G-modules M, isomorphic to $\mathbf{Z}$ as abelian groups, such that H acts trivially on M there are?

(c) Let H be an open normal subgroup of G with $[G : H] = 2$. Consider the G-module M, isomorphic to $\mathbf{Z}$ as an abelian group, and such that the action of G on M is defined by $g \cdot x = x$ if $g \in H$ and $g \cdot x = -x$ if $g \notin H$. Show that there is an exact sequence of G-modules:

$$0 \longrightarrow \mathbf{Z} \longrightarrow \mathbf{Z}[G/H] \longrightarrow M \longrightarrow 0,$$

then compute $H^1(G, M)$.

Exercise 4.10 Let G be a profinite group. Let A be a discrete G-module. Suppose that A is a free abelian group of finite type.
   (a) Show that if the action of G on A is trivial, then $H^1(G, A) = 0$.
   (b) Show that there exists an open normal subgroup U of G such that Inf : $H^1(G/U, A^U) \to H^1(G, A)$ is an isomorphism.
   (c) Deduce that $H^1(G, A)$ is finite. Does this property extend to the $H^i(G, A)$ for $i > 1$?

Exercise 4.11 Let G be a profinite group.
   (a) Show that a G-module A is cohomologically trivial if and only if for all open normal subgroups U of G, the G/U-module A^U is cohomologically trivial.
   (b) Show that an induced G-module is cohomologically trivial.
   (c) Extend Exercise 3.5 to the case where G is profinite and N is torsion-free of finite type.

Exercise 4.12 (*suggested by A. Pirutka*) Let k be a field of characteristic zero with Galois group $\Gamma_k = \mathrm{Gal}(\bar{k}/k)$. For any n-torsion Γ_k-module of finite type C, we denote by $C(-1)$ the Γ_k-module $\mathrm{Hom}(\mu_n, C)$.
   (a) Describe explicitly the Γ_k-module $\mu_n(-1)$.
   (b) Show that if $m > 0$ is an integer, then the Γ_k-module $\mu_n^{\otimes m}(-1)$ is isomorphic to $\mu_n^{\otimes (m-1)}$.
   (c) Let $\mu_n^{\otimes (-1)} = (\mathbf{Z}/n\mathbf{Z})(-1)$ and for all $m > 0$ define $\mu_n^{\otimes (-m)} = \mu_n^{\otimes (-1)} \otimes_{\mathbf{Z}} \cdots \otimes_{\mathbf{Z}} \mu_n^{\otimes (-1)}$ (m factors). Show that for all $p, q \in \mathbf{Z}$, the Γ_k-module $\mu_n^{\otimes (p+q)}$ is isomorphic to $\mu_n^{\otimes p} \otimes_{\mathbf{Z}} \mu_n^{\otimes q}$.
   (d) What is the Cartier dual of the Γ_k-module $\mu_n^{\otimes m}$?
   (e) Let G be the Galois group of the extension $k(\mu_n)/k$. Show that if m is a multiple of the exponent of G (for example if m is divisible by the value of the Euler totient function $\varphi(n)$ at n), then the Γ_k-module $\mu_n^{\otimes m}$ is isomorphic to $\mathbf{Z}/n\mathbf{Z}$.
   (f) Extend previous results to the case of a field k of characteristic $p > 0$ when p does not divide n.

Chapter 5
Cohomological Dimension

The notion of cohomological p-dimension of a profinite group is very important. Exercise 5.1 shows that it is mostly relevant in the case of an *infinite* group since the cohomological p-dimension of a finite group is either zero (if p does not divide its cardinality) or infinity (if p divides its cardinality).

5.1 Definitions, First Examples

Definition 5.1 A discrete G-module is *simple* if it is nonzero and admits no sub-G-module other than $\{0\}$ and itself.

Definition 5.2 Let G be a profinite group. For any prime number p, the *cohomological p-dimension of* G (denoted by $\mathrm{cd}_p(G)$) is the lower bound (in $\mathbf{N} \cup \{+\infty\}$) for the set of integers $n \in \mathbf{N}$ satisfying:

For any discrete *torsion* G-module A and any $q > n$, the p-primary component of $H^q(G, A)$ is zero (which is equivalent to the fact that the p-torsion subgroup $H^q(G, A)[p]$ is zero).

The *cohomological dimension of* G is $\mathrm{cd}(G) = \sup_p \mathrm{cd}_p(G)$.

Example 5.3 The group $G = \mathbf{Z}/2 = \mathrm{Gal}(\mathbf{C}/\mathbf{R})$ is of cohomological p-dimension 0 if $p \neq 2$ (by Corollary 1.49), but infinite if $p = 2$ by Theorem 2.16 and the fact that $\widehat{H}^0(G, \mathbf{Z}/2) = \widehat{H}^1(G, \mathbf{Z}/2) = \mathbf{Z}/2$ which is immediate. If p does not divide the order of G, we have $\mathrm{cd}_p(G) = 0$ by Corollaries 4.21 and 1.49.

Theorem 5.4 *Let G be a profinite group. Let p be a prime number and $n \in \mathbf{N}$. The following conditions are equivalent:*

(1) $\mathrm{cd}_p(G) \leq n$.
(2) For all $q > n$ and any discrete G-module A which is a p-primary abelian group, we have $H^q(G, A) = 0$.

© Springer Nature Switzerland AG 2020
D. Harari, *Galois Cohomology and Class Field Theory*, Universitext,
https://doi.org/10.1007/978-3-030-43901-9_5

*(3) We have $H^{n+1}(G, A) = 0$ for any discrete G-module A which is simple and
p-torsion.*

Note in passing that a simple and p-primary G module A is necessarily p-torsion
since $A[p]$ is a non-trivial sub-G-module of A.

Proof Write A as a direct sum of the $A\{p\}$. Then for $q \geq 1$ the group $H^q(G, A\{p\})$
is the p-primary component of $H^q(G, A)$ (by Proposition 4.18). The equivalence
between (1) and (2) follows. It is straightforward that (2) implies (3). Assume (3).
Let us first show that $H^{n+1}(G, A) = 0$ by induction on the cardinality of A when
A is finite and p-primary. It is trivial when $A = 0$, hence let us assume that $A \neq$
0. If A is simple, then $A[p]$ (which is nonzero as A is p-primary) is equal to A
and the assumption (3) gives $H^{n+1}(G, A) = 0$. Else, we have an exact sequence of
G-modules

$$0 \longrightarrow A_1 \longrightarrow A \longrightarrow A_2 \longrightarrow 0$$

with A_1 and A_2 of cardinality strictly smaller than that of A. It follows again that
$H^{n+1}(G, A) = 0$ via the long exact sequence and the induction assumption.

Now we again have $H^{n+1}(G, A) = 0$ (by Corollary 4.22), for any discrete
G-module A which is p-primary. This is because a sub-G-module of A of finite
type is then finite (it is of finite type and torsion over $\mathbf{Z}$).

We then prove (2) by induction on q embedding A in the induced module $I_G(A)$
(which indeed is p-primary since any continuous function from G to A is locally
constant, hence, by compactness of G takes only finitely many values), then applying
the induction assumption to the torsion p-primary module $I_G(A)/A$.   □

We deduce form Theorem 5.4 the following important criterion which allows us
to bound above the cohomological p-dimension of a pro-p-group:

Theorem 5.5 *Let G be a pro-p-group and $n \in \mathbf{N}$. Then $\mathrm{cd}_p(G) \leq n$ if and only if
$H^{n+1}(G, \mathbf{Z}/p) = 0$.*

Proof If $\mathrm{cd}_p(G) \leq n$, then $H^{n+1}(G, \mathbf{Z}/p) = 0$ by definition of the cohomological p-dimension. Thus, assume that $H^{n+1}(G, \mathbf{Z}/p) = 0$. By Theorem 5.4, we
are reduced to showing that if A is a discrete, simple, p-torsion G-module, then
$H^{n+1}(G, A) = 0$. To do this, we will in fact show that such an A is necessarily
isomorphic to $\mathbf{Z}/p$.

We already know that A is finite: indeed, the G-module M generated by a nonzero
element a of A is finite since it is of finite type over $\mathbf{Z}$ and torsion. But $M = A$ by
simplicity of A. Thus we can consider A as a G/U module, where U is an open
normal subgroup of G. Applying the result to G/U and to the simple G/U-module
A, we reduce ourselves to the case where G is a finite p-group. As $A \neq 0$, Lemma 3.1
implies that $A^G \neq 0$ and by simplicity of A, we obtain $A^G = A$. In other words, the
action of G on A is trivial.

Now we necessarily have $A = \mathbf{Z}/p$, since the abelian group generated by a
nonzero element of A is a sub-G-module of A, hence is equal to A (by simplicity of A) and isomorphic to $\mathbf{Z}/p$ (since A is p-torsion).   □

Example 5.6 Take $G = \mathbf{Z}_p = \varprojlim_n (\mathbf{Z}/p^n)$. We have $H^2(G, \mathbf{Z}/p) = 0$ by Proposition 4.18: Indeed, by Theorem 2.16, we have $H^2(\mathbf{Z}/p^n, \mathbf{Z}/p) = \widehat{H}^0(\mathbf{Z}/p^n, \mathbf{Z}/p) = \mathbf{Z}/p$, where the transition maps between the groups $H^2(\mathbf{Z}/p^n, \mathbf{Z}/p)$ correspond to multiplication by p (cf. Remark 4.29). Hence we have $\varinjlim_n H^2(\mathbf{Z}/p^n, \mathbf{Z}/p) = 0$. Theorem 5.5 then implies that $\mathrm{cd}_p(\mathbf{Z}_p) \leq 1$, and the equality comes from the fact that

$$H^1(\mathbf{Z}_p, \mathbf{Z}/p) = \mathrm{Hom}_c(\mathbf{Z}_p, \mathbf{Z}/p) = \mathrm{Hom}_c(\mathbf{Z}_p/p\mathbf{Z}_p, \mathbf{Z}/p) = \mathbf{Z}/p \neq 0.$$

When working with G-modules which are not necessarily torsion, we obtain the notion of strict cohomological dimension.

Definition 5.7 Let G be a profinite group, p a prime number, and $n \in \mathbf{N}$. The *strict cohomological p-dimension* of G is the lower bound for the set of $n \in \mathbf{N}$ such that for any discrete G-module A and any $q > n$, we have $H^q(G, A)\{p\} = 0$. We denote it by $\mathrm{scd}_p(G)$. The *strict cohomological dimension* of G is $\mathrm{scd}(G) = \sup_p \mathrm{scd}_p(G)$.

5.2 Properties of the Cohomological Dimension

We will now compare cd_p and scd_p, and their behaviour under passing to a subgroup or taking a quotient. We naturally have $\mathrm{scd}_p(G) \geq \mathrm{cd}_p(G)$, but we can say a lot more.

Proposition 5.8 *Let G be a profinite group. Then $\mathrm{scd}_p(G) \leq \mathrm{cd}_p(G) + 1$ for any prime number p. In particular, $\mathrm{scd}(G) \leq \mathrm{cd}(G) + 1$.*

Proof Let M be a G-module. Let $N = M[p]$ be the p-torsion submodule of M and $Q = M/pM$. Let n be the p-cohomological dimension of G. Let $I = pM$. Multiplication by p induces two exact sequences

$$0 \longrightarrow N \longrightarrow M \longrightarrow I \longrightarrow 0.$$

$$0 \longrightarrow I \longrightarrow M \longrightarrow Q \longrightarrow 0.$$

Let $q > n + 1$. As N and Q are p-primary torsion, we have $H^q(G, N) = H^{q-1}(G, Q) = 0$. Then the maps $H^q(G, M) \to H^q(G, I)$ and $H^q(G, I) \to H^q(G, M)$ induced by the two exact sequences above, are injective. Hence so is their composite, which is multiplication by p on $H^q(G, M)$. Lastly, $H^q(G, M)[p] = 0$ for all M, i.e., $\mathrm{scd}_p(G) \leq n + 1$. $\square$

Lemma 5.9 *Let G be a profinite group and A be a discrete G-module. Let p be a prime number and H a closed subgroup of G.*

(a) If p does not divide the supernatural number $[G : H]$, the map $\mathrm{Res} : H^q(G, A)$
     $\{p\} \to H^q(H, A)$ is injective for all $q > 0$.

(b) *If, in addition, H is open and $n = \mathrm{cd}_p(G)$ (resp. $n = \mathrm{scd}_p(G)$) is finite, then* Cores : $H^n(H, A)\{p\} \to H^n(G, A)\{p\}$ *is surjective for any discrete torsion G-module A (resp. for any discrete G-module A).*

Proof (a) If $[G : H]$ is finite, it is Proposition 3.2. We reduce the general case to it using Corollary 4.21 and the definition of the index of a closed subgroup.

(b) Let $I = I_G^H(A)$. We have a homomorphism $\pi : I \to A$ defined by

$$f \longmapsto \pi(f) := \sum_{g \in G/H} g \cdot f(g^{-1}).$$

This homomorphism is surjective since if $a \in A$, we can define a preimage f of a under π by setting $f(h) = h \cdot a$ for $h \in H$ and $f(g) = 0$ if $g \notin H$.

If A is torsion, I (and thus so is the kernel B of π) is torsion. On the other hand the induced homomorphism

$$H^n(G, I) = H^n(H, A) \longrightarrow H^n(G, A)$$

identifies with the corestriction (see Exercise 1.1: as usual, this follows from the fact that we obtain two natural transformations of cohomological functors that coincide in degree 0).

Assume now that $\mathrm{cd}_p(G) = n$. Since for a torsion module B, we have $H^{n+1}(G, B)\{p\} = 0$, we conclude using the long cohomology sequence (note that an exact sequence of *torsion* abelian groups remains exact when the functor $\{p\}$ is applied to it). The argument for scd_p is identical.   □

Proposition 5.10 *Let G be a profinite group and let H be a closed subgroup of G. Then for any prime number p, we have*

$$\mathrm{cd}_p(H) \leq \mathrm{cd}_p(G); \quad \mathrm{scd}_p(H) \leq \mathrm{scd}_p(G).$$

The equality holds if the index $[G : H]$ is prime to p, or if H is open and $\mathrm{cd}_p(G) < +\infty$.

Naturally, we have analogous results for $\mathrm{cd}(G)$ and $\mathrm{scd}(G)$. On the other hand, there is no analogous inequality for a quotient. For example the group $\mathbf{Z}_2$ has cohomological 2-dimension 1, but has a quotient isomorphic to $\mathbf{Z}/2$, whose cohomological 2-dimension is infinite.

Note also that the last assertion of the proposition is generally false if $\mathrm{cd}_p(G)$ is not assumed to be finite (take for example $G = \mathbf{Z}/2$, $H = 0$ and $p = 2$).

Proof Let us treat the case of cd_p (the arguments for scd_p are the same).

Let $A \in C_H$, then $I_G^H(A)$ (which is p-primary if A is p-primary) belongs to C_G and by Shapiro's lemma $H^q(G, I_G^H(A)) = H^q(H, A)$. This yields $\mathrm{cd}_p(H) \leq \mathrm{cd}_p(G)$. If we assume that $[G : H]$ is prime to p, then we have the equality by Lemma 5.9 (a).

Assume now that H is open and that $\mathrm{cd}_p(G) = n$ is finite and > 0 (the inequality $\mathrm{cd}_p(H) \geq \mathrm{cd}_p(G)$ is straightforward if $n = 0$). We can then find a discrete torsion G-module A such that $H^n(G, A)\{p\} \neq 0$. It suffices then to show that $H^n(H, A)\{p\} \neq 0$, which follows from Lemma 5.9 (b). $\qquad\square$

Corollary 5.11 *Let G_p be a p-Sylow of G. Then $\mathrm{cd}_p(G) = \mathrm{cd}_p(G_p) = \mathrm{cd}(G_p)$ and $\mathrm{scd}_p(G) = \mathrm{scd}_p(G_p) = \mathrm{scd}(G_p)$.*

Example 5.12 (a) By Corollary 5.11, we have $\mathrm{cd}_p(\widehat{\mathbf{Z}}) = \mathrm{cd}_p(\mathbf{Z}_p) = 1$. On the other hand, $H^2(\mathbf{Z}_p, \mathbf{Z}) \neq 0$ (indeed, $H^2(\mathbf{Z}_p, \mathbf{Z}) = \mathrm{Hom}_c(\mathbf{Z}_p, \mathbf{Q}/\mathbf{Z}) = \mathbf{Q}_p/\mathbf{Z}_p$), from which we deduce that $\mathrm{scd}_p(\widehat{\mathbf{Z}}) = \mathrm{scd}_p(\mathbf{Z}_p) = 2$.

(b) Later on, we will see that if k is a p-adic field and $G = \mathrm{Gal}(\bar{k}/k)$, then $\mathrm{cd}(G) = \mathrm{scd}(G) = 2$.

Proposition 5.13 *Let H be a normal closed subgroup of G. Then for any prime number p, we have*

$$\mathrm{cd}_p(G) \leq \mathrm{cd}_p(G/H) + \mathrm{cd}_p(H)$$

(the same holds for scd_p, $\mathrm{cd}(G)$, etc.).

Proof We will use the Hochschild–Serre spectral sequence (Theorem 1.44).

Let A be a discrete G-module of p-primary torsion. Let $n > \mathrm{cd}_p(G/H) + \mathrm{cd}_p(H)$. Then if $i + j = n$, we have either $i > \mathrm{cd}_p(G/H)$, or $j > \mathrm{cd}_p(H)$. In both cases, the group $E_2^{ij} = H^i(G/H, H^j(H, A))$ is trivial. Lastly, $H^n(G, A)$ (which admits a filtration whose successive quotients are subquotients of the E_2^{ij} for $i + j = n$) is trivial. $\qquad\square$

We have the following general criterion (due to Serre).

Proposition 5.14 *Let G be a profinite group of cohomological dimension n. Then the strict cohomological dimension of G is n if and only if for every open subgroup U of G, we have $H^{n+1}(U, \mathbf{Z}) = 0$.*

Naturally, we obtain an analogous result in the case of the cohomological p-dimension by restricting ourselves to the p-primary torsion of $H^{n+1}(U, \mathbf{Z})$.

Proof The condition is clearly necessary.

Conversely, assume the condition is satisfied. Then (by Shapiro's lemma) $H^{n+1}(G, A) = 0$ for any G-module A of the form $A = I_G^U(\mathbf{Z}^r) = \mathbf{Z}[G/U]^r$ with $r \geq 0$ and U an open normal subgroup of G. Let then M be a discrete finite type G-module.

Then there exists an open normal subgroup U of G acting trivially on M. It follows that we have an exact sequence

$$0 \longrightarrow B \longrightarrow \mathbf{Z}[G/U]^r \longrightarrow M \longrightarrow 0$$

which implies $H^{n+1}(G, M) = 0$ since $H^{n+2}(G, B) = 0$ by Proposition 5.8.

As any discrete G-module A is a union of discrete finite type G-modules, we obtain $H^{n+1}(G, A) = 0$. The result follows (again by Proposition 5.8). $\qquad\square$

Computation of the cohomological dimension of the absolute Galois group $\mathrm{Gal}(\bar{k}/k)$ of a field k is generally a difficult problem.

We have already seen that for $k = \mathbf{R}$, the cohomological p-dimension is 0 if $p \neq 2$ and infinity if $p = 2$. The Galois group of a finite field is isomorphic to $\widehat{\mathbf{Z}}$, whose cohomological dimension is 1.

We will see that for a finite extension of $\mathbf{Q}_p$, this dimension equals 2. The same is true for a totally imaginary finite extension of $\mathbf{Q}$.

5.3 Exercises

Exercise 5.1 Let G be a profinite group and p a prime number.

(a) Show that $\mathrm{cd}_p(G) = 0$ if and only if the order of G is prime to p (as a supernatural number).
(b) Show that the cohomological p-dimension of the finite group $\mathbf{Z}/p\mathbf{Z}$ is infinite.
(c) Show that if $\mathrm{cd}_p(G)$ is neither zero nor infinite, then the exponent of p in the order of G is infinite.
(d) Show that we can't have $\mathrm{scd}_p(G) = 1$.

Exercise 5.2 Let G be a profinite group and p a prime number such that $\mathrm{cd}_p(G) \leq n$ with $n \in \mathbf{N}$.

(a) Show that if A is a discrete p-divisible G-module, then for any $q > n$ the p-primary component $H^q(G, A)\{p\}$ is trivial.
(b) Deduce that if A is a discrete divisible G-module and $\mathrm{cd}(G) \leq n$, then $H^q(G, A) = 0$ for any $q > n$.

Exercise 5.3 Let G be the profinite group $\widehat{\mathbf{Z}} = \varprojlim_{n \in \mathbf{N}^*} \mathbf{Z}/n$.

(a) Let A be a discrete G-module. Let F be the automorphism of A induced by the canonical topological generator s of G (corresponding to $1 \in \widehat{\mathbf{Z}}$) and A' be the subgroup of A consisting of a such that there exists $n \in \mathbf{N}^*$ with

$$(1 + F + \cdots + F^{n-1})a = 0.$$

Show that $(F - 1)A$ is a subgroup of A' and that we have $H^1(G, A) = A'/(F - 1)A$.
(b) Show that if A is torsion, then $A' = A$.
(c) Compute the dual $\mathrm{Hom}_c(G, \mathbf{Q}/\mathbf{Z})$ of G.
(d) Show that if A is finite, then $H^2(G, A) = 0$ and recover the fact that $\mathrm{cd}(\widehat{\mathbf{Z}}) = 1$ (cf. Example 5.12, a).

Exercise 5.4 Let G be a profinite group of finite cohomological dimension. Show that G is torsion-free (that is, any element of G other than the identity is of infinite order).

Exercise 5.5 Let n be a strictly positive integer and p a prime number. Let G be a profinite group with $\mathrm{cd}_p(G) = n$. Show that $\mathrm{scd}_p(G) = n+1$ if and only if there exists an open subgroup H of G such that $H^n(H, \mathbf{Q}_p/\mathbf{Z}_p) \neq 0$.

Exercise 5.6 Let Γ be a profinite group and p a prime number. Assume that the cohomological p-dimension $\mathrm{cd}_p(\Gamma)$ is an integer $n > 0$. Let U be a normal open subgroup of Γ. Consider a p-primary torsion Γ-module A.

(a) Consider a resolution

$$0 \longrightarrow A \longrightarrow X^0 \longrightarrow \cdots \longrightarrow X^n \longrightarrow \cdots$$

by induced p-primary Γ-modules and let $A_n = \ker[X^n \to X^{n+1}]$. Show that A_n is a cohomologically trivial Γ-module.

(b) Consider the corestriction maps $\mathrm{Cores} : H^i(U, A) \to H^i(\Gamma, A)$ for $i \geq 0$. Show that they are obtained by taking the cohomology of the morphism of complexes $N_{\Gamma/U} : (X^\bullet)^U \to (X^\bullet)^\Gamma$ given by the norm of Γ/U. Then show that Cores induces a homomorphism (still denoted by Cores) $H^i(U, A)_{\Gamma/U} \to H^i(\Gamma, A)$ on the group of co-invariants $H^i(U, A)_{\Gamma/U}$ for the action of Γ/U on $H^i(U, A)$.

(c) Show that we have a commutative diagram, where N denotes the norm $N_{\Gamma/U}$:

$$
\begin{array}{ccccccc}
((X^{n-1})^U)_{\Gamma/U} & \longrightarrow & (A_n^U)_{\Gamma/U} & \longrightarrow & H^n(U, A)_{\Gamma/U} & \longrightarrow & 0 \\
\downarrow N & & \downarrow N & & \downarrow \mathrm{Cores} & & \\
(X^{n-1})^\Gamma & \longrightarrow & A_n^\Gamma & \longrightarrow & H^n(\Gamma, A) & \longrightarrow & 0.
\end{array}
$$

(d) Show that the left vertical arrow in this diagram is surjective.

(e) Show that the middle vertical arrow is injective. Deduce that the corestriction $H^n(U, A)_{\Gamma/U} \to H^n(\Gamma, A)$ is an isomorphism.

Chapter 6
First Notions of Galois Cohomology

In this chapter, k denotes a field and $\bar{k}$ its separable closure. We denote by Γ_k the profinite group $\Gamma_k := \mathrm{Gal}(\bar{k}/k)$. Our aim in this chapter is to introduce Galois cohomology, which is a special case of the cohomology of a profinite group, and to prove some of its general properties. In particular, we will encounter the *Brauer group* of a field, which will be used a lot in parts II and III.

6.1 Generalities

Let M be a discrete Γ_k-module. The cohomology groups $H^q(\Gamma_k, M)$ for $q \geqslant 0$ (as well as $\widehat{H}^0(\Gamma_k, M)$) have been defined in the previous chapters. Let now k_1 be an extension of k whose algebraic closure is denoted by $\bar{k}_1$ and let $j : \bar{k} \to \bar{k}_1$ be the field morphism extending the inclusion $i : k \to k_1$, then j defines a continuous homomorphism $f : \Gamma_{k_1} \longrightarrow \Gamma_k$. Thus we obtain homomorphisms

$$H^q(\Gamma_k, M) \longrightarrow H^q(\Gamma_{k_1}, M)$$

(cf. beginning of Sect. 1.5). If we change j, we change f by an inner automorphism of Γ_k. Therefore, Proposition 1.40 implies that these homomorphisms are in fact independent of the choice of j. In particular, two separable closures of k define canonically isomorphic $H^q(\Gamma_k, M)$. We can thus simply write $H^q(k, M)$ instead of $H^q(\Gamma_k, M)$. For any field extension k_1 of k, we have canonical homomorphisms $H^q(k, M) \to H^q(k_1, M)$.

Remark 6.1 If A is a *commutative group scheme* over k, we may consider the Galois cohomology groups $H^q(k, A) := H^q(k, A(\bar{k}))$. If k_1 is an extension of k, the above procedure yields canonical homomorphisms $H^q(k, A(\bar{k})) \to H^q(k_1, A(\bar{k}_1))$.

© Springer Nature Switzerland AG 2020
D. Harari, *Galois Cohomology and Class Field Theory*, Universitext,
https://doi.org/10.1007/978-3-030-43901-9_6

This applies for instance to the *additive group* $\mathbf{G}_a$ defined by $\mathbf{G}_a(k_1) = k_1$ and to the *multiplicative group* $\mathbf{G}_m$ defined by $\mathbf{G}_m(k_1) = k_1^*$.

The following proposition and its corollary show that the cohomology of the additive group is trivial.

Proposition 6.2 *Let L be a finite Galois extension of k. The*

$$\widehat{H}^q(\text{Gal}(L/k), L) = 0$$

for all $q \in \mathbf{Z}$.

Corollary 6.3 *We have $H^q(k, \bar{k}) = 0$ for all $q > 0$.*

Proof The corollary follows from the proposition by Corollary 4.21. The proposition follows from the fact that by the normal basis theorem (cf. [7], Sect. 10), the Gal(L/k)-module L is co-induced (isomorphic to $\mathbf{Z}[\text{Gal}(L/k)] \otimes_{\mathbf{Z}} k$) and hence induced since Gal(L/k) is finite.   □

Proposition 6.4 (Artin-Schreier) *Let k be a field of characteristic $p > 0$. Let Φ be the map from $\bar{k}$ to $\bar{k}$ defined by $\Phi(x) = x^p - x$. Then $k/\Phi(k) \simeq H^1(k, \mathbf{Z}/p)$ and $H^q(k, \mathbf{Z}/p) = 0$ for $q \geqslant 2$.*

Proof As $\bar{k}$ is of characteristic p, the map Φ is a morphism of Γ_k-modules. It is surjective as $\bar{k}$ is separably closed and for any $a \in \bar{k}$, the polynomial $X^p - X - a$ is separable (its derivative is -1). The kernel of Φ is the prime subfield of $\bar{k}$, hence isomorphic to the Γ_k-module $\mathbf{Z}/p$ (with the trivial action of Γ_k) and we have the exact sequence of Γ_k-modules:

$$0 \longrightarrow \mathbf{Z}/p \longrightarrow \bar{k} \xrightarrow{\ \Phi\ } \bar{k} \longrightarrow 0.$$

We conclude using Corollary 6.3 and the long exact cohomology sequence.   □

6.2   Hilbert's Theorem 90 and Applications

We consider the natural action of the Galois group Γ_k on the abelian group $\bar{k}^*$, giving $\bar{k}^*$ the structure of a discrete Γ_k-module.

Theorem 6.5 (Hilbert 90) *Let L be a finite Galois extension of k. Let G be the Galois group $G = \text{Gal}(L/k)$. Then*

$$H^1(G, L^*) = 0 \quad and \quad H^1(k, \bar{k}^*) = 0.$$

Proof The second statement is deduced from the first one using Corollary 4.21. Let $s \mapsto a_s$ be a cocycle in $Z^1(G, L^*)$. By Dedekind's theorem on linear independence of morphisms ([7], Sect. 7, no 5), we can find an element c of L^* such that the element

$$b := \sum_{t \in G} a_t t(c)$$

is nonzero. We then have for all s in G:

$$s(b) = \sum_{t \in G} s(a_t) \cdot (st)(c) = \sum_{t \in G} a_s^{-1} a_{st} \cdot (st)(c) = a_s^{-1} b$$

hence $a_s = s(b^{-1})/b^{-1}$, showing that $s \mapsto a_s$ is a coboundary. □

Corollary 6.6 ("Kummer theory") *Let n be an integer, that we assume to be invertible in k. Let μ_n be the multiplicative group of nth roots of unity in $\bar{k}$. Then*

$$k^*/k^{*n} \simeq H^1(k, \mu_n).$$

Proof It follows from the long exact cohomology sequence associated to

$$1 \longrightarrow \mu_n \longrightarrow \bar{k}^* \xrightarrow{\cdot n} \bar{k}^* \longrightarrow 1$$

and from Hilbert's theorem 90. □

In the next paragraph, we will see that unlike the Γ_k-module $\bar{k}$, the cohomology of $\bar{k}^*$ is not always trivial.

Remark 6.7 For any discrete Γ_k-module M, the set $H^1(k, M)$ can also be interpreted as the set of isomorphism classes of *principal homogeneous spaces* (or *torsors*) of M, cf. [49], Part. I, Sect. 5.2. In this language, the $\mathbf{Z}/p\mathbf{Z}$-torsor associated to an element $a \in k/\Phi(k)$ as in Proposition 6.4 is given by the equation $\Phi(x) = a$ and the μ_n torsor associated to an element $b \in k^*/k^{*n}$ as in the Corollary 6.6 is given by the equation $x^n = b$.

6.3 Brauer Group of a Field

Definition 6.8 Let k be a field with absolute Galois group $\Gamma_k = \mathrm{Gal}(\bar{k}/k)$. The *Brauer group* of k is the cohomology group $H^2(\Gamma_k, \bar{k}^*)$. We denote it by $\mathrm{Br}\, k$.

Thus $\mathrm{Br}\, k$ is the inductive limit of the groups $\mathrm{Br}(L/k) := H^2(\mathrm{Gal}(L/k), L^*)$ (for L/k a finite Galois extensions).

Note also that if K is an extension of k, we have a homomorphism $\operatorname{Br} k \to \operatorname{Br} K$ induced by the natural morphism $i : \Gamma_K \to \Gamma_k$ and the inclusion $\bar{k}^* \to \bar{K}^*$. Indeed, even though i is only defined up to conjugation, the same arguments as in paragraph Sect. 6.1 show that the homomorphism $\operatorname{Br} k \to \operatorname{Br} K$ is well defined.

Remark 6.9 The Brauer group of a field k can also be defined without using Galois cohomology, using the equivalence classes of finite-dimensional central simple algebras over k. See, for example, Sect. 2.4 of [16] or Chap. X (Sect. 5) of [47]. We will not be using this characterisation in the book, even though it may sometimes be useful (see, for example, the remark at the end of Exercise 6.1 and the preamble to Chap. 8).

Proposition 6.10 *Let L be a finite Galois extension of k. Then*

$$\operatorname{Br}(L/k) = \ker[\operatorname{Br} k \longrightarrow \operatorname{Br} L].$$

Thus $\operatorname{Br} k$ is the union of the $\operatorname{Br}(L/k)$ for L/k a finite Galois extension.

Proof This follows from Theorem 6.5 (Hilbert 90) and Corollary 1.45 (when the group G is profinite and H is a closed subgroup of G): take $q = 2$, $G = \Gamma_k$, $H = \Gamma_L = \operatorname{Gal}(\bar{k}/L)$ and $A = \bar{k}^*$. □

Remark 6.11 Previous proposition immediately extends to the case where L is a Galois extension (not necessarily finite) of k.

Proposition 6.12 *Let n be an integer which is invertible in k. Then*

$$H^2(k, \mu_n) = (\operatorname{Br} k)[n].$$

In particular, if we further have $\mu_n \subset k$, then $H^2(k, \mathbf{Z}/n) \simeq (\operatorname{Br} k)[n]$ (via the choice of a primitive nth root of unity in k).

Proof This follows from the long exact cohomology sequence associated to the exact sequence

$$1 \longrightarrow \mu_n \longrightarrow \bar{k}^* \xrightarrow{\;\cdot n\;} \bar{k}^* \longrightarrow 1$$

using the fact that $H^1(k, \bar{k}^*) = 0$ (Hilbert 90). □

Example 6.13 (a) The Brauer group of a separably closed field is by definition trivial. As we will see later, the same holds true for a finite field.

(b) The Brauer group of the field $\mathbf{R}$ is $\mathbf{Z}/2$ since $H^2(\Gamma_{\mathbf{R}}, \mathbf{C}^*)$ is isomorphic to $\widehat{H}^0(\Gamma_{\mathbf{R}}, \mathbf{C}^*) = \mathbf{R}^*/\mathbf{R}^*_+$ by Theorem 2.16 (as $\Gamma_{\mathbf{R}}$ is cyclic).

(c) We will later see that the Brauer group of a p-adic field is $\mathbf{Q}/\mathbf{Z}$.

6.4 Cohomological Dimension of a Field

Let k be a field with absolute Galois group $\Gamma_k = \mathrm{Gal}(\bar{k}/k)$. For each prime number p, we have the notion of cohomological p-dimension of the profinite group Γ_k defined in Chap. 5. We will use it to define the notion of cohomological p-dimension of a field of characteristic different from p, or perfect of characteristic p.

Definition 6.14 Let k be a field and p a prime number. Assume that the characteristic of k is different from p, or that k is perfect of characteristic p. The cohomological *p-dimension* (resp. the *cohomological dimension* if k is of characteristic zero or perfect of > 0 characteristic) of k is by definition that of the absolute Galois group Γ_k. We denote it by $\mathrm{cd}_p(k)$ (resp. $\mathrm{cd}(k)$).

Naturally, we have an analogous definition for the strict cohomological dimension.

In this book, we will not be concerned with the cohomological p-dimension of an imperfect field of characteristic p. The reason being that the previous notion is in a certain sense "not the good one" in this case (even if certain texts such as [16], [42] or [51] retain the convention $\mathrm{cd}_p(k) = \mathrm{cd}_p(\Gamma_k)$ in this case). In particular, we have the following statement (while we do not necessarily want to say that any field of characteristic p is of cohomological p-dimension $\leqslant 1$, cf. remark 6.18 below):

Proposition 6.15 *Let k be a field of characteristic $p > 0$. Then we have $\mathrm{cd}_p(\Gamma_k) \leqslant 1$ (and therefore $\mathrm{scd}_p(\Gamma_k) \leqslant 2$).*

Proof Let G_p be a p-Sylow of Γ_k. By Galois theory, we have $G_p = \mathrm{Gal}(\bar{k}/K)$, where K is an extension of k contained in $\bar{k}$. Corollary 5.11 says that $\mathrm{cd}_p(\Gamma_k) = \mathrm{cd}_p(G_p)$.

By Theorem 5.5, it is enough to show that $H^2(K, \mathbf{Z}/p) = 0$. But this follows from Proposition 6.4, as K is of characteristic p. $\qquad\square$

Remark 6.16 For imperfect fields of characteristic p, several substitutes to the notion of cohomological p-dimension exist. We can replace Galois cohomology by *flat cohomology* (cf. part III of [39]) and define cohomological p-dimension using p-group schemes. We can also use the groups H_p^r defined by Kato ([27]) and define the separable cohomological p-dimension as in [50] Sect. 10, or [2].

The fields k of cohomological dimension $\leqslant 1$ are particularly important. It is remarkable that this property can be characterised using only the Brauer group of finite extensions of k.

Theorem 6.17 *Let k be a field. Let p be a prime number different from the characteristic of k. Then the following statements are equivalent:*

(i) We have $\mathrm{cd}_p(k) \leqslant 1$.

(ii) For any algebraic separable extension K of k, the p-torsion $(\mathrm{Br}\, K)[p]$ of $\mathrm{Br}\, K$ is trivial.

(iii) For any finite separable extension K of k, the p-torsion $(\mathrm{Br}\, K)[p]$ of $\mathrm{Br}\, K$ is trivial.

If k is of characteristic zero, we can of course replace everywhere cd_p by cd and $(\mathrm{Br}\,K)[p]$ by $\mathrm{Br}\,K$.

Proof Assume that (i) holds.

Let K be an algebraic separable extension of k. Then the absolute Galois group Γ_K is isomorphic to a closed subgroup of Γ_k, hence $\mathrm{cd}_p(K) \leqslant 1$ by Proposition 5.10. This implies $(\mathrm{Br}\,K)[p] = H^2(K, \mu_p) = 0$. The implication (ii) $\implies$ (iii) is trivial.

Assume that (iii) holds. Let G_p be a p-Sylow of Γ_k and $K_p \subset \bar{k}$ its fixed field. Then K_p contains the group μ_p of pth roots of unity in $\bar{k}$. This is because the degree $[K_p(\mu_p) : K_p]$ divides p and $p - 1$. Hence $H^2(K_p, \mathbf{Z}/p) = H^2(K_p, \mu_p)$. Let $K' \subset K_p$ be an arbitrary finite separable extension of k containing μ_p, then the property (iii) implies $H^2(K', \mu_p) = 0$. As K_p is the union of the extensions K' as above, we obtain $H^2(K_p, \mu_p) = 0$ by remark 4.27 applied to open subgroups $\mathrm{Gal}(\bar{k}/K')$ of Γ_k, whose intersection is $\mathrm{Gal}(\bar{k}/K_p)$. Lastly $H^2(K_p, \mathbf{Z}/p) = 0$, hence $\mathrm{cd}_p(k) = \mathrm{cd}_p(K_p) \leqslant 1$ by Corollary 5.11 and Theorem 5.5. $\square$

Remark 6.18 If k is a perfect field of characteristic p, we have seen that $\mathrm{cd}_p(k) \leqslant 1$ (Proposition 6.15), and we also have $(\mathrm{Br}\,K)[p] = 0$ for any algebraic separable extension K of k: indeed, K is still perfect, hence $x \mapsto x^p$ is an isomorphism of $\bar{K}^*$ to itself, which implies $(\mathrm{Br}\,K)[p] = 0$. Hence Theorem 6.17 remains true for a perfect field of characteristic p.

On the other hand, we can have $(\mathrm{Br}\,k)[p] \neq 0$ for an imperfect field k of characteristic p. Later on we will see for example that the p-torsion of the Brauer group of a local field of characteristic p is isomorphic to $\mathbf{Z}/p$.

This is why some authors, like Serre ([49], Part. II, Sect. 3), define, for any imperfect field k of characteristic p, the property $\dim(k) \leqslant 1$ by combining the conditions $\mathrm{cd}(\Gamma_k) \leqslant 1$ and $(\mathrm{Br}\,K)[p] = 0$ for any algebraic extension (or any separable algebraic extension, which comes down to the same) K of k.

6.5 C_1 Fields

The most frequent examples of fields of cohomological dimension 1 are the fields C_1, defined by the following very concrete property.

Definition 6.19 A field k is called C_1 if any homogeneous polynomial $f \in k[X_1, \ldots, X_n]$ of degree $d < n$ has at least one non-trivial zero.

Example 6.20 (a) A finite field is C_1 (Chevalley's theorem, [16], Th. 6.2.6.).

(b) If k is an algebraically closed field, then $k(t)$ (more generally, any extension of k of transcendence degree 1) is C_1 (Tsen's theorem, [16], Th. 6.2.8.). The same holds true for $k((t))$ (this result is due to Lang, [16], Th. 6.2.1.)

(c) Lang ([30]) has also proved that the maximal unramified extension of a p-adic field is C_1. This result holds more generally for the field of fractions K of a henselian excellent discrete valuation ring (it is automatically excellent if $\mathrm{Car}\,K = 0$) with algebraically closed residual field.

Lemma 6.21 *Let k be a C_1 field and let k_1 be an algebraic extension of k. Then k_1 is C_1.*

Proof Let F be a homogeneous polynomial of degree d in n variables with coefficients in k_1, with $d < n$. We want to show that it admits a non-trivial zero. As the coefficients of F are algebraic over k, there exists a finite extension of k which contains them all and we can thus assume that k_1 is a finite extension of k, whose degree we denote by m.

Set $f(x) = N_{k_1/k}(F(x))$. Then f is a homogeneous polynomial of degree dm in nm variables with coefficients in k (take a basis $(e_1, \ldots, e_m)$ of k_1 over k and decompose $x \in k_1^n$ in this basis). As k is C_1, the polynomial f has a non-trivial zero. Thus we have $x \in k_1^n$ such that $f(x) = 0$, which implies $F(x) = 0$. $\square$

Theorem 6.22 *Let k be a C_1 field.*

(a) *If k is of characteristic zero, or perfect of characteristic $p > 0$, we have $\mathrm{cd}(k) \leqslant 1$. In particular $\mathrm{Br}\, k = 0$.*

(b) *If k is imperfect of characteristic $p > 0$, we have $\mathrm{cd}(\Gamma_k) \leqslant 1$ and $\mathrm{Br}\, K = 0$ for any algebraic extension K of k (in particular k is of cohomological dimension $\leqslant 1$ in the "strong" sense of [49]).*

Note that the converse is false: Ax has constructed a field of cohomological dimension 1 that is not C_1, cf. [49], Exer. p. 90.

Proof Let K be an algebraic extension (that in the case (b) we do not assume separable) of k. Let L be a finite Galois extension of degree d of K. Let $a \in K^*$. Let N be the norm map from L to K. As K is C_1 by the previous lemma, the equation

$$N(x) = ax_0^d$$

for $x \in L$, $x_0 \in K$, has a non-trivial solution (x, x_0) (it is a polynomial equation of degree d in $d + 1$ variables over K). We have $x_0 \neq 0$ (otherwise $N(x) = 0$ hence $x = 0$). It follows that $N(x/x_0) = a$. Lastly, the norm $N_{L/K} : L^* \to K^*$ is surjective. Combining this with the Hilbert 90 theorem, we obtain $\widehat{H}^n(G, L^*) = 0$ for $n = 0, 1$, where $G := \mathrm{Gal}(L/K)$. It also holds for any intermediate extension L/K' of the extension L/K by Lemma 6.21. Theorem 3.7 implies that the G-module L^* is cohomologically trivial. In particular, $\mathrm{Br}(L/K) = H^2(G, L^*)$ is trivial. Passing to the limit, we obtain $\mathrm{Br}\, K = 0$. The result follows from Theorem 6.17 combined with the fact that $\mathrm{cd}_p(\Gamma_k) \leqslant 1$ in characteristic p is automatic. $\square$

Remark 6.23 Note that in (b), we do not need to assume that K is a separable extension of k. That is due to the fact that property C_1 extends to any algebraic extension.

Corollary 6.24 *A finite field is of cohomological dimension $\leqslant 1$. In particular, its Brauer group is trivial.*

This corollary can also be obtained by noticing that the Galois group of a finite field is isomorphic to $\widehat{\mathbf{Z}}$.

Remark 6.25 It is possible to prove Theorem 6.22 without using the difficult Theorem 3.7. One can show directly (using the interpretation of the Brauer group in terms of central simple algebras) that the Brauer group of a C_1 field is trivial. This is the method used in [16], Prop. 6.2.3.

6.6 Exercises

Exercise 6.1 Let k be a perfect field (e.g. of characteristic zero).

(1) Show that the following properties are equivalent:

    (a) $\mathrm{cd}(k) \leqslant 1$.

    (b) For any finite extension K of k and any finite extension L of K, the norm $N : L^* \to K^*$ is surjective.

(2) Show that if these properties are satisfied, then (b) continues to hold for any algebraic extension (not necessarily finite) K of k.

(3) Show that for these properties to hold, it is enough to assume (b) for finite extensions L of K which are Galois and of prime degree.

Remark. If we assume (b) only for finite extensions L of K which are Galois and of prime degree, we can still deduce (a) without using the difficult Theorem 3.7 (cf. [16], Th. 6.1.8). However, this approach does not imply (b) in its full generality and does not prove the triviality of the p-primary torsion of the Brauer group of an imperfect C_1 field of characteristic p.

Exercise 6.2 Let k be a field (not necessarily perfect). Let p be a prime number. In this exercise, the absolute Galois group of a field F will be denoted by G_F.

For any algebraic quasi-Galois field extension L/K [1] (not necessarily separable), we will still denote by $\mathrm{Gal}(L/K)$ the group of automorphisms of L inducing the identity on K. One can use properties of quasi-Galois extensions found in no.3 of [7], Sect. 9.

(a) Fix an algebraic closure k^{alg} of k and a separable closure F of $k^{\mathrm{alg}}(t)$. Show that G_k is isomorphic to $\mathrm{Gal}(k^{\mathrm{alg}}(t)/k(t))$. What is the cohomological dimension of $\mathrm{Gal}(F/k^{\mathrm{alg}}(t))$?

(b) Compare $G_{k(t)}$ and $\mathrm{Gal}(F/k(t))$.

(c) Show that $\mathrm{cd}_p(G_{k(t)}) \leqslant \mathrm{cd}_p(G_k) + 1$.

(d) Let k' be an algebraic extension of k. Compare $\mathrm{cd}_p(G_k)$ and $\mathrm{cd}_p(G_{k'})$.

(e) Deduce that if K is an extension k of transcendence degree N, then

$$\mathrm{cd}_p(G_K) \leqslant N + \mathrm{cd}_p(G_k).$$

---

[1] One also says *normal* instead of quasi-Galois.

(f) What conclusions can we obtain when k is separably closed?

Exercise 6.3 Let p be a prime number and k a field of characteristic $\neq p$, with separable closure $\bar{k}$. Let $n \in \mathbf{N}^*$. Prove the equivalence of the following properties:
(i) $\mathrm{cd}_p(k) \leqslant n$;
  (ii) for every algebraic separable extension $K \subset \bar{k}$ of k, we have

$$H^{n+1}(K, \bar{k}^*)\{p\} = 0$$

and $H^n(K, \bar{k}^*)\{p\}$ is p-divisible;
  (iii) same statement as in (ii) but restricted to extensions K/k which are in addition finite and of degree prime to p. (You may first translate (ii) in terms of the Galois module μ_p).

Exercise 6.4 Let k be a C_1 field of characteristic $p > 0$. Show that $[k : k^p]$ is equal to 1 or p.

Exercise 6.5 Let k be a perfect field with algebraic closure $\bar{k}$ and absolute Galois group $\Gamma_k = \mathrm{Gal}(\bar{k}/k)$. A k-*torus* T is an algebraic group over k that becomes isomorphic to $\mathbf{G}_m^r$ (for a certain integer $r \geqslant 0$) over $\bar{k}$, where $\mathbf{G}_m$ is the multiplicative group. In particular, there is a finite Galois extension L of k such that T becomes isomorphic to $\mathbf{G}_m^r$ over L. This implies that the Γ_L-module $T(\bar{k})$ consisting of the $\bar{k}$-points of T is isomorphic to $(\bar{k}^*)^r$ with the natural action of $\Gamma_L := \mathrm{Gal}(\bar{k}/L)$.

In what follows, we will denote by $H^i(k, T)$ the Galois cohomology groups $H^i(k, T(\bar{k}))$.

(a) Let $n > 0$. Show that we have an exact sequence of Γ_k-modules:

$$0 \longrightarrow T(\bar{k})[n] \longrightarrow T(\bar{k}) \xrightarrow{\cdot n} T(\bar{k}) \longrightarrow 0.$$

(b) Show that if $d := [L : k]$, then $H^1(k, T)$ is d-torsion.
(c) Assume that k is a perfect field of cohomological dimension $\leqslant 1$. Show that $H^1(k, T) = 0$ (show first that this group is divisible).

Part II
Local Fields

This part concerns itself with the cohomological class field theory of local fields, which are p-adic fields (finite extension of $\mathbf{Q_p}$) and the fields of Laurent series over a finite field.

After a chapter that recalls facts about local fields, we study in detail the Brauer group of a local field, which, by Tate–Nakayama theorem, allows us to prove in Chap. 9 the main properties of the *reciprocity map*. From this we deduce the structure of the Galois group of a p-adic field. Chapter 10 is devoted to the local Tate duality theorem and some of its applications. Lastly, in Chap. 11, we describe the construction of Lubin–Tate which allows us to find some of the previous results in a more precise form and extend the theorem on the structure of the abelian Galois group to the case of a local field of strictly positive characteristic.

There exist different approaches to class field theory. Let us in particular mention that of J. Neukirch [40] that works equally well in both local and global cases: it does not use cohomology groups of degree 2 or the Tate–Nakayama theorem to construct the reciprocity map. It is defined explicitly using the *class field axiom* (whose local version, we will see in Proposition 8.2 and the global one in Theorem 13.23). Since these cohomological notions will be essential to us in Part IV, we have chosen here to freely use results proved in Part I (which are of independent interest) that also provide us with a very useful computation of the Brauer group. This viewpoint will be very fruitful in Part III.

Chapter 7
Basic Facts About Local Fields

In this chapter we recall some standard results on local fields. For more details, one can consult the first two parts of [47].

7.1 Discrete Valuation Rings

Definition 7.1 *A discrete valuation ring* is a commutative principal ideal domain [1] A, which has a single nonzero prime ideal $\mathfrak{m}$ ($\mathfrak{m}$ is then the unique maximal ideal of A). The field $A/\mathfrak{m}$ is called the *residue field* of A. Any generator π of $\mathfrak{m}$ is called a *uniformiser* of A.

From this definition it follows that the group A^* of invertible elements of A coincides with $A - \mathfrak{m}$. If we fix a uniformiser π, then up to multiplication by an invertible element, the only irreducible element in A is π.

Thus, any nonzero element x in A can be uniquely written as $x = u \cdot \pi^n$ with $n \in \mathbf{N}$ and $u \in A^*$. The integer $v(x) := n$ is called the *valuation* of x. Convening that $v(0) = +\infty$, we obtain the usual properties of valuations:

$$v(x + y) \geqslant \min(v(x), v(y)), \quad v(xy) = v(x) + v(y),$$

from which we deduce that $v(x + y) = \min(v(x), v(y))$ if $v(x) \neq v(y)$.

We extend the valuation to $K = \mathrm{Frac}\, A$ by setting $v(x/y) = v(x) - v(y)$, so that $v : K^* \to \mathbf{Z}$ becomes a surjective homomorphism whose kernel is A^*.

We say that A is the *ring of integers* of K. Note also that if the fraction field K of A is of characteristic $p > 0$, then so is the residue field κ, and if κ is of characteristic zero, it is also the case of K. On the other hand it may happen (the case of *unequal*

---

[1] Recall that, by definition this implies in particular that A is an integral domain.

© Springer Nature Switzerland AG 2020
D. Harari, *Galois Cohomology and Class Field Theory*, Universitext,
https://doi.org/10.1007/978-3-030-43901-9_7

characteristic) that K is of characteristic zero and κ is of characteristic $p > 0$ (it is the case of the examples (b) and (c) below).

Example 7.2 (a) If k is a field, the ring $k[[t]]$ of formal series with coefficients in k is a discrete valuation ring, the valuation of a nonzero formal power series $\sum_{n \geqslant 0} a_n t^n$ is the smallest n such that $a_n \neq 0$. The residue field of $k[[t]]$ is k and its fraction field is the field $k((t))$ of Laurent series, i.e., of formal sums $\sum_{n \in \mathbf{Z}} a_n t^n$ such that only finitely many a_n with $n < 0$ are nonzero.

(b) Let p be a prime number. The subring $\mathbf{Z}_{(p)}$ of $\mathbf{Q}$, consisting of x/y with $x, y \in \mathbf{Z}$ and y not divisible by p, is a discrete valuation ring with field of fraction $\mathbf{Q}$ and residue field $\mathbf{Z}/p\mathbf{Z}$. The valuation (called *p-adic valuation*) of a nonzero $x \in \mathbf{Z}$ is the largest integer n such that p^n divides x.

(c) Let p be a prime number. The ring $\mathbf{Z}_p = \varprojlim_{n \in \mathbf{N}^*} \mathbf{Z}/p^n\mathbf{Z}$ of *p-adic integers* is a discrete valuation ring, with residue field $\mathbf{Z}/p\mathbf{Z}$. The group of invertible elements $\mathbf{Z}_p^*$ consists of elements $x = (x_n)$ with x_1 (which is in $\mathbf{Z}/p\mathbf{Z}$) nonzero, the maximal ideal of $\mathbf{Z}_p$ is $p\mathbf{Z}_p = \mathbf{Z}_p - \mathbf{Z}_p^*$. The valuation of $x \neq 0$ is 0 if $x \in \mathbf{Z}_p^*$ and is the largest n such that $x_n = 0$ otherwise. We denote by $\mathbf{Q}_p = \operatorname{Frac} \mathbf{Z}_p$ the *field of p-adic numbers*.

7.2 Complete Field for a Discrete Valuation

Let A be a discrete valuation ring with the field of fractions K and valuation $v :$ $K^* \to \mathbf{Z}$. Let a be a real number with $0 < a < 1$. Define an *ultrametric absolute value* $x \mapsto |x|$ on K by setting

$$|x| = a^{v(x)}$$

(by convention $|0| = 0$). For all x, y in K, the three following properties characterise such an absolute value:

$$|xy| = |x||y|,$$
$$|x + y| \leqslant \max(|x|, |y|),$$
$$|x| = 0 \iff x = 0.$$

This absolute value defines a metric (even ultrametric) space topology on K, associated to the distance $d(x, y) = |x - y|$. Note that a different a defines an equivalent absolute value (and hence distance). This metric space is totally disconnected.

Proposition 7.3 *For the topology defined above, the field K is locally compact if and only it is complete and its residue field is finite. In this case, the ring of integers A of K is compact (it is thus a profinite abelian group).*

For a proof, see [47], Chap. II, Prop. 1.

Definition 7.4 A *local field*[2] is a complete field with respect to a discrete valuation, with a finite residue field.

Note that if K is a local field with uniformiser π, then its ring of integers A has a basis of open neighbourhoods of 0 consisting of open subgroups $\pi^n A$ for $n \geqslant 0$. We will make the structure of local fields more precise later.

Example 7.5 (a) Let k be a field. Then $k((t))$ is complete for the valuation defined in the example 7.2 a). It is locally compact if and only if k is finite.

(b) Let p be a prime number. The field $\mathbf{Q}$ is not complete for the p-adic valuation. Its completion is the field $\mathbf{Q}_p$, which is locally compact.

Remark 7.6 When K is complete for a discrete valuation v and finite residue field κ, it will be convenient (in particular to obtain the *product formula* for global fields) to choose as absolute value $x \mapsto q^{-v(x)}$, where q is the cardinality of κ. This absolute value is called *normalised*.

7.3 Extensions of Complete Fields

In this paragraph, we denote by K a complete field with respect to a discrete valuation v, ring of integers A and residue field κ. We denote by $\mathfrak{p}$ the prime ideal of A. The following theorem (whose proof can be found in [47], Chap. II, Sect. 2) summarise the properties of finite extensions of such a field.

Theorem 7.7 *Let L be a finite extension of K and let B be the integral closure of A in L. Then:*

(a) *B is a discrete valuation ring, and it is a free A-module of rank $n := [L : K]$.*

(b) *The field L is complete for the topology induced by B, and there exists a unique valuation w of L inducing [3] on K the topology induced by v.*

(c) *Let $\mathfrak{p}_B$ be a prime ideal of B. Let $\mathfrak{p}B = \mathfrak{p}_B^e$ with $e > 0$ and let f be the degree of the field extension $(B/\mathfrak{p}_B)$ of κ. Then $ef = n$.*

(d) *Two elements of L conjugated over K have the same valuation. The valuation $w : L \to \mathbf{Z} \cup \{+\infty\}$ of L is defined by the formula*

$$w(x) = \frac{1}{f} \, v(N_{L/K}(x))$$

---

[2] This definition is not universal, some authors consider $\mathbf{R}$ and $\mathbf{C}$ to be local fields, or do not require the residue field to be finite. The definition we adopt here is more common and better suited for number theoretic question.

[3] Sometimes one says w *extending* v, but it is important to note that in the case of a ramified extension, one can not ask that w takes values in $\mathbf{Z} \cup \{+\infty\}$ and that at the same time the restriction of w to K is equal to v, see Remark 7.9 below.

for all x of L, where $N_{L/K}$ denotes the norm from L to K. In other words, the absolute value on K extends to L using the formula

$$|x|_L = |N_{L/K}(x)|_K. \tag{7.1}$$

Definition 7.8 With the notations of the Theorem 7.7, the integer $e \geqslant 1$ is called the *ramification index* of the extension L, and the integer $f \geqslant 1$ its *residual degree*. An extension is called *unramified* if $e = 1$, and in addition $(B/\mathfrak{p}_B)$ is a separable extension of κ. It is called *totally ramified* if $f = 1$, tamely ramified if the characteristic of κ does not divide e.

Remark 7.9 If L is an unramified extension of K, then the restriction of the valuation $w : L \to \mathbf{Z} \cup \{+\infty\}$ from L to K is exactly v (in the general case, it is $e \cdot v$, where e is the ramification index).

Example 7.10 (a) If a is an element of $\mathbf{Z}_p^*$ which is not a square modulo p and $p \neq 2$, then $\mathbf{Q}_p(\sqrt{a})$ is an unramified quadratic extension of $\mathbf{Q}_p$.
(b) The extension $\mathbf{Q}_p(\sqrt{p})$ is totally ramified. It is tamely ramified if and only if $p \neq 2$.
(c) Let k be a field. The finite and separable extensions of $k((t))$ are the $k'((t))$ with k' finite separable extension of k.

Example 7.11 Let A be a complete discrete valuation ring with field of fractions K and residue field κ. Let L be an unramified extension of K of degree n and let B be the integral closure of A in L. Then there exists a monic polynomial $f \in A[X]$ (of degree $[L : K]$), whose reduction in $\kappa[X]$ is separable, and such that B is isomorphic to $A[X]/f$ and L to $K[X]/f$. This follows from [47], Chap. I, Prop. 16 and from the fact that every finite separable extension of the residue field of A is generated by one element by the primitive element theorem. Conversely, a field extension L of K of this type is unramified (loc. cit., Cor. 2). For example, the unramified quadratic extension of $\mathbf{Q}_2$ is $\mathbf{Q}_2(\sqrt{5}) = \mathbf{Q}_2[X]/(X^2 + 3X + 1)$.

Example 7.12 Let R be a discrete valuation ring with maximal ideal $\mathfrak{m}$ and field of fractions F. An *Eisenstein polynomial* is a polynomial in $R[X]$ of the form

$$P = X^n + a_{n-1}X^{n-1} + \cdots + a_0,$$

with $a_i \in \mathfrak{m}$ for all $i \in \{0, \ldots, n-1\}$ and $a_0 \notin \mathfrak{m}^2$. Such a polynomial is irreducible ([47], Chap. I, Corollary to Proposition 17). If R is complete, the field $L = F[X]/(P)$ is a totally ramified extension of F (loc. cit.). Conversely, all totally ramified extensions are obtained in this fashion (loc. cit., Prop. 18).

7.4   Galois Theory of a Complete Field for a Discrete Valuation

In this paragraph, we consider a field K complete for a discrete valuation, with ring of integers A and residue field κ. Assume for simplicity that κ is perfect (for example of characteristic zero or finite). The following result (Theorem 2 of [47], Chap. III, Sect. 5) relates unramified extensions and extensions of the residue field.

Theorem 7.13 *(a) Let κ' be a finite extension of κ. Then there exists a finite unramified extension K', unique up to isomorphism whose corresponding residual extension is isomorphic to κ'/κ. The extension K'/K is Galois if and only if κ'/κ is.*

(b) Let $\bar{\kappa}$ be an algebraic closure of κ. Let K_{nr} be the inductive limit of all unramified extensions corresponding to finite subextensions of $\bar{\kappa}$. Then the field K_{nr} is a Galois extension (generally infinite) of K, and we have $\mathrm{Gal}(K_{nr}/K) \simeq \mathrm{Gal}(\bar{\kappa}/\kappa)$.

Definition 7.14 We say that K_{nr} is the *maximal unramified extension* of K. Fix a separable closure $\bar{K}$ of K, then we can see K_{nr} as a subextension of K. The subgroup $\mathrm{Gal}(\bar{K}/K_{nr})$ of $\mathrm{Gal}(\bar{K}/K)$ is called the *inertia group* of K.

Note that by Remark 7.9, the valuation of K extends uniquely to a discrete valuation v of K_{nr}. Note though that K_{nr} is not necessarily complete for this valuation (it remains *henselian*, in the sense that its ring of integers satisfies the conclusion of Theorem 7.16 below), see Exercise 7.3.

Example 7.15 (a) Let k be a perfect field. Then the maximal unramified extension of $k((t))$ is the union of the $k'((t))$ for k' a finite extension of k (warning: this is not the same as $\bar{k}((t))$, which is generally larger). If k is assumed to be algebraically closed, all the extensions of $k((t))$ are totally ramified. If in addition k is of characteristic zero, then the algebraic closure of $k((t))$ is the union of the $k((t^{1/n}))$ for $n > 0$ ([47], Chap. IV, Prop. 2).

(b) Assume that K is a local field, i.e., the residue field κ is finite. Then for all $n > 0$, there exists a unique extension of degree n of κ and its Galois group is cyclic. It follows that there is a unique unramified extension of K of degree n, it is Galois with Galois group $\mathbf{Z}/n\mathbf{Z}$. We thus have $\mathrm{Gal}(K_{nr}/K) \simeq \mathrm{Gal}(\bar{\kappa}/\kappa) = \widehat{\mathbf{Z}}$. The group $\mathrm{Gal}(\bar{\kappa}/\kappa)$ has a canonical topological generator F, called the *Frobenius*, which can also be viewed as an element of $\mathrm{Gal}(K_{nr}/K)$. It corresponds to the automorphism $x \mapsto x^q$ of $\bar{\kappa}$, where q is the cardinality of κ.

If L is a finite Galois extension of a local field K, then the group $\mathrm{Gal}(L/K)$ is solvable ([47], Chap. IV, Cor. 5). In the case where K is a finite field extension of $\mathbf{Q}_p$, we have that ([47], Chap. IV, Cor. 4 or [42], Prop. 7.5.2) the inertia subgroup I of $\mathrm{Gal}(\bar{K}/K)$ has a unique p-Sylow I_p which is normal in I, and the quotient $V := I/I_p$ is isomorphic to $\mathbf{Z}'_p := \prod_{\ell \neq p} \mathbf{Z}_\ell$. More precisely, we have $I_p = \mathrm{Gal}(\bar{K}/K_{mr})$, where K_{mr} is the maximal tamely ramified extension of K.

7.5   Structure Theorem; Filtration of the Group of Units

We begin by the following important result proved (under more general hypotheses) in [47], Chap. II, Prop. 7:

Theorem 7.16 (Hensel lemma) *Let K be a complete field for a discrete valuation with ring of integers A and residue field κ. Let $f \in A[X]$ be a polynomial with reduction $\bar{f} \in \kappa[X]$ modulo the maximal ideal of A. Then any simple root of $\bar{f}$ in κ lifts uniquely to a root of f in A.*

The following theorem (proved in a much more general setting in [47], Chap. II, Th. 2 and Th. 4) describes the structure of local fields depending on whether their characteristic is 0 or $p > 0$.

Theorem 7.17 *Let K be a complete field for a discrete valuation whose residue field κ is of characteristic p. Then:*

(a) *If K is of characteristic zero, then K is a finite extension of the field $\mathbf{Q}_p$. In this case we call K a p-adic field.*

(b) *IF K of characteristic p, it is isomorphic to the field of Laurent series $\kappa((t))$.*

Let K be a local field with ring of integers $\mathcal{O}_K$ and residue field κ. Let $U_K = \mathcal{O}_K^*$ be the *group of units* of K. It is the multiplicative group of elements with zero valuation, so that the multiplicative group K^* is isomorphic (via the choice of a uniformiser) to $\mathbf{Z} \times U_K$. We define a filtration of U_K by setting for each $i \geqslant 1$:

$$U_K^i = \{x \in K, v(1-x) \geqslant i\}.$$

The groups $U_K, U_K^1, \ldots, U_K^i, \ldots$ form a decreasing sequence of subgroups of U_K whose intersection is $\{1\}$. Elements of U_K^1 are called *principal units*. By definition of the topology induced by the valuation, the group U_K is profinite and the U_K^i form a basis of neighbourhood of $\{1\}$. The group U_K^1 is the projective limit of the U_K^1/U_K^i for $i \geqslant 1$.

Theorem 7.18 *With the notations below:*

(a) *We have $U_K/U_K^1 \simeq \kappa^*$ and for $i \geqslant 1$, the group U_K^i/U_K^{i+1} is isomorphic to the additive group κ. In particular U_K^1 is a pro-p-group.*

(b) *Assume that K is a p-adic field with $[K : \mathbf{Q}_p] = n$. Then $U_K^m \subset K^{*p}$ for m large enough and the abelian group U_K^1 is isomorphic to a direct product $\mathbf{Z}_p^n \times F \simeq \mathcal{O}_K \times F$, where F is a finite cyclic group whose order is a power of p.*

(c) *Under the assumptions and notations of (b), the group U_K is isomorphic to $U_K^1 \times \kappa^*$, and the group K^* to $\mathbf{Z} \times \mathbf{Z}_p^n \times F \times \kappa^*$.*

For a) (which holds true without the assumption that the residue field is finite), see [47], Chap. IV, Prop. 6. Part (b) (which is classical when $K = \mathbf{Q}_p$) is a straightforward consequence of loc. cit., Chap. XIV, Prop. 9 and 10.

$$y_N = \prod_{i=1}^{f} (1 + \alpha_i \pi^n)^{a_i p^\nu},$$

and where the $a_i \in \{0, \ldots, p-1\}$ have been chosen so that $y_N \equiv 1 + x_N \bmod \mathfrak{m}^{N+1}$ (we still set $N = np^\nu$ where n is not divisible by p). Show that additionally such a y_N is unique and that

$$1 + x_1 = \prod_{N=1}^{\infty} y_N.$$

(d) Show that any element of U_K^1 is uniquely written as an infinite product of factors of the form $(1 + \alpha_i \pi^n)^b$ with $1 \leqslant i \leqslant f$, n a positive integer prime to p, and $b \in \mathbf{Z}_p$.

(e) Deduce that U_K^1 is isomorphic (as $\mathbf{Z}_p$-module or as a profinite group) to $\mathbf{Z}_p^{\mathbf{N}}$.

Chapter 8
Brauer Group of a Local Field

Throughout this chapter, K is a *local field*, that is, a complete field for a discrete valuation v with finite residue field κ of characteristic p. Recall (see Theorem 7.17) that when K is of characteristic zero, it is a p-adic field ($=$ finite extension of $\mathbf{Q}_p$); else K is of characteristic p and it is then isomorphic to the field of Laurent series $\kappa((t))$.

We start developing the *local class field theory* whose aim is to describe abelian extensions of a local field K, their Galois groups and their link to finite index open subgroups of K^*. The first step, which is the main objective of this chapter, is to calculate the Brauer group Br K. The key point is to show that Br $K = \mathrm{Br}(K_{\mathrm{nr}}/K)$, for example by showing that the Brauer group of the maximal unramified extension K_{nr} is trivial (or even that K_{nr} is of cohomological dimension 1). One can establish this directly if one knows the (rather difficult) theorem of Lang stating that K_{nr} is C_1 ([30]) or else one can show it using the same technique as in Theorem 6.22 (see also Exercise 6.1) by norm computations in local fields ([47], Sect. XII.1). Another option is to use the characterisation of the Brauer group in terms of central simple algebras ([47], Sect. XII.2). We will follow the method of [42] (see also Serre's article in [9], Chap. VI), which is slightly less general in as far as it uses the assumption that κ is finite, but has an advantage of only using what has been previously covered in this book.

8.1 Local Class Field Axiom

Recall (Sect. 7.5) that we have a filtration of the group of units $U_K = \mathcal{O}_K^*$ by subgroups $U_K^i, i \geqslant 1$.

© Springer Nature Switzerland AG 2020
D. Harari, *Galois Cohomology and Class Field Theory*, Universitext,
https://doi.org/10.1007/978-3-030-43901-9_8

Lemma 8.1 *Let L be a finite Galois extension of K with Galois group G. Then there exists a sub-G-module V_1 of U_L^1 which is of finite index (in U_L or U_L^1) and cohomologically trivial.*

Proof By the normal basis theorem ([7], Sect. 10), there exists $\alpha \in L$ such that the family $(g \cdot \alpha)_{g \in G}$ is a basis of the K-vector space L. For $a \in K^*$ of sufficiently large valuation, we then have $M \subset \mathcal{O}_L$, where M is the $\mathcal{O}_K$-module generated by $(ag \cdot \alpha)_{g \in G}$. We note that the G-module M is isomorphic to $\mathcal{O}_K[G]$ and is an open subgroup of the profinite group $\mathcal{O}_L$ (indeed it is defined by a condition of the type: $x \in M$ if and only if each of the coordinates x_g in the basis $(g \cdot \alpha)$ has valuation greater or equal to that of a). In particular, M is of finite index in $\mathcal{O}_L$, which yields an integer $m \geqslant 1$ such that $M \supset \pi^m \mathcal{O}_L$, where π is a uniformiser of K. We then define a sub-G-module V_i of U_L^1 by $V_i = 1 + \pi^{m+i} M$ (the fact that it is indeed a multiplicative subgroup follows from the property $M \supset \pi^m \mathcal{O}_L$).

Each V_i / V_{i+1} is isomorphic to $M/\pi M$ via

$$v_i \longmapsto \pi^{-m-i}(v_i - 1) \quad (\pi M)$$

which implies that these are finite and cohomologically trivial G-modules (isomorphic to the induced modules $(\mathcal{O}_K/\pi)[G]$). By induction, it follows that the same is true for the V_1/V_i for $i \geqslant 1$. As the cohomology of a finite group commutes with projective limits (indexed by the integers) by Proposition 1.31, we deduce that the projective limit V_1 of the V_1/V_i is also a cohomologically trivial G-module. Furthermore, V_1 clearly has a finite index in U_L^1.   □

Proposition 8.2 ("The axiom of the local class field theory") *Let L be a finite Galois extension of K, with cyclic Galois group G. Then $H^1(G, L^*) = 0$ and $\widehat{H}^0(G, L^*)$ is of cardinality $[L : K]$.*

Proof The first assertion follows from Hilbert 90. We now apply the preceding lemma. The Herbrand quotient $h(G, V_1)$ is 1 since V_1 is cohomologically trivial, and $h(G, U_L/V_1) = 1$ by Theorem 2.20, (b). We deduce that $h(G, U_L) = 1$ by Theorem 2.20, (a). As the G-module $\mathbf{Z}$ (with trivial action) is isomorphic to the quotient L^*/U_L, we obtain $h(G, L^*) = h(G, \mathbf{Z}) = [L : K]$ as $H^1(G, \mathbf{Z}) = 0$ and $\widehat{H}^0(G, \mathbf{Z}) = \mathbf{Z}/[L : K]\mathbf{Z}$. As $H^1(G, L^*) = 0$, it shows that the cardinality $\widehat{H}^0(G, L^*)$ is $[L : K]$, as desired.   □

8.2 Computation of the Brauer Group

We start with an important property of cohomology of the group of units.

Proposition 8.3 *Let L be a finite Galois extension of K with Galois group $G = \mathrm{Gal}(L/K)$. Assume that the extension L/K is unramified. Then U_L and U_L^1 are cohomologically trivial G-modules. In particular, $\widehat{H}^0(G, U_L) = U_K/N_{L/K}U_L$ is the trivial group.*

Proof Let λ be the residue field of L. As the extension L/K is unramified, we also have $G = \mathrm{Gal}(\lambda/\kappa)$. Besides we have a filtration of the group of units of L:

$$U_L \supset U_L^1 \supset \cdots \supset U_L^i \supset \cdots$$

with an isomorphism of G-modules $U_L^i/U_L^{i+1} \simeq \lambda$ for any $i > 0$ (by the Theorem 7.18, a). It follows that each U_L^i/U_L^{i+1} is a cohomologically trivial G-module by Proposition 6.2, and by induction on i we immediately deduce that the same holds true for U_L^1/U_L^i and each $i > 0$. We deduce (using Proposition 1.31) that $U_L^1 = \varprojlim_{i>0}(U_L^1/U_L^i)$ is a cohomologically trivial G-module.

On the other hand for every subgroup H of G (corresponding to a finite extension κ' of κ), we have $H^1(H, \lambda^*) = 0$ by Hilbert 90, and $H^2(H, \lambda^*) = 0$ by the triviality of the Brauer group of the finite field κ'. The 2-periodicity of the cohomology of the cyclic group H (Theorem 2.16) implies that λ^* is a cohomologically trivial G-module. It is then also the case for U_L via the exact sequence

$$0 \longrightarrow U_L^1 \longrightarrow U_L \longrightarrow \lambda^* \longrightarrow 0.$$

$\square$

For a generalisation of the statement about U_L^1, see Exercise 8.5.

Let now K_{nr} be the maximal unramified extension of K. The Galois group $\widetilde{\Gamma}_K = \mathrm{Gal}(K_{\mathrm{nr}}/K) \simeq \mathrm{Gal}(\bar{\kappa}/\kappa)$ (that we will simply denote by $\widetilde{\Gamma}$ if there is no risk of confusion) is isomorphic to $\widehat{\mathbf{Z}}$. It is topologically generated by the Frobenius F (Example 7.15, b). We have isomorphisms:

$$H^2(\widetilde{\Gamma}, K_{\mathrm{nr}}^*) \xrightarrow{\ \beta\ } H^2(\widetilde{\Gamma}, \mathbf{Z}) \xrightarrow{\ \delta^{-1}\ } H^1(\widetilde{\Gamma}, \mathbf{Q}/\mathbf{Z}) \xrightarrow{\ \gamma\ } \mathbf{Q}/\mathbf{Z}.$$

Here β is induced by the exact sequence of $\widetilde{\Gamma}$-modules (cf. Remark 7.9)

$$0 \longrightarrow U_{K_{\mathrm{nr}}} \longrightarrow K_{\mathrm{nr}}^* \xrightarrow{\ v\ } \mathbf{Z} \longrightarrow 0,$$

where $U_{K_{\mathrm{nr}}} \subset K_{\mathrm{nr}}^*$ is the subgroup of units, i.e., the group of invertible elements of the ring of integers of K_{nr}: indeed, $H^i(\widetilde{\Gamma}, U_{K_{\mathrm{nr}}}) = 0$ for any $i > 0$ by Proposition 8.3, passing to the limit by Proposition 4.18. The isomorphism δ comes from the triviality of the cohomology of $\mathbf{Q}$ (Corollary 1.51). Lastly γ is obtained by sending every character $\chi \in H^1(\widetilde{\Gamma}, \mathbf{Q}/\mathbf{Z})$ to $\chi(F)$.

Proposition 8.4 Let $\mathrm{inv}_K : H^2(\widetilde{\Gamma}_K, K_{\mathrm{nr}}^*) \to \mathbf{Q}/\mathbf{Z}$ be the composite of the above isomorphisms. Let L be a finite extension of K. Then, we have the commutative diagram

$$\begin{array}{ccc} H^2(\widetilde{\Gamma}_K, K_{\mathrm{nr}}^*) & \xrightarrow{\ \mathrm{inv}_K\ } & \mathbf{Q}/\mathbf{Z} \\ {\scriptstyle \mathrm{Res}} \downarrow & & \downarrow {\scriptstyle \cdot [L:K]} \\ H^2(\widetilde{\Gamma}_L, L_{\mathrm{nr}}^*) & \xrightarrow{\ \mathrm{inv}_L\ } & \mathbf{Q}/\mathbf{Z}. \end{array}$$

Proof Let e be the ramification index of L/K and let f be the residual degree. We have $[L:K] = ef$ (Theorem 7.7). As the isomorphism $\beta_K : H^2(\widetilde{\Gamma}_K, K_{\mathrm{nr}}^*) \to H^2(\widetilde{\Gamma}_K, \mathbf{Z})$ is induced by the valuation (same for $\widetilde{\Gamma}_L$), we have (with the Remark 7.9) $\beta_L \circ \mathrm{Res} = e . \mathrm{Res} \circ \beta_K$. The compatibility of Res with the long exact sequences yields, with similar notations, $\delta_L^{-1} \circ \mathrm{Res} = \mathrm{Res} \circ \delta_K^{-1}$. Lastly, we have $\gamma_L \circ \mathrm{Res} = f . \mathrm{Res} \circ \gamma_K$ as the image of the Frobenius of $\widetilde{\Gamma}_L$ in $\widetilde{\Gamma}_K$ is the fth power of the Frobenius of $\widetilde{\Gamma}_K$. The result then follows from the definition of inv_K. $\square$

Remark 8.5 Note that we do not need the assumption that L is a separable extension of K in the above statement.

We know that $H^2(\widetilde{\Gamma}_K, K_{\mathrm{nr}}^*) = \mathrm{Br}(K_{\mathrm{nr}}/K)$ is a subgroup of the Brauer group Br K. The next very important statement shows that it is in fact the whole of Br K. Our method only uses general results on the cohomology of finite groups combined with the basic properties of local fields. On the other hand it relies on Lemma 8.1 and Proposition 8.2, which use in an essential way that the residue field of K is finite, while other methods work for any complete field K with a discrete valuation and perfect residue field.

Theorem 8.6 *We have* $\mathrm{Br}(K_{\mathrm{nr}}/K) = \mathrm{Br}\, K$.

Proof Let L be a finite Galois extension of K with Galois group G. Let $n = [L:K]$. The proof uses two lemmas:

Lemma 8.7 *The group* $H^2(G, L^*)$ *is finite and its cardinality divides* n.

Proof If G is cyclic, this follows immediately from Proposition 8.2 and Theorem 2.16. If now G is an ℓ-group with ℓ prime, its centre is non-trivial and thus contains a normal subgroup $H = \mathrm{Gal}(L/K_1)$ of cardinality ℓ.

We obtain the result by induction on the cardinality of G via the exact sequence (which follows from Proposition 6.10)

$$0 \longrightarrow H^2(\mathrm{Gal}(K_1/K), K_1^*) \longrightarrow H^2(G, L^*) \longrightarrow H^2(\mathrm{Gal}(L/K_1), L^*).$$

In the general case, let S be the set of prime numbers dividing n and consider, for $\ell \in S$, a ℓ-Sylow G_ℓ of G. As $H^2(G, L^*) = \bigoplus_{\ell \in S} H^2(G, L^*)\{\ell\}$, the Lemma 3.2 says that the restriction

$$H^2(G, L^*) \longrightarrow \bigoplus_{\ell \in S} H^2(G_\ell, L^*)$$

is injective. The case of ℓ-groups (applied to the finite extension of K associated with each G_ℓ by the Galois correspondence) gives that the cardinality of $H^2(G, L^*)$ is finite and divides $\prod_{\ell \in S} \#G_\ell = n$. $\square$

Lemma 8.8 *Let K_n be the unramified extension of K of degree n. Then we have*

$$H^2(G, L^*) = H^2(\mathrm{Gal}(K_n/K), K_n^*),$$

where we have identified the two groups with their images in Br K.

Proof By Lemma 8.7, the cardinality of $H^2(G, L^*)$ divides $n = [L : K] = [K_n : K]$, which is the cardinality of $H^2(\mathrm{Gal}(K_n/K), K_n^*)$ by Proposition 8.2 and Theorem 2.16 since K_n/K is cyclic (cf. Example 7.15, b). It is thus enough to prove that $H^2(\mathrm{Gal}(K_n/K), K_n^*) \subset H^2(G, L^*)$. By Proposition 8.4, we have a commutative diagram whose arrows inv_K and inv_L are isomorphisms and the first line is exact:

$$
\begin{array}{ccccccc}
0 & \longrightarrow & H^2(G, L^*) & \longrightarrow & \mathrm{Br}\,K & \xrightarrow{\ \mathrm{Res}\ } & \mathrm{Br}\,L \\
 & & & & \uparrow & & \uparrow \\
 & & & & H^2(\widetilde{\Gamma}_K, K_{\mathrm{nr}}^*) & \xrightarrow{\ \mathrm{Res}\ } & H^2(\widetilde{\Gamma}_L, L_{\mathrm{nr}}^*) \\
 & & & & \mathrm{inv}_K \downarrow & & \downarrow \mathrm{inv}_L \\
 & & & & \mathbf{Q}/\mathbf{Z} & \xrightarrow{\ \cdot n\ } & \mathbf{Q}/\mathbf{Z}.
\end{array}
$$

Let $a \in H^2(\mathrm{Gal}(K_n/K), K_n^*)$. Then we have $na = 0$ and $a \in H^2(\widetilde{\Gamma}_K, K_{\mathrm{nr}}^*)$ since K_n/K is unramified. The diagram then implies that the restriction of a to Br L is zero, i.e., $a \in H^2(G, L^*)$. $\square$

End of Proof of the Theorem 8.6 As Br K is the inductive limit (over finite Galois extensions L of K) of the $H^2(\mathrm{Gal}(L/K), L^*)$, the preceding lemma implies that Br K is contained in the inductive limit (over $n > 0$) of the $H^2(\mathrm{Gal}(K_n/K), K_n^*)$, hence in $\mathrm{Br}(K_{\mathrm{nr}}/K)$. $\square$

We deduce

Theorem 8.9 *Let K be a local field. Then we have an isomorphism*

$$\mathrm{inv}_K : \mathrm{Br}\,K \longrightarrow \mathbf{Q}/\mathbf{Z}.$$

If L is a finite extension of K, the restriction Br $K \to$ Br L *corresponds to the multiplication by $[L : K]$ on $\mathbf{Q}/\mathbf{Z}$. If furthermore L/K is separable, the corestriction* Br $L \to$ Br K *corresponds to the identity of $\mathbf{Q}/\mathbf{Z}$.*

Proof This follows from Theorem 8.6, and Proposition 8.4, and from the formula of Theorem 1.48, which says that Cores ∘ Res is the multiplication by $[L : K]$. $\square$

Corollary 8.10 *Under the assumptions of Theorem 8.9, an element a of* Br K *has image zero in* Br L *if and only if na = 0, where n := [L : K].*

If we furthermore assume that L/K is Galois, the image of $\mathrm{Br}(L/K) = H^2(\mathrm{Gal}(L/K), L^*)$ *by* inv_K *is the subgroup* $(\frac{1}{n}\mathbf{Z})/\mathbf{Z}$ *of* $\mathbf{Q}/\mathbf{Z}$.

8.3 Cohomological Dimension; Finiteness Theorem

Let G be a profinite group. Recall that we have defined the order of G and the index of a closed subgroup of G as supernatural numbers (Definition 4.6).

We deduce from Theorem 8.6 the cohomological dimension of a local field.

Theorem 8.11 *Let K be a local field. Let ℓ be a prime number. We denote by K_{nr} the maximal unramified extension of K.*

(a) *Let L be an algebraic separable extension of K with absolute Galois group Γ_L. If ℓ^∞ divides $[L : K]$, then $\mathrm{cd}_\ell(\Gamma_L) \leqslant 1$ and $(\mathrm{Br}\, L)\{\ell\} = 0$.*

(b) *The absolute Galois group I of K_{nr} (which is the inertia group of K) satisfies $\mathrm{cd}(I) \leqslant 1$, and we have* Br $K_{\mathrm{nr}} = 0$.

(c) *If ℓ is different from the characteristic of K, we have $\mathrm{cd}_\ell(K) = 2$. In particular, $\mathrm{cd}(K) = 2$.*

Proof (a) Note that if ℓ is the characteristic of K, the property $\mathrm{cd}_\ell(\Gamma_L) \leqslant 1$ is automatic by Proposition 6.15. It is then enough to prove that $(\mathrm{Br}\, L_1)\{\ell\} = 0$ for any algebraic separable extension L_1 of L (by Theorem 6.17), and we are immediately reduced to the case $L_1 = L$ since ℓ^∞ then divides $[L_1 : K]$. Observe that Br L is the inductive limit of the Br K' for K' finite extension of K contained in L, the transition maps being the restrictions (this follows from Remark 4.27). Let K' be such an extension and let $\alpha \in$ Br K' be a ℓ^m-torsion element with $m > 0$. Then there exists an intermediate extension K_1 of L/K', finite over K', and whose degree over K' is divisible by ℓ^m since ℓ^∞ divides $[L : K']$. The restriction of α in Br K_1 is zero by Corollary 8.10, hence so is its image in Br L. We have shown that $(\mathrm{Br}\, L)\{\ell\} = 0$, as desired.

(b) We can apply (a) to $L = K_{\mathrm{nr}}$, since for any prime ℓ prime, the order of $\mathrm{Gal}(K_{\mathrm{nr}}/K) \simeq \hat{\mathbf{Z}}$ is divisible by ℓ^∞.

(c) Let $\Gamma_K = \mathrm{Gal}(\bar{K}/K)$ and $I = \mathrm{Gal}(\bar{K}/K_{\mathrm{nr}})$. We just saw that $\mathrm{cd}_\ell(I) \leqslant 1$ and in addition $\mathrm{cd}_\ell(\Gamma_K/I) \leqslant 1$ since Γ_K/I is the absolute Galois group of the residue field κ, assumed finite, hence of cohomological dimension $\leqslant 1$ by Corollary 6.24. Thus we obtain $\mathrm{cd}_\ell(K) \leqslant 2$ by Proposition 5.13. On the other hand for ℓ different from the characteristic of K, we have $H^2(K, \mu_\ell) = (\mathrm{Br}\, K)[\ell] \neq 0$, hence $\mathrm{cd}_\ell(K) = 2$.   □

Remark 8.12 If K is of characteristic $p > 0$ and with the absolute Galois group Γ_K, we have $\mathrm{cd}_p(\Gamma_K) = 1$ by Proposition 6.15 but the Theorem 8.9 shows that *we do* *not have* $(\mathrm{Br}\,K)[p] = 0$. Hence we do not have $\dim(K) \leqslant 1$ in the "strong " sense of Serre ([49], Part. II, Sect. 3). On the other hand the assertion (a) implies that an algebraic separable extension of K such that p^∞ divides $[L:K]$ (e.g. K_{nr}) satisfies this stronger property. We will see later (Theorem 10.6 and Remark 10.7) that for $\ell \neq \mathrm{Car}\,K$, we have $\mathrm{scd}_\ell(K) = 2$.

Proposition 8.13 *Let K be a p-adic field. Let $n > 0$.*

(1) The group $H^1(K, \mu_n) \simeq K^*/K^{*^n}$ is finite.
(2) We have $H^2(K, \mu_n) \simeq \mathbf{Z}/n\mathbf{Z}$.

Proof (1) We have already seen (via the Kummer exact sequence and Hilbert 90) the isomorphism $H^1(K, \mu_n) \simeq K^*/K^{*^n}$. The fact that these groups are finite follows from Theorem 7.18, (b) and the isomorphism $K^* \simeq \mathbf{Z} \times \kappa^* \times U_K^1$ (Theorem 7.18, a), where κ is the residue field of K.
(2) As $\mathrm{Br}\,K \simeq \mathbf{Q}/\mathbf{Z}$, we have $H^2(K, \mu_n) = (\mathrm{Br}\,K)[n] \simeq \mathbf{Z}/n\mathbf{Z}$. $\square$

Remark 8.14 If $K = k((t))$, with k finite of characteristic $p > 0$, it is no longer true that K^*/K^{*^p} and $H^1(K, \mathbf{Z}/p)$ are finite. On the other hand the Proposition 8.13 is still true for n not divisible by p since the nth powers form an open subgroup of the profinite group U_K (cf. Remark 7.20). Similarly the following corollary is still valid if the order of M is not divisible by p.

Corollary 8.15 *Let K be a p-adic field and M a finite Γ_K-module. Then $H^r(\Gamma_K, M)$ is finite for any $r \geqslant 0$.*

Proof Let n be the order of M. By what we have seen, $H^r(K, \mu_n)$ is finite for $r = 0, 1, 2$, and zero for $r \geqslant 3$. As M is finite, we can find a finite Galois extension L/K such that the action of Γ_L on μ_n and on M is trivial. In particular the Γ_L-module M is isomorphic to a direct sum of the μ_{n_i}. As, by the structure of the cohomology of μ_n, the groups $H^q(\Gamma_L, M)$ are finite for any $q \geqslant 0$, the Hochschild–Serre spectral sequence (Theorem 1.44)

$$H^p(\mathrm{Gal}(L/K), H^q(\Gamma_L, M)) \implies H^{p+q}(\Gamma_K, M)$$

allows us to conclude that the $H^r(\Gamma_K, M)$ are finite. $\square$

8.4 Exercises

Exercise 8.1 Let p be a prime number. Let $K = \mathbf{F}_p((t))$ and $\bar{K}$ a separable closure of K. We denote by Γ_K the absolute Galois group of K and Γ_p a p-Sylow of Γ_K. Let $L \subset \bar{K}$ be the fixed field of Γ_p and $\Gamma_L = \mathrm{Gal}(\bar{K}/L)$ its absolute Galois group.

(a) Show that $cd(\Gamma_L) \leqslant 1$.
(b) Do we have $\mathrm{Br}\, L = 0$?

Exercise 8.2 Let ℓ be a prime number. Let K be a finite extension of $\mathbf{Q}_l$ with Galois group $G = \mathrm{Gal}(\bar{K}/K)$. Fix a prime number p (possibly equal to ℓ).

(a) Let L be an algebraic extension of K. Assume that $[L : K]$ is divisible by p^∞. What is $(\mathrm{Br}\, L)\{p\}$?
(b) Let $G_K(p) = G/I$ be the largest quotient of G which is a pro-p-group: thus we have $G_K(p) = \mathrm{Gal}(K(p)/K)$, where $K(p)$ is an algebraic extension of K and $I = \mathrm{Gal}(\bar{K}/K(p))$. Show that $cd_p(I) \leqslant 1$.
(c) Show that every homomorphism from I to a pro-p-group is trivial. Deduce that if A is a p-primary torsion $G_K(p)$-module, we have $H^1(I, A) = 0$.
(d) Let A be a p-primary torsion $G_K(p)$-module. Show that for any integer $i \geqslant 0$, the inflation homomorphism $H^i(G_K(p), A) \to H^i(G, A)$ is an isomorphism.

Exercise 8.3 Let K be a p-adic field with absolute Galois group Γ_K. Let M be a finite type Γ_K-module. Show that $H^1(K, M)$ is finite.

Exercise 8.4 Let G be a profinite group. Let M be a discrete and divisible G-module. For any abelian group A, we denote by $A[n]$ the n-torsion subgroup of A and by A/n the quotient of A by the subgroup of elements of the form ny with $y \in A$.

(a) Show that for any $n > 0$ and any $i > 0$, we have an exact sequence

$$0 \longrightarrow H^{i-1}(G, M)/n \longrightarrow H^i(G, M[n]) \longrightarrow H^i(G, M)[n] \longrightarrow 0.$$

(b) In the rest of this exercise we suppose that $G = \mathrm{Gal}(\bar{K}/K)$ is the absolute Galois group of a p-adic field K, and that M is isomorphic as an abelian group to $(\bar{K}^*)^m$ with $m \in \mathbf{N}^*$. Show that for any integer $n > 0$, the group $H^i(G, M)[n]$ is finite.
(c) Assume furthermore that there exists a finite Galois extension L of K such that M is isomorphic as a $\mathrm{Gal}(\bar{K}/L)$-module to $(\bar{K}^*)^m$. Show that $H^1(G, M)$ is finite.

Exercise 8.5 (a) Let G be a finite group. Let M be a G-module filtered by a decreasing sequence $(M_n)_{n \geqslant 1}$ of submodules

$$M = M_1 \supset M_2 \supset \cdots \supset M_n \supset \cdots$$

such that the canonical map $M \to \varprojlim_n (M/M_n)$ is an isomorphism. Let q be an integer such that $H^q(G, M_n/M_{n+1}) = 0$ for every $n \geqslant 1$. Show that $H^q(G, M) = 0$ (hint: one can work directly with the cocycles).

(b) Let K be a complete field for a discrete valuation with perfect residue field κ. Let L be a finite unramified Galois extension of K, let $G = \mathrm{Gal}(L/K)$. Let U_L^1 be the multiplicative group consisting of those $x \in L$ such that $v_L(1 - x) > 0$, where v_L is the valuation of L. Show that $H^q(G, U_L^1) = 0$ for any $q > 0$.

Exercise 8.6 Let K be a local field of characteristic $p > 0$. Show that the group $H^1(K, \mathbf{Z}/p\mathbf{Z})$ is infinite. (If one considers the "fppf" cohomology, the group $H^1(K, \mu_p)$ is isomorphic to K^*/K^{*p}, which is also infinite: the sheaf μ_p is nontrivial and fits into an exact sequence $0 \to \mu_p \to \mathbf{G}_m \xrightarrow{\cdot p} \mathbf{G}_m \to 0$, where $\mathbf{G}_m$ is the multiplicative group, cf. [39], Part. III; while in Galois cohomology, we have $\mu_p(\bar{K}) = 0$ in characteristic p.)

Chapter 9
Local Class Field Theory: The Reciprocity Map

Throughout this chapter K is a local field. We start our study of the abelian Galois group $\mathrm{Gal}(K^{\mathrm{ab}}/K)$ and complete it in the next two chapters.

General results on the reciprocity map will in particular be used in all the proofs of the *existence theorem*, whose first proof will be given at the end of the present chapter in the case of a p-adic field.

9.1 Definition and Main Properties

Definition 9.1 Let L be a finite Galois extension of K with group G. Let $n :=$ $[L : K]$. We call *fundamental class* of the extension L/K the unique element $u_{L/K}$ of $\mathrm{Br}(L/K) = H^2(G, L^*)$ such that $\mathrm{inv}_K(u_{L/K}) = 1/n \in \mathbf{Q}/\mathbf{Z}$.

Recall that we have an isomorphism $\mathrm{inv}_K : \mathrm{Br}\, K \to \mathbf{Q}/\mathbf{Z}$ (Theorem 8.9) and that the group $\mathrm{Br}(L/K)$ is precisely the n-torsion subgroup of $\mathrm{Br}\, K$ (Corollary 8.10), which justifies the definition. We can thus apply Theorem 3.14 (Tate–Nakayama) to the G-module $A = L^*$ and the element $u_{L/K}$ of $H^2(G, L^*)$. Let indeed $G_p = \mathrm{Gal}(L/K_p)$ be a p-Sylow of G. We have $H^1(G_p, L^*) = 0$ by Hilbert 90, and $H^2(G_p, L^*) = \mathrm{Br}(L/K_p)$ is of order $\#G_p$. It is generated by the restriction u_{L/K_p} of u to $H^2(G_p, L^*)$ (by Theorem 8.9). By taking $n = -2$ and applying Proposition 2.4, we obtain the following result, where $N_{L/K}$ (or simply N when there is no possible confusion) denotes the norm of the extension L/K:

Theorem 9.2 *Let L be a finite Galois extension of a local field K. Then the cup-product with $u_{L/K}$ defines an isomorphism*

$$\theta_{L/K} : G^{\mathrm{ab}} \longrightarrow K^*/NL^*,$$

© Springer Nature Switzerland AG 2020
D. Harari, *Galois Cohomology and Class Field Theory*, Universitext,
https://doi.org/10.1007/978-3-030-43901-9_9

where $G := \mathrm{Gal}(L/K)$. The isomorphism

$$\omega_{L/K} := \theta_{L/K}^{-1} : K^*/NL^* \longrightarrow G^{\mathrm{ab}}$$

is called the reciprocity isomorphism *associated to the extension L/K.*

For simplicity, we will also denote by $\omega_{L/K}$ the surjection $K^* \to G^{\mathrm{ab}}$ induced by the reciprocity isomorphism. The following proposition establishes a link between this isomorphism and the cup-product.

Proposition 9.3 *Let L be a finite abelian extension of a local field K. Let $G = \mathrm{Gal}(L/K)$ and let $\chi : G \longrightarrow \mathbf{Q}/\mathbf{Z}$ be a character of G. Denote by d_χ its image in $H^2(G, \mathbf{Z})$ via the coboundary associated to the exact sequence*

$$0 \longrightarrow \mathbf{Z} \longrightarrow \mathbf{Q} \longrightarrow \mathbf{Q}/\mathbf{Z} \longrightarrow 0.$$

Let $a \in K^*$ with image $\bar{a} \in \widehat{H}^0(G, L^*) = K^*/NL^*$. Then

$$\chi(\omega_{L/K}(\bar{a})) = \mathrm{inv}_K(\bar{a} \cup d_\chi),$$

where $\mathrm{inv}_K : \mathrm{Br}(L/K) \to \mathbf{Q}/\mathbf{Z}$ is the local invariant.

Note that here we do not need to worry about in which order we take the cup-product (cf. Proposition 2.27, b) as $\bar{a} \cup d_\chi = d_\chi \cup \bar{a}$ since the cohomology classes we consider are in even degree.

Proof Let $u = u_{L/K} \in H^2(G, L^*)$ be the fundamental class. Let $s = \omega_{L/K}(\bar{a}) \in G = \widehat{H}^{-2}(G, \mathbf{Z})$ (Recall that G is abelian). Let $n := [L : K]$. By definition of the reciprocity isomorphism, we have

$$u \cup s = \bar{a} \in \widehat{H}^0(G, L^*)$$

hence, by the associativity of the cup-product (Proposition 2.27, a) and compatibility with the coboundaries (Proposition 2.23)

$$\bar{a} \cup d_\chi = u \cup (s \cup d_\chi) = u \cup (d(s \cup \chi)),$$

where $s \cup \chi \in \widehat{H}^{-1}(G, \mathbf{Q}/\mathbf{Z})$ and we still denote by d the coboundary $\widehat{H}^{-1}(G, \mathbf{Q}/\mathbf{Z}) \to \widehat{H}^0(G, \mathbf{Z})$. As the action of G on $\mathbf{Q}/\mathbf{Z}$ is trivial, the group $\widehat{H}^{-1}(G, \mathbf{Q}/\mathbf{Z})$ is simply the subgroup $\frac{1}{n}\mathbf{Z}/\mathbf{Z}$ of $\mathbf{Q}/\mathbf{Z}$ and by Proposition 2.32, the cup-product $s \cup \chi$ identifies with $\chi(s)$. Let then $\chi(s) = r/n$ with $r \in \mathbf{Z}$. It is straightforward that $d(r/n) = r \in \widehat{H}^0(G, \mathbf{Z}) = \mathbf{Z}/n\mathbf{Z}$, hence

$$u \cup (d(s \cup \chi)) = u \cup r = r \cdot u \in \mathrm{Br}(L/K) = H^2(G, L^*).$$

On the other hand we have the equality $\mathrm{inv}_K(r \cdot u) = r/n$ by definition of u and hence finally $\mathrm{inv}_K(r \cdot u) = \chi(s)$, or in other words

$$\text{inv}_K(\bar{a} \cup d_\chi) = \chi(s),$$

as desired. □

Corollary 9.4 *Let L be a finite Galois extension of K. Let M be a finite Galois extension of K containing L. We have a commutative diagram:*

$$
\begin{array}{ccc}
K^*/N_{M/K}M^* & \xrightarrow{\ \omega_{M/K}\ } & \text{Gal}(M/K)^{\text{ab}} \\
\downarrow & & \downarrow \\
K^*/N_{L/K}L^* & \xrightarrow{\ \omega_{L/K}\ } & \text{Gal}(L/K)^{\text{ab}}
\end{array}
$$

where the vertical arrows are the canonical surjections.

Proof We are immediately reduced to the case where L and M are abelian extensions of K (by if necessary replacing them by the maximal abelian extensions of K contained in L and M, respectively). Let then $a \in K^*$ with image $\bar{a}_L$ in $K^*/N_{L/K}L^*$ and $\bar{a}_M$ in $K^*/N_{M/K}M^*$. Let χ an arbitrary character of $\text{Gal}(L/K)$. We still denote by χ the character it induces on $\text{Gal}(M/K)$ (of which $\text{Gal}(L/K)$ is a quotient). We then have, by Proposition 9.3 and by the functoriality of the cup-product:

$$\chi(\omega_{L/K}(\bar{a}_L)) = \text{inv}_K(\bar{a}_L \cup d_\chi) = \text{inv}_K(\bar{a}_M \cup d_\chi) = \chi(\omega_{M/K}(\bar{a}_M)).$$

This proves the desired compatibility as the character χ is arbitrary.

The above corollary allows us to pass to the limit. This justifies the following definition. □

Definition 9.5 The homomorphism

$$\omega : K^* \longrightarrow \Gamma_K^{\text{ab}}$$

obtained from the $\omega_{L/K}$ by passing to the limit over the finite abelian extensions L of K, is called the *reciprocity map*. It has a dense image and induces an isomorphism of $\varprojlim_L K^*/NL^*$ with Γ_K^{ab}.

Here Γ_K^{ab} is the abelianisation of Γ_K as a profinite group (it is the quotient of Γ_K by the closure of its derived subgroup in the usual sense). The density of the image of ω follows from the surjectivity of the $\omega_{L/K}$ and from the fact that the groups $\text{Gal}(K^{\text{ab}}/L)$ (for L/K finite abelian) form a basis of open sets of $\{1\}$ in Γ_K^{ab}.

We will see in Sect. 9.2 that, for a p-adic field, the reciprocity map $\omega : K^* \to \Gamma_K^{\text{ab}}$ induces an isomorphism of the profinite completion of K^* with Γ_K^{ab}, and hence this profinite completion is isomorphic to $\varprojlim_L K^*/NL^*$ (the limit being taken over the finite abelian extensions L of K). However this requires the knowledge of the existence theorem, whose first proof in the case of a p-adic fields will be given at the end of this chapter. For now, we just deduce from the above results that $\varprojlim_L K^*/NL^*$ is a quotient of the profinite completion $\widehat{K}^*$ of K^*.

Corollary 9.6 *Let K be a local field with Galois group $\Gamma_K = \mathrm{Gal}(\bar{K}/K)$. Let χ be a character of Γ_K^{ab} (or, what comes down to the same, of Γ_K). Let $b \in K^*$. Then:*
   (a) we have

$$\mathrm{inv}_K(b \cup \chi) = \chi(\omega(b)),$$

where, in the cup-product, χ is seen as an element of $H^2(K, \mathbf{Z})$ and b as an element of $H^0(K, \bar{K}^*)$.
   (b) Let $n > 0$. Denote by Γ_K^{ab}/n the quotient of the group Γ_K^{ab} (the group law denoted additively) by its subgroup $n\Gamma_K^{ab}$. Let a be the class of b in $H^1(K, \mu_n) = K^*/K^{*n}$. Assume that χ is in the n-torsion subgroup $H^1(K, \mathbf{Z}/n\mathbf{Z})$ of $H^1(K, \mathbf{Q}/\mathbf{Z}) = H^2(K, \mathbf{Z})$, so that χ induces a character (still denoted by χ) of Γ_K^{ab}/n. Let $\omega_n : K^*/K^{*n} \to \Gamma_K^{ab}/n$ be the map induced by ω. Then we have

$$\mathrm{inv}_K(\chi \cup a) = \chi(\omega_n(b)),$$

where $\chi \cup a$ is seen in $H^2(K, \mu_n) = {}_n\,\mathrm{Br}\,K$.

Proof (a) follows from the Proposition 9.3 by passing to the limit over the finite abelian extensions L of K. We obtain (b) from (a) and from the compatibility of the cup-product, up to a sign change (Proposition 2.28), with the exact sequences

$$0 \longrightarrow \mu_n \longrightarrow \bar{K}^* \longrightarrow \bar{K}^* \longrightarrow 0; \quad 0 \longrightarrow \mathbf{Z} \longrightarrow \mathbf{Z} \longrightarrow \mathbf{Z}/n\mathbf{Z} \longrightarrow 0. \qquad \square$$

The next proposition provides another compatibility of the reciprocity maps and an application to the subgroups of K^* defined by the norms, that we will revisit in Sect. 9.2.

Proposition 9.7 *Let K be a local field with separable closure $\bar{K}$. Let $E \subset \bar{K}$ be a finite separable extension of K. Let $\Gamma_K = \mathrm{Gal}(\bar{K}/K)$ and $\Gamma_E = \mathrm{Gal}(\bar{K}/E)$ be the absolute Galois groups of K and E, respectively.*
   (a) we have a commutative diagram:

$$
\begin{array}{ccc}
E^* & \xrightarrow{\;\omega_E\;} & \Gamma_E^{ab} \\
{\scriptstyle N_{E/K}}\downarrow & & \downarrow{\scriptstyle i} \\
K^* & \xrightarrow{\;\omega_K\;} & \Gamma_K^{ab}
\end{array}
$$

where i is induced by the inclusion $\Gamma_E \to \Gamma_K$.
   (b) Let L be the maximal abelian extension of K contained in E. Then $N_{E/K}E^* = N_{L/K}L^*$.

Proof (a) Let F be a finite Galois extension of K containing E. Observe that the map $i_F : \mathrm{Gal}(F/E)^{\mathrm{ab}} \to \mathrm{Gal}(F/K)^{\mathrm{ab}}$ induced by the inclusion $\mathrm{Gal}(F/E) \to \mathrm{Gal}(F/K)$ corresponds to the corestriction

$$H_1(\mathrm{Gal}(F/E), \mathbf{Z}) = \widehat{H}^{-2}(\mathrm{Gal}(F/E), \mathbf{Z})$$
$$\longrightarrow H_1(\mathrm{Gal}(F/K), \mathbf{Z}) = \widehat{H}^{-2}(\mathrm{Gal}(F/K), \mathbf{Z}),$$

by Proposition 2.4 and the definition of the corestriction in homology given at the beginning of Sect. 2.2. As the norm

$$H^0(\mathrm{Gal}(F/E), F^*) = E^* \longrightarrow H^0(\mathrm{Gal}(F/K), F^*) = K^*$$

also corresponds to the corestriction, the compatibility of the cup-product to corestrictions (Proposition 2.27, e) and the definition of $\theta_{F/E} = \omega_{F/E}^{-1}$ yield a commutative diagram:

$$
\begin{array}{ccc}
E^* & \xrightarrow{\;\;\omega_{F/E}\;\;} & \mathrm{Gal}(F/E)^{\mathrm{ab}} \\
{\scriptstyle N_{E/K}}\Big\downarrow & & \Big\downarrow{\scriptstyle i_F} \\
K^* & \xrightarrow{\;\;\omega_{F/K}\;\;} & \mathrm{Gal}(F/K)^{\mathrm{ab}}.
\end{array}
\tag{9.1}
$$

We obtain the result by passing to the limit over the finite Galois extensions F of K containing E.

(b) The inclusion $N_{E/K}E^* \subset N_{L/K}L^*$ follows from the transitivity of the norms. Let us show the reverse inclusion (that would easily follow from Corollary 9.4 if we had further assumed E/K to be Galois). Let F be the Galois closure of E over K, let $G = \mathrm{Gal}(F/K)$ and $H = \mathrm{Gal}(F/E)$. We thus have

$$K \subset L \subset E \subset F.$$

Let G' be the derived subgroup of G. By definition of L, Galois theory says that $\mathrm{Gal}(F/L)$ is the smallest normal subgroup G_1 of G such that $H \subset G_1$, and G/G_1 is abelian. This last condition is equivalent to $G' \subset G_1$. In other words, we have $\mathrm{Gal}(F/L) = G'.H$.

Let $a \in N_{L/K}L^*$. Corollary 9.4 applied to extensions $K \subset L \subset F$ says that $\omega_{F/K}(a)$ is in the image of the canonical map $H^{\mathrm{ab}} = H/H' \to G/G'$. The diagram (9.1) and the surjectivity of $\omega_{F/E}$ imply the existence of a $b \in E^*$ such that

$$\omega_{F/K}(N_{E/K}(b)) = \omega_{F/K}(a).$$

As the cokernel of $\omega_{F/K}$ is $N_{F/K}F^*$, we can find $b' \in F^*$ such that

$$a = N_{E/K}(b).N_{F/K}(b'),$$

which shows that $a \in N_{E/K} E^*$ because $N_{F/K}(b') \in N_{E/K} E^*$ by the transitivity of the norms.   □

To finish this paragraph, we will prove a lemma about the reciprocity map which will prove useful in the sequel (particularly for the Lubin–Tate theory).

Lemma 9.8 *Let K be a local field with Galois group $\Gamma_K = \mathrm{Gal}(\bar{K}/K)$. Let $\omega_K : K^* \to \Gamma_K^{\mathrm{ab}}$ be the reciprocity map. Then:*

(a) If K' is a finite unramified (hence cyclic) extension of K, we have, for any x in K^*,

$$\omega_{K'/K}(x) = F_K^{v(x)}, \tag{9.2}$$

where F_K is the canonical generator of $\mathrm{Gal}(K'/K)$.

(b) The image by the reciprocity map ω_K of the group of units $U_K = \mathcal{O}_K^*$ is exactly the abelian inertia subgroup $I_K^{\mathrm{ab}} = \mathrm{Gal}(K^{\mathrm{ab}}/K_{\mathrm{nr}})$ of Γ_K^{ab}.

We will later see (Corollary 9.15) that ω_K is in fact injective, and hence induces an isomorphism of U_K with I_K^{ab}.

Proof (a) follows easily from the fact that for any character χ of $\mathrm{Gal}(K'/K)$, we have

$$\chi(\omega_{K'/K}(x)) = \mathrm{inv}_K(x \cup \chi)$$

(Proposition 9.3) and from the definition of inv_K given just before the Proposition 8.4.

To prove (b), it is enough to verify it at a finite level. More precisely, let L be a finite abelian extension with group G over K, we need to show that the image of U_K by the reciprocity map $\omega_{L/K} : K^* \to G$ is exactly the inertia subgroup I of G. Let $I = \mathrm{Gal}(L/K')$, where K' is the maximal unramified extension of K contained in L. By identifying $\mathrm{Gal}(K'/K)$ with the Galois group of the residual extension $\mathrm{Gal}(\kappa'/\kappa)$, we deduce from (a) that for $x \in U_K$, the image of x by $\omega_{K'/K}$ is trivial, which means that $\omega_{L/K}(x)$ is in $I = \mathrm{Gal}(L/K')$.

Conversely, let $t \in I$. As $\omega_{L/K}$ is surjective (with kernel $N_{L/K} L^*$), we can write $t = \omega_{L/K}(a)$ with $a \in K^*$. Let $m = [K' : K] = [\kappa' : \kappa]$ (it is the residual degree of the extension L/K). As the image of t in $\mathrm{Gal}(K'/K)$ is trivial, the formula (9.2) implies that m divides $v(a)$. We know that for any $b \in L^*$, we have $v(N_{L/K}(b)) = m v(b)$ by Theorem 7.7 (d). Choose b in L^* with valuation $v(a)/m$. We then obtain that a and $N_{L/K}(b)$ have the same valuation. Let then $u = a/N_{L/K}(b)$, then $u \in U_K$ and $t = \omega_{L/K}(a) = \omega_{L/K}(u)$, which shows that t is in the image of U_K by $\omega_{L/K}$. □

9.2 The Existence Theorem: Preliminary Lemmas and the Case of a p-adic Field

We start with a lemma, which is the first step in all proofs of the local existence theorem. It will be generalised later to class formations (Proposition 16.31), which can for example be useful for the global existence theorem (Theorem 15.9).

Lemma 9.9 *Let K be a local field.*

(a) Let U be a subgroup of K^*. Assume that U contains a subgroup V of the form $V = N_{L/K} L^*$, where L is a finite abelian extension of K. Then U is itself of the form $U = N_{F/K} F^*$ for some finite abelian extension F of K contained in L.

(b) Fix a separable closure $\bar{K}$ of K. Let L and M be two finite abelian extensions of K, contained in $\bar{K}$, and such that $N_{L/K} L^* = N_{M/K} M^*$. Then $L = M$.

(c) Let F be a finite abelian extension of K. Then $N_{F/K} F^*$ is an open finite index subgroup of K^*.

Note that we already know (Corollary 7.19) that for K a p-adic field, every finite index subgroup of K^* is open. On the other hand, Proposition 9.7, (b) implies that any subgroup of K^* of the form $N_{E/K} E^*$ (where E is a finite separable extension of K) can also be written as $N_{L/K} L^*$ with L/K finite abelian.

Proof (a) the image of U by the reciprocity map $\omega_{L/K} : K^* \to \mathrm{Gal}(L/K)$ is of the form $\mathrm{Gal}(L/F)$, where F is an intermediate extension between K and L. As by assumption U contains $V = \mathrm{Ker}\, \omega_{L/K}$, we have

$$U = \omega_{L/K}^{-1}(\mathrm{Gal}(L/F)).$$

We deduce that $U = N_{F/K} F^*$ using the commutative diagram

$$
\begin{array}{ccc}
K^* & \xrightarrow{\;\omega_{L/K}\;} & \mathrm{Gal}(L/K) \\
\| & & \downarrow \\
K^* & \xrightarrow{\;\omega_{F/K}\;} & \mathrm{Gal}(F/K)
\end{array}
$$

and the fact that the kernel of $\omega_{F/K} : K^* \to \mathrm{Gal}(F/K)$ is $N_{F/K} F^*$.

(b) Let $K^{\mathrm{ab}} \subset \bar{K}$ be the maximal abelian extension of K. The preimages of $\mathrm{Gal}(K^{\mathrm{ab}}/L)$ and $\mathrm{Gal}(K^{\mathrm{ab}}/M)$ by the reciprocity map (whose image is dense) $\omega : K^* \to \mathrm{Gal}(K^{\mathrm{ab}}/K)$ coincide. We deduce that $\mathrm{Gal}(K^{\mathrm{ab}}/L) = \mathrm{Gal}(K^{\mathrm{ab}}/M)$, as if we had for example an element $x \in \mathrm{Gal}(K^{\mathrm{ab}}/L)$ with $x \notin \mathrm{Gal}(K^{\mathrm{ab}}/M)$, we would be able to find an open neighbourhood V of x in $\mathrm{Gal}(K^{\mathrm{ab}}/K)$ which does not meet the closed subgroup $\mathrm{Gal}(K^{\mathrm{ab}}/M)$. As V meets $\mathrm{Im}\, \omega$ by density, we would obtain a contradiction. We then conclude that $L = M$ by Galois theory.

(c) We know that the reciprocity map induces an isomorphism between $K^*/N_{F/K} F^*$ and $\mathrm{Gal}(F/K)$, which shows that $N_{F/K} F^*$ is of finite index in K^*. On the other hand the norm $N_{F/K} : F^* \to K^*$ is continuous (if we fix a basis of the K-vector space F, it is given by a polynomial in the coordinates) and *proper* (the preimage of a compact is compact): indeed, the valuation in K of $N_{F/K}(x)$ is $d.v_F(x)$ (where v_F is the valuation in F and d is the residual degree of F/K) by Theorem 7.7 (d), which shows that the preimage of U_K by $N_{F/K}$ (which is closed in F^* by continuity of $N_{F/K}$) is contained in U_F (which is compact), hence is compact itself. But any compact of K^* is contained in a union of finitely many translates of

U_K since U_K is an open subgroup of K^*, which shows that the preimage of a compact by $N_{F/K}$ is compact.

The properness of $N_{F/K}$ then implies ([6], Sect. 10, Proposition 7) that $N_{F/K}$ is closed, hence $N_{F/K} F^*$ is of finite index and closed in K^*, i.e., open of finite index.   □

The crucial step in the proof of the existence theorem that we will give here in the case of a p-adic field uses *Kummer extensions* (that we will encounter again in the global case, cf. Sect. 13.3); these are extensions obtained by extracting the nth roots. It is convenient to formulate the results in terms of the *local symbols*. More precisely, let k be a field with absolute Galois group Γ_k. Let n be an integer > 0 not divisible by Car k, and such that k *contains an nth primitive root of unity* ζ. Let $a \in k^*$. The choice of ζ (that we fix once and for all in the remainder of this paragraph) allows us to define a character χ_a of $\Gamma_k = \mathrm{Gal}(\bar{k}/k)$ associated to a, by identifying the image $\bar{a}$ of a in $k^*/k^{*n} = H^1(k, \mu_n)$ with an element of $H^1(k, \mathbf{Z}/n)$. In particular, χ_a is a character of Γ_k whose kernel is $\mathrm{Gal}(\bar{k}/k(a^{1/n}))$, and $\mathrm{Gal}(k(a^{1/n})/k)$ is a cyclic group of order dividing n. Thus χ_a is an n-torsion element of the character group of Γ_k.

Definition 9.10 Let a, b in k^*. We define the *symbol* $(a, b) \in (\mathrm{Br}\, k)[n]$ to be the cup-product $b \cup \chi_a$ of $b \in H^0(k, \bar{k}^*) = k^*$ with $\chi_a \in H^2(k, \mathbf{Z}) \simeq H^1(k, \mathbf{Q}/\mathbf{Z})$. It is a bilinear map of the multiplicative group $k^* \times k^*$ to the additive group $\mathrm{Br}\, k$.

Example 9.11 For $n = 2$, the symbol (a, b) is the classical Hilbert symbol, that can be thought of as taking values in $\mathbf{Z}/2\mathbf{Z}$ or in $\{\pm 1\}$ when k is a p-adic field or the field $\mathbf{R}$. In the case where $k = \mathbf{Q}_p$ with p prime, we have explicit formulas to calculate these symbols. These are obtained by local calculations following quite easily from the Hensel lemma and its refinement, cf. [46], Chap. III, Theorem 1. The result is as follows: for an odd prime p and an integer x not divisible by p (or equivalently $x \in \mathbf{F}_p^*$), define the *Legendre symbol* $(\frac{x}{p})$ by $(\frac{x}{p}) = 1$ if x is a square modulo p, and $(\frac{x}{p}) = -1$ otherwise. For x odd, denote also by $\varepsilon(x)$ the class of $(x - 1)/2$ modulo 2 and by $\omega(x)$ the class of $(x^2 - 1)/8$ modulo 2. Then, if we write $a = p^\alpha u$ and $b = p^\beta v$ with $\alpha, \beta \in \mathbf{Z}$ and $u, v \in \mathbf{Z}_p^*$, we have, by considering the Hilbert symbol (a, b) as taking values in $\{\pm 1\}$:

$$(a, b) = (-1)^{\alpha\beta\varepsilon(p)} \left(\frac{u}{p}\right)^\beta \left(\frac{v}{p}\right)^\alpha$$

if $p \neq 2$ and

$$(a, b) = (-1)^{\varepsilon(u)\varepsilon(v) + \alpha\omega(v) + \beta\omega(u)}$$

if $p = 2$. For (a, b) in $\mathbf{R}^*$, naturally (a, b) is non-trivial if and only if a and b are both < 0.

Proposition 9.12 *Let k be a field containing a primitive nth root of unity, with n not divisible by Car k. Let $a, b \in k^*$.*

(a) *If b is a norm of the extension $k(a^{1/n})/k$, then $(a, b) = 0$.*
(b) *we have $(a, -a) = 0$ and $(a, b) = -(b, a)$; Furthermore, we have $(a, 1 - a) = 0$*
   if $a \neq 1$.
(c) *In addition, assume that k is a p-adic field. If an element b of k^* satisfies $(a, b) = 0$ for all a of k^*, then $b \in k^{*n}$.*

Proof (a) Let $L = k(a^{1/n})$. Let $G = \mathrm{Gal}(L/k)$. Then χ_a identifies with an element of $H^2(G, \mathbf{Z})$, so that (Proposition 2.27, d) $\chi_a \cup b$ is also the cup-product of χ_a with the image of b in $\widehat{H}^0(G, L^*) = k^*/N_{L/k}L^*$. The result follows.

(b) Let $x \in k$ be such that $x^n - a \neq 0$. Let α be an nth root of a, so that $k(\alpha) = k(a^{1/n})$. Then

$$x^n - a = \prod_{i=0}^{n-1} (x - \zeta^i \alpha).$$

Let $d \geq 1$ be the largest divisor of n such that a has a dth root in k. Let $m = n/d$. Then the extension $k(a^{1/n})/k$ is cyclic of degree m and the conjugates of $(x - \zeta^i \alpha)$ are the $(x - \zeta^j \alpha)$ with j congruent to i modulo d. Thus

$$x^n - a = \prod_{i=0}^{d-1} N_{k(a^{1/n})/k}(x - \zeta^i \alpha),$$

which shows that $x^n - a$ is a norm of the extension $k(a^{1/n})/k$. By (a), we have $(a, x^n - a) = 0$. We obtain that $(a, -a) = 0$ by letting $x = 0$, and $(a, 1 - a) = 0$ by letting $x = 1$. We finally have:

$$(a, b) + (b, a) = ((a, b) + (a, -a)) + ((b, a) + (b, -b))$$
$$= (a, -ab) + (b, -ba) = (ab, -ab) = 0.$$

(c) By (b), we have $(b, a) = 0$, hence $\chi_b \cup a = 0$. By Corollary 9.6, we have $\chi_b(\omega_k(a)) = 0$ for any a in k^*. This implies that $\chi_b = 0$ (by density of the image of the reciprocity map $\omega_k : k^* \to \Gamma_k^{ab}$), or that the class of b in $k^*/k^{*n} \simeq H^1(k, \mathbf{Z}/n)$ is trivial. $\qquad\square$

Theorem 9.13 (existence theorem) *Let K be a p-adic field. Fix its algebraic closure $\bar{K}$. Consider a finite index subgroup U of K^*. Then there exists a unique finite abelian extension $L \subset \bar{K}$ of K such that $U = N_{L/K}L^*$.*

Proof The uniqueness is the (b) of the Lemma 9.9. Let us prove the existence. Let U be a subgroup of finite index n in K^*. Then U contains the subgroup K^{*n} as any element x of K^*/U satisfies $x^n = 1$. By Lemma 9.9, it is thus enough to show that K^{*n} is of the form $N_{L/K}L^*$ for some finite abelian extension L of K. To do this, we can assume that K contains a primitive nth root of unity ζ: indeed, if F is a finite extension of K containing ζ and if $F^{*n} = N_{E/F}E^*$ with E a finite extension of F, then

$$K^{*n} \supset N_{F/K}(F^{*n}) = N_{E/K}E^*,$$

which is sufficient, by Proposition 9.7 (b) and Lemma 9.9 (a).

Assume that K contains a primitive root of unity. Let $\Gamma = \mathrm{Gal}(K^{\mathrm{ab}}/K)$ be the Galois group of the maximal abelian extension of K, then the subgroup $n\Gamma$ of Γ is closed (since it is compact), and the dual $\mathrm{Hom}_c(\Gamma/n, \mathbf{Q}/\mathbf{Z})$ of the profinite group Γ/n is $H^1(K, \mathbf{Z}/n)$ which is finite by Corollary 8.15. It follows that the group Γ/n is itself finite, and by Galois theory, we can write $\Gamma/n = \mathrm{Gal}(L/K)$, where L is a finite abelian extension of K. Consider the map

$$\omega_n : K^* \longrightarrow \Gamma/n$$

induced by the reciprocity map ω_K. Its kernel contains K^{*n}. Conversely, let $b \in \mathrm{Ker}\,\omega_n$. Let $a \in K^*$, we have $(a, b) = b \cup \chi_a$. As $\chi_a(\omega_K(b)) = 0$ since by assumption $\omega_K(b)$ is in $n\Gamma$ and χ_a is n-torsion, Corollary 9.6 implies that $(a, b) = 0$. As this holds for any a, Proposition 9.12 (c) implies that $b \in K^{*n}$. Finally K^{*n} is the kernel of the reciprocity map $K^* \to \mathrm{Gal}(L/K)$, and hence $K^{*n} = N_{L/K}K^*$ (Theorem 9.2).   □

Remark 9.14 Conversely we know that all the subgroups of the form $N_{L/K}L^*$ for L/K are open and of finite index (Lemma 9.9, c).

Corollary 9.15 *Let K be a p-adic field. The reciprocity map $\omega : K^* \to \mathrm{Gal}(K^{\mathrm{ab}}/K)$ induces an isomorphism from the profinite completion $(K^*)^\wedge$ of K^* to $\mathrm{Gal}(K^{\mathrm{ab}}/K)$. In particular, this group is isomorphic to the direct product $\widehat{\mathbf{Z}} \times U_K$. Furthermore, ω is injective. In other words, the universal norm group $\bigcap_L N_{L/K}L^*$ (the intersection is taken over the finite abelian extensions L of K) is reduced to $\{1\}$.*

Proof We know (cf. Definition 9.5) that ω induces an isomorphism of $\varprojlim_L K^*/N_{L/K}L^*$ with $\mathrm{Gal}(K^{\mathrm{ab}}/K)$. The existence theorem identifies $\varprojlim_L K^*/N_{L/K}L^*$ to the profinite completion $(K^*)^\wedge$ of K^*. We deduce (as K^* is isomorphic to the direct product $\mathbf{Z} \times U_K$, which injects into its profinite completion $\widehat{\mathbf{Z}} \times U_K$) that the canonical arrow $K^* \to \varprojlim_L K^*/N_{L/K}L^*$ is injective, hence so is ω.   □

Corollary 9.16 *The reciprocity map $\omega : K^* \to \mathrm{Gal}(K^{\mathrm{ab}}/K)$ induces an isomorphism between the group of units U_K and the abelian inertia group I_K^{ab}.*

Proof Combine Corollary 9.15 and Lemma 9.8.   □

Remark 9.17 The method used in this paragraph to prove the existence theorem based on Kummer extensions does not work for a characteristic $p > 0$ local field if one considers open subgroups of K^* of index divisible by p. Instead, one can use the *Artin–Schreier* extensions, see [47], Chap. XIV, Sects. 5 and 6.

9.3 Exercises

Exercise 9.1 We use the assumptions and notations of Proposition 9.7.

(a) Show that the index $[K^* : NE^*]$ divides $[E : K]$, and is equal to it if and only if E is an abelian extension of K.
(b) Do conclusions of Proposition 9.7 (b) and part (a) of this exercise still hold if E is no longer assumed to be separable over K?
   (For a generalisation of the results of Proposition 9.7 and of this exercise to the class formations, see Proposition 16.29.)

Exercise 9.2 Let K be a local field. Let L be a non-trivial finite Galois extension of K. Using the structure of the absolute Galois group of K (cf. Example 7.15), show that the group $K^*/N_{L/K}L^*$ is of cardinality at least 2. Would this result still hold if the extension L/K is only assumed to be finite separable?

Exercise 9.3 Let K be a local field. Let E be a finite Galois extension of K and $L \subset E$ a subextension. Show that we have a commutative diagram:

$$
\begin{array}{ccc}
K^* & \xrightarrow{\ i\ } & L^* \\
{\scriptstyle \omega_{E/K}}\downarrow & & \downarrow{\scriptstyle \omega_{E/L}} \\
\mathrm{Gal}(E/K)^{ab} & \xrightarrow{\ v\ } & \mathrm{Gal}(E/L)^{ab}
\end{array}
$$

where i is the inclusion, $\omega_{E/K}$ and $\omega_{E/L}$ are reciprocity maps and $v : \mathrm{Gal}(E/K)^{ab} \to \mathrm{Gal}(E/L)^{ab}$ is the transfer map (cf. Definition 2.11) associated to the inclusion $\mathrm{Gal}(E/L) \to \mathrm{Gal}(E/K)$.

(Compare this to Proposition 9.7, a). This exercise will also be generalised to class formations, see Proposition 16.30).

Exercise 9.4 Under the assumptions and notations of Proposition 9.12, show the converse of (a): if $(a, b) = 0$, then b is a norm of the extension $k(a^{1/n})/k$.

Exercise 9.5 Let k be a field with separable closure $\bar{k}$. Let $n > 0$ be an integer not divisible by $\mathrm{Car}\, k$. Assume that k contains a primitive nth root of 1. Fix such a root ζ, which identifies k^*/k^{*n} (resp. $H^2(k, \mu_n)$) with $H^1(k, \mathbf{Z}/n)$ (resp. $H^2(k, \mathbf{Z}/n)$). For a, b in k^*, denote by φ_a, φ_b their respective images in $H^1(k, \mathbf{Z}/n)$. We can thus consider their cup-product $\varphi_a \cup \varphi_b \in H^2(k, \mathbf{Z}/n)$. Let i be the injection $H^2(k, \mathbf{Z}/n) \longrightarrow \mathrm{Br}\, k$ induced by the exact sequence

$$0 \longrightarrow \mathbf{Z}/n \longrightarrow \bar{k}^* \xrightarrow{\ \cdot\, n\ } \bar{k}^* \longrightarrow 0.$$

Show that $i(\varphi_a \cup \varphi_b)$ coincides with the symbol (a, b).

Exercise 9.6 Let K be a p-adic field. Let $n > 0$ such that K contains a primitive nth root of 1. Let $a \in K^*$. Show that $(a, b) = 0$ for any $b \in U_K$ if and only if $K(a^{1/n})$ is an unramified extension of K.

Chapter 10
The Tate Local Duality Theorem

In this chapter, we denote by K a p-adic field. Our aim here is to prove the Tate local duality theorem. Combining this duality with the results of Chap. 9, we will derive a slightly different proof of the existence theorem for p-adic fields. In Chap. 11, the existence theorem will be proved by another method (Lubin–Tate theory), which works as well for local fields of characteristic p.

The first four paragraphs of this chapter are independent of Chap. 9 and in particular do not appeal to the Tate–Nakayama theorem.

10.1 The Dualising Module

Let G be a profinite group of cohomological dimension $n < +\infty$. Recall that if B is a discrete torsion abelian group (e.g. a finite abelian group), we have $B^* = \mathrm{Hom}(B, \mathbf{Q}/\mathbf{Z})$. The functor $B \mapsto B^*$ (from the category of torsion discrete abelian groups to the category of abelian groups) is exact.

We consider the functor $A \mapsto H^n(G, A)^*$ from the category C_G^f of finite discrete G-modules to the category of abelian groups. It turns out that there is a general result, which ensures, under relatively weak hypotheses, that this functor is representable by a discrete torsion G-module. More precisely:

Theorem 10.1 *Let G be a profinite group of cohomological dimension $n < +\infty$. Assume that for every $A \in C_G^f$, the group $H^n(G, A)$ is finite. Then there exists a discrete torsion G-module I such that the functors $\mathrm{Hom}_G(., I)$ and $H^n(G, .)^*$ (from C_G^f to Ab) are isomorphic. We then say that I is the dualising module of the profinite group G.*

Proof It is a consequence of a general theorem of homological algebra, Theorem A.26. Indeed, the category C_G^f is clearly noetherian and the assumption

© Springer Nature Switzerland AG 2020
D. Harari, *Galois Cohomology and Class Field Theory*, Universitext,
https://doi.org/10.1007/978-3-030-43901-9_10

$cd(G) = n$ implies immediately that the functor $H^n(G, .)$ is right exact, hence that the contravariant functor $H^n(G, .)^*$ is left exact. Thus there exists an inductive system (I_j) of finite discrete G-modules such that the functors $A \mapsto H^n(G, A)^*$ and $A \mapsto \varinjlim_j \mathrm{Hom}_G(A, I_j)$ (defined on C_G^f) are isomorphic. Let $I = \varinjlim_j I_j$; then I is a discrete torsion G-module and the condition that A is finite allows us to identify $\varinjlim_j \mathrm{Hom}_G(A, I_j)$ and $\mathrm{Hom}_G(A, I)$. $\square$

In other words, we have $\mathrm{Hom}_G(A, I) \simeq H^n(G, A)^*$ (the isomorphism is functorial) for any finite discrete G-module A, or, to put it differently, a perfect pairing of finite groups

$$\langle , \rangle_A : \mathrm{Hom}_G(A, I) \times H^n(G, A) \longrightarrow \mathbf{Q}/\mathbf{Z}.$$

Note that if A is a discrete torsion G-module (not necessarily finite), we obtain by passing to the limit that the discrete group $H^n(G, A)$ is the dual of the profinite group $\mathrm{Hom}_G(A, I)$ (endowed with the topology of simple convergence), which allows us to define $\langle , \rangle_A$ for such an A, for example for I. We also have a theorem and analogous definitions with the cohomological p-dimension (restricting ourselves to torsion p-primary modules).

For any G-module A, denote by A' the G-module $A' = \mathrm{Hom}_{\mathbf{Z}}(A, I)$. We thus have $\mathrm{Hom}_G(A, I) = H^0(G, A')$. We have the following compatibility[1] of the duality pairing with the the cup-product:

Proposition 10.2 *Let $i : H^n(G, I) \to \mathbf{Q}/\mathbf{Z}$ be the homomorphism defined by $\langle \mathrm{Id}_I, .\rangle_I : x \mapsto \langle \mathrm{Id}_I, x\rangle_I$. Then for any finite discrete G-module A, the pairing $\langle , \rangle_A$ is the composite of the cup-product*

$$H^0(G, A') \times H^n(G, A) \longrightarrow H^n(G, I)$$

with i.

Proof Let $\rho : A \to I$ be any element of $H^0(G, A') = \mathrm{Hom}_G(A, I)$. This element induces a map

$$\rho_* : H^n(G, A) \longrightarrow H^n(G, I),$$

and also for any G-module M a homomorphism

$$\rho^* : \mathrm{Hom}_G(I, M) \longrightarrow \mathrm{Hom}_G(A, M).$$

This in particular applies to $M = I$. By functoriality of the duality, we have a commutative diagram:

---

[1] Thanks to Demarche, who pointed out to me the necessity of this verification to prove Theorem 10.9.

$$
\begin{array}{ccc}
\mathrm{Hom}_G(A, I) \times H^n(G, A) & \xrightarrow{\langle\,,\,\rangle_A} & \mathbf{Q}/\mathbf{Z} \\
{\scriptstyle \rho^*}\big\uparrow & \big\downarrow{\scriptstyle \rho_*} & \big\| \\
\mathrm{Hom}_G(I, I) \times H^n(G, I) & \xrightarrow{\langle\,,\,\rangle_I} & \mathbf{Q}/\mathbf{Z}.
\end{array}
$$

We conclude by observing that $\rho^*(\mathrm{Id}_I) = \rho$ and that for any $\alpha \in H^n(G, A)$, the cup-product $\rho \cup \alpha$ is $\rho_*(\alpha)$ by Remark 2.22. □

Remark 10.3 From the above properties it immediately follows that the pair (I, i) is unique up to isomorphism. Thus we can speak of *the* dualising module (rather than simply a dualising module).

The existence of the dualising module does not in fact require the assumption of finiteness of $H^n(G, A)$. It is possible to describe it explicitly without a priori assuming its existence (and hence without using Theorem A.26), but it requires a lot of work (see [49], Part. I, Ann. 1, or [42], Sect. III.4). We will be calculating the dualising module of a p-adic field and deduce an application to the strict cohomological dimension of a p-adic field and to the Tate duality theorem. As usual, the results will be valid for a local field of characteristic p, provided we restrict ourselves to p-torsion-free modules.

Proposition 10.4 *Let G be a profinite group of cohomological dimension $n \in \mathbf{N}^*$, satisfying the assumptions of Theorem 10.1. Let U an open subgroup of G. Let I be the dualising module of G. Then:*

(a) the dualising module of U is I viewed as a U-module.
(b) the homomorphism $H^n(U, A) \to H^n(G, A)$ obtained by dualising the canonical injection $\mathrm{Hom}_G(A, I) \to \mathrm{Hom}_U(A, I)$ is the corestriction.

Proof

(a) we have $\mathrm{cd}(U) = \mathrm{cd}(G)$ by Proposition 5.10. The result then follows from Shapiro's lemma and Proposition 1.15 (which extends in a straightforward way to the case of an open subgroup of a profinite group), by taking $M = I_G^U(A)$ (where A is a finite U-module) in the property characterising the dualising module.

(b) The corestriction $H^n(U, A) \to H^n(G, A)$ is obtained (see the Proof of Lemma 5.9b, or Exercise 1.1) by applying the functor $H^n(G, .)$ to the morphism of G-modules

$$
\pi : I_G^U(A) \longrightarrow A, \quad f \longmapsto \sum_{g \in G/U} g \cdot f(g^{-1}).
$$

Using (a) and the definition of the dualising module, we obtain that the dual of the corestriction is the homomorphism

$$
\mathrm{Hom}_G(A, I) \longrightarrow \mathrm{Hom}_G(I_G^U(A), I)
$$

induced by π, which is identified, by Proposition 1.15, to the canonical injection
$\mathrm{Hom}_G(A, I) \to \mathrm{Hom}_U(A, I)$. $\square$

Let now K be a p-adic field. We have seen (Theorem 8.11) that its Galois group
$\Gamma := \Gamma_K$ is of cohomological dimension 2 and that $H^2(\Gamma, A)$ is finite for all finite
Γ-module A (Corollary 8.15). We can thus apply previous results with $n = 2$. The
next proposition calculates the dualising module of Γ.

Proposition 10.5 *The dualising module I of Γ is canonically isomorphic to the
Γ-module $\mu = \mathbf{Q}/\mathbf{Z}(1)$ of roots of unity of $\bar{K}^*$.*

Proof Let $n > 0$, denote by I_n the kernel of the multiplication by n on I. For any
open subgroup U of Γ, I is also the dualising module of U (Proposition 10.4, a)
hence, by applying the definition of the dualising module:

$$\mathrm{Hom}_U(\mu_n, I_n) = \mathrm{Hom}_U(\mu_n, I) \simeq H^2(U, \mu_n)^* \simeq \mathbf{Z}/n\mathbf{Z},$$

where the isomorphism comes from Proposition 8.13. Furthermore these isomorphisms are compatible with each other for various U by Proposition 10.4 (b) and
Theorem 8.9. Thus the group $\mathrm{Hom}_U(\mu_n, I_n)$ is independent of U; we obtain that
$\mathrm{Hom}(\mu_n, I_n) \simeq \mathbf{Z}/n\mathbf{Z}$, and in addition the fact that Γ acts trivially on this group (by
taking $U = \Gamma$) and that the isomorphisms for various n are compatible with each
other. Choose a generator f_n of $\mathrm{Hom}_\Gamma(\mu_n, I_n)$ corresponding to $\bar{1} \in \mathbf{Z}/n\mathbf{Z}$. The
homomorphism f_n is injective because it is of order n. It is also surjective, otherwise
there would exist an x in I_n which is not in its image, hence a homomorphism from
μ_n to I_n which is not in the subgroup generated by f_n. Define f by $f_{|\mu_n} = f_n$. We
thus obtain an isomorphism of Γ-modules from μ to I, using the functorial property
of the dualising module and the fact that in the isomorphism $H^2(K, \mu_n) \simeq \mathbf{Z}/n\mathbf{Z}$ of
Proposition 8.13, the map $H^2(K, \mu_n) \to H^2(K, \mu_m)$ induces the canonical injection
from $\mathbf{Z}/n\mathbf{Z}$ to $\mathbf{Z}/m\mathbf{Z}$ if m divides n. $\square$

We deduce:

Theorem 10.6 *Let K be a p-adic field. Then it is of strict cohomological
dimension 2.*

Proof We use Proposition 5.14. Let U be an open subgroup of Γ. We then have
$H^3(U, \mathbf{Z}) = H^2(U, \mathbf{Q}/\mathbf{Z})$ (Example 1.52). By Proposition 10.5, this last group is
the dual of $\mathrm{Hom}_U(\mathbf{Q}/\mathbf{Z}, \mu)$. Let L/K be the finite extension corresponding to U.
Then the image of a homomorphism $f : \mathbf{Q}/\mathbf{Z} \to \mu$ of U-modules is divisible and
contained in $H^0(L, \mu)$. But L (which is a p-adic field) contains only finitely many
roots of unity by Theorem 7.18, hence $H^0(L, \mu)$ is finite. We conclude that $\mathrm{Im}\, f = 0$,
that is, $\mathrm{Hom}_U(\mathbf{Q}/\mathbf{Z}, \mu) = 0$, which finishes the proof. $\square$

Remark 10.7 The same proof shows that if K is a local field of characteristic $p > 0$,
then $\mathrm{scd}_\ell(K) = 2$ si $\ell \neq p$.

We conclude this paragraph with a topological remark:

Remark 10.8 The dual $A^* = \mathrm{Hom}_c(A, \mathbf{Q}/\mathbf{Z})$ of a topological group A behaves well (it coincides with the classical Pontryagin dual) if A is profinite or discrete and torsion (cf. Example 4.3).

However, for arbitrary abelian locally compact and completely disconnected groups, the situation is more complicated. For example with our definition, the dual of $\mathbf{Z}$ is $\mathbf{Q}/\mathbf{Z}$ but that of $\mathbf{Q}/\mathbf{Z}$ is $\widehat{\mathbf{Z}}$ (which shows that to call A^* the dual of A may be dangerous!) Also, if

$$0 \longrightarrow A \longrightarrow B \longrightarrow C \longrightarrow 0$$

is a short exact sequence (the morphisms are assumed to be continuous) of locally compact and completely disconnected abelian groups, the dual sequence

$$0 \longrightarrow C^* \longrightarrow B^* \longrightarrow A^* \longrightarrow 0$$

is exact (cf. [19], Lemma 2.4). Note though that this does not imply that the dual of an injective continuous morphism between such groups is surjective. For example $\mathbf{Z} \to \mathbf{Z}_p$ is injective, but the dual morphism $\mathbf{Q}_p/\mathbf{Z}_p \to \mathbf{Q}/\mathbf{Z}$ is not surjective (the problem is that the quotient $\mathbf{Z}_p/\mathbf{Z}$ is not Hausdorff, or that the morphism under consideration is not *strict*. Thus we do not have a short exact sequence all of whose three terms remain in the category under consideration).

10.2 The Local Duality Theorem

Let k be a field with absolute Galois group $\Gamma_k = \mathrm{Gal}(\bar{k}/k)$. Let $\mu \subset \bar{k}^*$ be the multiplicative group consisting of all the roots of unity. For any finite Γ_k-module M whose torsion is not divisible by Car k, we denote by

$$M' := \mathrm{Hom}_{\mathbf{Z}}(M, \bar{k}^*) = \mathrm{Hom}(M, \mu)$$

its *Cartier dual*, which is a finite Γ_k-module for the action defined in Example 4.16, (d). If M is n-torsion, we also have $M' = \mathrm{Hom}(M, \mu_n)$. For a finite M, we can identify $(M')'$ to M, but one should remember that it is not the case in general when M is only assumed to be n-torsion (cf. Exercise 10.8).

We have a pairing of Γ_k-modules

$$M \times M' \longrightarrow \mu \subset \bar{k}^*$$

which defines a cup-product.

Theorem 10.9 (Tate) *Let K be a p-adic field with Galois group Γ, and M a finite Γ-module. Then for $i = 0, 1, 2$, the cup-product*

$$H^i(\Gamma, M) \times H^{2-i}(\Gamma, M') \longrightarrow H^2(K, \bar{K}^*) = \mathbf{Q}/\mathbf{Z}$$

is a perfect duality of finite groups.

Proof The finiteness of all the groups has already been seen (Corollary 8.15). For $i = 2$, the theorem states that the groups $H^2(\Gamma, M)$ and $H^0(\Gamma, M')$ are in perfect duality via the bilinear map $(x, a) \mapsto x \cup a$ from $H^2(\Gamma, M) \times H^0(\Gamma, M')$ to $H^2(\Gamma, \mu) = \mathbf{Q}/\mathbf{Z}$. This follows from Propositions 10.5 and 10.2. The case $i = 0$ is symmetrical since $(M')' = M$. To deal with the case $i = 1$, it is thus enough to show that the homomorphism

$$H^1(\Gamma, M) \longrightarrow H^1(\Gamma, M')^*$$

induced by the cup-product is injective (since by applying the same result to M', we will in addition obtain that the cardinality of the left-hand side group is at least that of the right-hand side group). To do this, write an exact sequence

$$0 \longrightarrow M \longrightarrow B \longrightarrow C \longrightarrow 0$$

such that B and C are finite and the map $H^1(\Gamma, M) \to H^1(\Gamma, B)$ is zero. To find such a B (cf. also exercise 4.7), we start by embedding M into the induced module $I_\Gamma(M)$ (which is torsion), then we note that $H^1(\Gamma, M)$ is finite (Corollary 8.15) and that $0 = H^1(\Gamma, I_\Gamma(M))$ is the inductive limit of the $H^1(\Gamma, F)$ for F finite contained in $I_\Gamma(M)$, by Proposition 4.18.

By properties of compatibility of the cup-product with respect to exact sequences (Proposition 2.28), we have an exact (up to sign) commutative diagram:

$$
\begin{array}{ccccccc}
H^0(\Gamma, B) & \longrightarrow & H^0(\Gamma, C) & \longrightarrow & H^1(\Gamma, M) & \longrightarrow & 0 \\
\downarrow & & \downarrow & & \downarrow & & \\
H^2(\Gamma, B')^* & \longrightarrow & H^2(\Gamma, C')^* & \longrightarrow & H^1(\Gamma, M')^*. & &
\end{array}
$$

By the case $i = 0$, the first two vertical arrows are isomorphisms (note the importance of having chosen B and C to be finite, cf. Exercise 10.8). We deduce the injectivity of the third arrow by diagram chasing.   □

Remark 10.10 For local fields of characteristic p, the theorem remains true (with the same proof) provided one restricts to M whose torsion is prime to p. Otherwise an analogous result (due to Shatz) is much more difficult (see [51], Chap. VI, or [39], Sect. III.6), in particular because $H^1(\Gamma, M)$ is not in general finite.

Also, there is a generalisation of Theorem 10.9 to the case where A is only assumed to be of finite type over $\mathbf{Z}$ (in this case the situation is no longer symmetrical, the Γ-module A' is then the group of $\bar{K}$-points of a *group of multiplicative type* over K). See [49], Part. II, Sect. 5.8, as well as Theorem 16.26 and Exercise 16.3.

10.3 The Euler–Poincaré Characteristic

Let K be a p-adic field with absolute Galois group $\Gamma = \mathrm{Gal}(\bar{K}/K)$. Let A be a finite Γ-module. Let $h^i(A)$ be the cardinality of the finite group $H^i(K, A) = H^i(\Gamma, A)$.

Definition 10.11 The *Euler–Poincaré characteristic* of A is the strictly positive rational number

$$\chi(A) := \frac{h^0(A) \cdot h^2(A)}{h^1(A)}.$$

If $0 \to A \to B \to C \to 0$ is an exact sequence of finite Γ-modules, then we obtain $\chi(B) = \chi(A).\chi(C)$ using the long cohomology exact sequence. A theorem of Tate ([49], Theorem 5 p. 109) yields the equality

$$\chi(A) = 1/[\mathcal{O}_K : a\mathcal{O}_K],$$

where a is the cardinality of A (in particular $\chi(A)$ depends only on a). The proof of this theorem is long and uses subtle results from representation theory of finite groups. We will thus only treat a simpler special case, which will suffice for applications to unramified cohomology. The refined version of Tate's theorem is on the other hand indispensable if one wishes to prove the other of his duality theorems concerning abelian varieties over local fields ([39], Part. I, Chap. 3).

Proposition 10.12 *Assume that the cardinality a of A is prime to p. Then $\chi(A) = 1$.*

Proof Let I_K be the inertia group $\mathrm{Gal}(\bar{K}/K_{\mathrm{nr}})$. The quotient Γ/I_K is isomorphic to $\hat{\mathbf{Z}}$ and I_K has a unique p-Sylow I_p which is normal in I_K, such that the quotient $V := I_K/I_p$ is isomorphic to $\mathbf{Z}'_p := \prod_{\ell \neq p} \mathbf{Z}_\ell$ (cf. Example 7.15, b).

Lemma 10.13 *The group $H^i(I_K, A)$ is finite for any $i \geqslant 0$ and trivial if $i \geqslant 2$.*

Proof The case $i = 0$ is straightforward. We know (Theorem 8.11, b) that I_K is of cohomological dimension $\leqslant 1$ hence $H^i(I_K, A) = 0$ for $i \geqslant 2$. On the other hand $H^i(I_p, A) = 0$ for $i \geqslant 1$ by Theorem 1.48, since I_p is a pro-p-group and the cardinality of A is prime to p. Thus $H^1(I_K, A) = H^1(V, A^{I_p})$ via the restriction–inflation exact sequence and we are reduced to showing that $H^1(V, A^{I_p})$ is finite. By decomposing the finite V-module A^{I_p} as a finite direct sum of its ℓ-primary components (for ℓ prime different from p), we see (always using Theorem 1.48, and the inflation–restriction exact sequence applied to the subgroup $\mathbf{Z}_l$ of V) that it remains to check that $H^1(\mathbf{Z}_\ell, B)$ is finite for any finite ℓ-primary module B. Let W be an open subgroup of $\mathbf{Z}_\ell$ acting trivially on B. Then by the the inflation–restriction sequence it is enough to see that $H^1(W, B)$ is finite (as $\mathbf{Z}_\ell/W$ is finite, hence so is $H^1(\mathbf{Z}_\ell/W, B)$) or in other words that $\mathrm{Hom}_c(W, B)$ is finite. But if B is of cardinality ℓ^m, then $\mathrm{Hom}_c(W, B) = \mathrm{Hom}(W/\ell^m W, B)$ is finite since $W/\ell^m W$ is finite (the open subgroup W of the profinite group $\mathbf{Z}_\ell$ is of finite index in $\mathbf{Z}_\ell$ and $\ell^m \mathbf{Z}_\ell$ is of finite index in $\mathbf{Z}_\ell$; it is enough then to apply the five lemma). $\square$

Let us take up again the Proof of Proposition 10.12. The exact sequence of the first terms of the Hochschild–Serre spectral sequence

$$H^i(\Gamma/I_K, H^j(I_K, A)) \implies H^{i+j}(\Gamma, A)$$

yields (taking into account the above results and the fact that cd $\widehat{\mathbf{Z}} = 1$):

$$H^0(K, A) = H^0(\widehat{\mathbf{Z}}, H^0(I_K, A)); \quad H^2(K, A) = H^1(\widehat{\mathbf{Z}}, H^1(I_K, A))$$

and an exact sequence

$$0 \longrightarrow H^1(\widehat{\mathbf{Z}}, H^0(I_K, A)) \longrightarrow H^1(K, A) \longrightarrow H^0(\widehat{\mathbf{Z}}, H^1(I_K, A)) \longrightarrow 0.$$

To conclude, it is enough to apply the following lemma (cf. also Exercise 5.3) to the finite $\widehat{\mathbf{Z}}$-modules $H^0(I_K, A)$ and $H^1(I_K, A)$.

Lemma 10.14 *Let M be a finite $\widehat{\mathbf{Z}}$-module. Then $H^0(\widehat{\mathbf{Z}}, M)$ and $H^1(\widehat{\mathbf{Z}}, M)$ are finite of the same cardinality.*

Proof of the Lemma The canonical topological generator 1 of $\widehat{\mathbf{Z}}$ induces an automorphism F of M. Let $s = mn$, where n and m are two integers chosen so that the mth power F^m of F acts trivially on M and $M = M[n]$. Then we can see M as a $\mathbf{Z}/s$-module and the norm $N_{\mathbf{Z}/s} : M \to M$ is the zero map since for any x in M, we have

$$N_{\mathbf{Z}/s}(x) = (1 + F + \cdots + F^{mn-1})x = n(1 + F + \cdots + F^{m-1}) \cdot x = 0.$$

Then $H^0(\widehat{\mathbf{Z}}, M)$ is the kernel of the endomorphism $F - 1$ of M, while $H^1(\widehat{\mathbf{Z}}, M)$ is the inductive limit over i of the $H^1(\mathbf{Z}/i, M^{F^i})$, where M^{F^i} is the submodule of M consisting of elements fixed by F^i. By definition of an inductive limit, we can restrict ourselves to i which are multiples of s, so that $H^1(\widehat{\mathbf{Z}}, M)$ is simply the inductive limit for i multiple of s of the $H^1(\mathbf{Z}/i, M)$. But $H^1(\mathbf{Z}/i, M) = \widehat{H}^{-1}(\mathbf{Z}/i, M)$ (Theorem 2.16) is the cokernel of $F - 1$ since $N_{\mathbf{Z}/i}$ is zero for i multiple of s, whence the result. $\square$

10.4   Unramified Cohomology

Definition 10.15 Let A be a Γ-module. We call A *unramified* if the group $I_K = \mathrm{Gal}(\bar{K}/K_{nr})$ acts trivially on A. In this case, we define the groups $H^i_{nr}(K, A) := H^i(\mathrm{Gal}(K_{nr}/K), A)$.

Proposition 10.16 *Let A be a finite and unramified Γ-module. Then we have the equalities $H^0_{nr}(K, A) = H^0(K, A)$ and $H^i_{nr}(K, A) = 0$ for $i \geqslant 2$. The group*

$H^1_{\mathrm{nr}}(K, A)$ *identifies with a subgroup of* $H^1(K, A)$ *and its cardinality is that of* $H^0(K, A)$.

Proof The assertion about H^0 is straightforward. The assertion about the H^i for $i \geqslant 2$ follows from the fact that $\mathrm{cd}(\widehat{\mathbf{Z}}) = 1$. Lastly, the assertion about H^1 comes from the restriction–inflation exact sequence and Lemma 10.14 applied to A. $\qquad\square$

Theorem 10.17 *Let A be a finite unramified Γ-module, of order prime to p. Then its dual A' has the same properties. Furthermore, in the duality between $H^1(K, A)$ and $H^1(K, A')$, each of the subgroups $H^1_{\mathrm{nr}}(K, A)$ and $H^1_{\mathrm{nr}}(K, A')$ is the orthogonal of the other.*

Proof Let μ be the Γ-module of roots of unity in $\bar{K}^*$ (it is the dualising module of Γ) and let $\bar{\mu}$ be the submodule consisting of elements of order prime to p. As the nth roots of unity for n prime to p are in K_{nr}, the Γ-module $\bar{\mu}$ is unramified, which immediately implies that $A' = \mathrm{Hom}(A, \bar{\mu})$ is unramified. The cup-product

$$H^1_{\mathrm{nr}}(K, A) \times H^1_{\mathrm{nr}}(K, A') \longrightarrow H^2(k, \mu)$$

factors through $H^2_{\mathrm{nr}}(k, \bar{\mu})$ which is zero, which implies that $H^1_{\mathrm{nr}}(K, A)$ and $H^1_{\mathrm{nr}}(K, A')$ are orthogonal. To conclude, it is enough to show that the cardinality $h^1(A')$ of $H^1(K, A')$ is the product $h^1_{\mathrm{nr}}(A) \cdot h^1_{\mathrm{nr}}(A')$ of cardinalities of $H^1_{\mathrm{nr}}(K, A)$ and $H^1_{\mathrm{nr}}(K, A')$. Indeed, this will imply that the homomorphism

$$H^1_{\mathrm{nr}}(K, A) \longrightarrow (H^1(K, A')/H^1_{\mathrm{nr}}(K, A'))^*$$

induced by the local duality is an isomorphism (we know already that this homomorphism is injective by Theorem 10.9). But $h^1_{\mathrm{nr}}(A) = h^0(A)$ and $h^1_{\mathrm{nr}}(A') = h^0(A') = h^2(A)$ by Proposition 10.16 and Theorem 10.9. The last theorem also implies that $h^1(A) = h^1(A')$. The result then follows from Proposition 10.12. $\qquad\square$

Remark 10.18 Proposition 10.12, Definition 10.15, and Theorem 10.17 extend immediately, *mutatis mutandis*, to the case of a local field characteristic $p > 0$ and a module A *of cardinality prime to p*. Furthermore, Milne proved these results under much weaker assumptions, see [39], Chap. III.1 and III.7.

Example 10.19 Let K be a p-adic field with absolute Galois group Γ and inertia group $I_K = \mathrm{Gal}(\bar{K}/K_{\mathrm{nr}})$. Let $\mathcal{O}^{\mathrm{nr}}_K$ be the ring of integers of K_{nr}. Then, for n prime to p, the Γ-module μ_n is unramified. Hensel lemma (applied to finite unramified extensions of K) provides an exact sequence

$$0 \longrightarrow \mu_n \longrightarrow \mathcal{O}^{\mathrm{nr}*}_K \xrightarrow{\ \cdot n\ } \mathcal{O}^{\mathrm{nr}*}_K \longrightarrow 0,$$

hence we deduce an exact sequence of $\Gamma/I_K = \mathrm{Gal}(K_{\mathrm{nr}}/K)$-modules:

$$0 \longrightarrow \mathcal{O}^*_K/\mathcal{O}^{*n}_K \longrightarrow H^1((\Gamma/I_K), \mu_n) \longrightarrow H^1((\Gamma/I_K), \mathcal{O}^{\mathrm{nr}*}_K),$$

with $H^1((\Gamma/I_K), \mu_n) = H^1_{\mathrm{nr}}(K, \mu_n)$. By passing to the limit in Proposition 8.3, we obtain $H^1((\Gamma/I_K), \mathcal{O}_K^{\mathrm{nr}*}) = 0$, hence finally

$$H^1_{\mathrm{nr}}(K, \mu_n) \simeq \mathcal{O}_K^* / \mathcal{O}_K^{*n}.$$

The result also holds (with the same proof) for a local field of characteristic p.

10.5 From the Duality Theorem to the Existence Theorem

In this paragraph, we will see a proof of the existence theorem for p-adic fields. The final argument does not use Kummer extensions as in Sect. 9.2, but rests on the local Tate duality (of which we have given a proof resting on the calculation of the dualising module).

We first recover, using the local duality a statement which had been seen as a consequence of the existence theorem (Corollary 9.15).

Theorem 10.20 *Let K be a p-adic field with absolute Galois group Γ. The reciprocity map $\omega_K : K^* \to \Gamma^{\mathrm{ab}}$ induces an isomorphism*

$$(K^*)^\wedge = \varprojlim_{n \in \mathbf{N}^*} (K^*/K^{*n}) \simeq \Gamma^{\mathrm{ab}}.$$

Thus Γ^{ab} is isomorphic (not canonically) to $\widehat{\mathbf{Z}} \times U_K$.

Note that the profinite completion $(K^*)^\wedge$ of K^* is the projective limit of the K^*/K^{*n} since they are finite (Proposition 8.13) and furthermore the subgroups K^{*n} form a cofinal family among the finite index subgroups of K^* (indeed, if U and V are two such subgroups, then $U \cap V$ is also one, hence $U \cap V$ contains K^{*n} for a certain $n > 0$).

Proof Let $n > 0$. The reciprocity map induces a map

$$\omega_n : K^*/K^{*n} = H^1(K, \mu_n) \longrightarrow \Gamma^{\mathrm{ab}}/n = H^1(K, \mathbf{Z}/n)^*$$

which, by Corollary 9.6 (b), is the map induced by the cup-product

$$H^1(K, \mu_n) \times H^1(K, \mathbf{Z}/n) \longrightarrow H^2(K, \mu_n) \simeq \mathbf{Z}/n \hookrightarrow \mathbf{Q}/\mathbf{Z},$$

that is, the isomorphism given by the local Tate duality. We conclude by passing to the projective limit on n. $\square$

Remark 10.21 Theorem 10.20 also holds for a finite extension of $\mathbf{F}_p((t))$ provided we limit ourselves to the completion of K^* for the topology defined by the *open* subgroups of finite index.

This can for example be deduced from the Shatz duality theorem or from the results of the next chapter.

Theorem 10.22 (existence theorem) *Let K be a p-adic field with fixed algebraic closure $\bar{K}$. We consider a subgroup U of finite index of K^*. Then there exists a unique finite abelian extension $L \subset \bar{K}$ of K such that $U = N_{L/K} L^*$.*

Proof As we have seen at the beginning of the first proof (cf. Theorem 9.13), it is enough to show that if $n > 0$ is an integer, then K^{*n} is of the form $N_{L/K} L^*$ for some finite abelian extension L of K. As the subgroup $n\Gamma^{ab}$ of Γ^{ab} is compact (image of a compact set by a continuous map), it is a closed subgroup of Γ^{ab}. Theorem 10.20 says that the reciprocity map induces an isomorphism of the finite group K^*/K^{*n} with Γ^{ab}/n. This implies that $n\Gamma^{ab}$ is closed of finite index in Γ^{ab}. It is thus (by infinite Galois theory) of the form $\mathrm{Gal}(\bar{K}/L)$, where L is a finite abelian extension of K, hence $\Gamma^{ab}/n = \mathrm{Gal}(L/K)$. This implies that K^{*n} is the kernel of reciprocity map $\omega_{L/K} : K^* \to \mathrm{Gal}(L/K)$, and hence $K^{*n} = N_{L/K} L^*$ by Theorem 9.2. $\qquad\square$

Remark 10.23 The difference from the method of Chap. 9 is just the way we are showing that K^{*n} is the kernel of a reciprocity map $\omega_{L/K} : K^* \to \mathrm{Gal}(L/K)$ (it involves the local duality theorem). We also know that all subgroups $N_{L/K} L^*$ are closed (Lemma 9.9, c).

Note also that if n is divisible by p, this method does not work in the case of a local field of characteristic p. In the next chapter we will see a uniform method allowing to obtain the existence theorem in this case.

10.6 Exercises

Exercise 10.1 Using Exercise 5.5, recover the fact that a p-adic field K satisfies $\mathrm{scd}(K) = 2$.

Exercise 10.2 Let K be a p-adic field with absolute Galois group Γ. Let M a Γ-module which is free when seen as an abelian group. Assume that M is unramified.

(a) Show that $H^1_{nr}(K, M)$ is equal to $H^1(K, M)$.
(b) Give an example where $H^2_{nr}(K, M)$ is non-trivial and show that we have an exact sequence
$$0 \longrightarrow H^2_{nr}(K, M) \longrightarrow H^2(K, M) \longrightarrow (I_K^*)^r,$$
where I_K is the inertia group of K, I_K^* denotes the dual of the profinite group I_K, and r is the rank of M as a $\mathbf{Z}$-module.

Exercise 10.3 Let p an odd prime number.

(a) Show that the only solution to the equation $x^p = 1$ in $\mathbf{Q}_p$ is $x = 1$.
(b) Show that $H^2(\mathbf{Q}_p, \mathbf{Z}/p\mathbf{Z}) = 0$. Does this result remain valid if we replace $\mathbf{Z}/p\mathbf{Z}$ by the Galois module μ_p of pth roots of unity in $\overline{\mathbf{Q}_p}$?

(c) Give an example of a p-adic field K such that $H^2(K, \mathbf{Z}/p\mathbf{Z})$ is non-trivial.

(d) Let K be a p-adic field. Can $\varinjlim_{n \geqslant 1} H^2(K, \mathbf{Z}/p^n\mathbf{Z})$ be non-trivial?

Exercise 10.4 Let p be a prime number and let K be a p-adic field with absolute Galois group $G_K = \mathrm{Gal}(\bar{K}/K)$. Let $K(p)$ be the *maximal p-extension* of K: by definition this means that the Galois group $G_K(p) := \mathrm{Gal}(K(p)/K)$ is the largest quotient of G_K which is a pro-p-group.

We will admit the following result (cf. Exercise 8.2): for any p-primary torsion $G_K(p)$-module A and any $i \geqslant 0$, the inflation homomorphism

$$H^i(G_K(p), A) \longrightarrow H^i(G_K, A)$$

is an isomorphism.

(a) To begin with, assume that K does not contain a primitive pth root of 1. Show that $H^2(G_K(p), \mathbf{Z}/p\mathbf{Z}) = 0$, and deduce the p-cohomological dimension of $G_K(p)$.

Compute, depending on the integer $N = [K : \mathbf{Q}_p]$, the dimension of the $\mathbf{Z}/p\mathbf{Z}$-vector space $H^1(G_K(p), \mathbf{Z}/p\mathbf{Z})$ (first compute F/pF, where F is the torsion subgroup of the group of units U_K of K).

(b) We suppose now that the field K contains all the pth roots of 1. Show that the cohomological p-dimension of $G_K(p)$ is 2. What is (depending on N) the dimension of the $\mathbf{Z}/p\mathbf{Z}$-vector space $H^1(G_K(p), \mathbf{Z}/p\mathbf{Z})$?

Exercise 10.5 Let K be a p-adic field. Let A be an *abelian variety* over K. For $n > 0$, let A_n be the subgroup of $A(\bar{K})$ consisting of the elements of n-torsion. Recall the following facts: the multiplication by n on $A(\bar{K})$ is surjective; the Galois module A_n is the dual (this follows from the properties of the Weil pairing) of A'_n, where A' is the *dual abelian variety* of A. The transpose of the inclusion $A_m \to A_n$ for m dividing n is the multiplication by n/m from A'_n to A'_m. Furthermore $A(K) = H^0(K, A(\bar{K}))$ is a p-adic compact Lie group (which in particular implies that it is isomorphic—as a topological group—to the product of a finite group and of $\mathbf{Z}_p^N$ for some N).

(a) Show that $H^2(K, A) = \varinjlim_n H^2(K, A_n)$.

(b) Show that the torsion subgroup of $A'(K)$ is finite.

(c) Deduce that $H^2(K, A) = 0$.

Exercise 10.6 Let G be the profinite group $\hat{\mathbf{Z}}$, which is of cohomological dimension 1. Compute the dualising module of G (use Exercise 5.3).

Exercise 10.7 Let G be a profinite group with $\mathrm{cd}(G) = n$, satisfying the assumptions of Theorem 10.1. Let I be the dualising module of G and let A be a finite G-module. Show that the group $\mathrm{Hom}_{\mathbf{Z}}(A, I)$ is the inductive limit of the duals of the $H^n(H, A)$ for H ranging through the set of open subgroups of G, the transition maps being the transposes of the corestrictions.

Exercise 10.8 (*suggested by Riou*) Let K be a p-adic field with absolute Galois group Γ. Let ℓ be a prime number. We consider the ℓ-torsion Γ-module $A = \bigoplus_{\mathbf{N}} \mathbf{Z}/\ell\mathbf{Z}$ (infinite countable direct sum of the $\mathbf{Z}/\ell\mathbf{Z}$ with a trivial action of Γ).

(a) What is the dual $A' = \mathrm{Hom}(A, \mu_\ell)$ of A?
(b) Show that for any integer $n > 0$, there exists a Γ-module B and a split exact sequence of Γ-modules

$$0 \longrightarrow \prod_{i=1}^{n} \mu_l \longrightarrow A' \longrightarrow B \longrightarrow 0.$$

(c) Show that the canonical injection $A \to (A')'$ is not an isomorphism.
(d) Show that $H^2(\Gamma, A')$ is a $\mathbf{Z}/\ell\mathbf{Z}$-vector space of infinite dimension and that A is countable.
(e) Show that the map $H^0(\Gamma, A) \to H^2(\Gamma, A')^*$ induced by the cup-product $H^0(\Gamma, A) \times H^2(\Gamma, A') \to H^2(\Gamma, \mu_\ell) \simeq \mathbf{Z}/\ell\mathbf{Z}$ is *not* an isomorphism, despite the fact that A' is ℓ-torsion.

Chapter 11
Local Class Field Theory: Lubin–Tate Theory

The aim of this chapter (which can be read independently of the preceding one) is to give an explicit description, using *formal groups*, of the maximal abelian extension of a local field. This, in particular, will allow us to give an easy proof (and valid for any characteristic) of the existence theorem (already proved in the p-adic case in Sects. 9.2 and 10.5 using slightly different approaches). Furthermore we will obtain explicit formulas to compute local symbols $(a, L/K) := \omega_{L/K}(a)$ when $a \in K^*$ and L is an abelian extension of a local field K. These formulas will also be useful for the global class field theory.

In what follows, K denotes a local field (i.e., complete for a discrete valuation, with a finite residue field) with ring of integers $\mathcal{O}_K$. Set also $U_K = \mathcal{O}_K^*$ and let $\mathfrak{m}_K$ be the maximal ideal of $\mathcal{O}_K$.

11.1 Formal Groups

The multiplicative group law on $U_K^1 := 1 + \mathfrak{m}_K$ corresponds to the additive group law on $\mathfrak{m}_K$ given by $(X, Y) \mapsto X + Y + XY$. This observation justifies the following definition.

Definition 11.1 Let A be a commutative ring. Let $F \in A[[X, Y]]$ be a formal series in two variables. We say that F is a *commutative formal group law* if the following properties hold:
  (a) (associativity) $F(X, F(Y, Z)) = F(F(X, Y), Z)$;
  (b) (identity) $F(0, Y) = Y$ and $F(X, 0) = X$;
  (c) (symmetry) there exists a unique $G(X) \in A[[X]]$ such that:

$$F(X, G(X)) = 0\,;$$

  (d) (commutativity) $F(X, Y) = F(Y, X)$;
  (e) $F(X, Y) \equiv X + Y \pmod{\deg 2}$.

© Springer Nature Switzerland AG 2020
D. Harari, *Galois Cohomology and Class Field Theory*, Universitext,
https://doi.org/10.1007/978-3-030-43901-9_11

(Two formal series are congruent modulo deg n if they have the same terms of degree $< n$. One can check that (c) and (e) can be deduced from the three other properties. Recall also that if n is a positive integer, we can substitute a family of n formal series without constant terms in any formal series in n variables.)

If we set $A = \mathcal{O}_K$, we can then define $x \star y := F(x, y)$ for x, y in $\mathfrak{m}_K$, which turns $\mathfrak{m}_K$ into a commutative group that we will denote $F(\mathfrak{m}_K)$. Similarly, we can define the group $F(\mathfrak{m}_L)$ (which can also be seen "concretely" as a subset of L) for any finite extension L of K, and even for any algebraic extension by passing to the limit. The case of $1 + \mathfrak{m}_K$ corresponds to the formal group law $F(X, Y) = X + Y + XY$.

Remark 11.2 The above definition corresponds to formal group laws of dimension 1. More generally, we have a formal group law *of dimension n* (i.e., involving a family of n formal series in $2n$ variables) associated to any commutative group scheme of dimension n: the law is defined from coordinates, by taking the origin as the neutral element. This in particular applies to abelian schemes, for example elliptic curves over a field, to which are associated formal group laws in the sense of Definition 11.1.

Let now κ be the residue field of the local field K and let q be the cardinality of κ. Fix a uniformiser π of K. Let $\mathcal{F}_\pi$ be the set of formal series $f \in A[[X]]$ such that $f(X) \equiv \pi X \bmod \deg 2$ and $f(X) \equiv X^q \bmod \pi$ (the second condition means that $f(X) - X^q$ is divisible by π in $A[[X]]$). We can take for example $f(X) = \pi X + X^q$, or $f(X) = (1 + X)^p - 1$ over $K = \mathbf{Q}_p$.

Theorem 11.3 *Let $f \in \mathcal{F}_\pi$. Then there exists a unique formal group law $F_f \in A[[X, Y]]$ such that*

$$f(F_f(X, Y)) = F_f(f(X), f(Y))$$

(in other words, f is an endomorphism for the law F_f). We say that F_f is the Lubin–Tate formal group law associated to f and π.

The proof is based on the following proposition.

Proposition 11.4 *Let $f, g \in \mathcal{F}_\pi$. Let $n \in \mathbf{N}^*$ and let $\phi_1(X_1, \ldots, X_n)$ a linear form in n variables with coefficients in A. Then there exists a unique formal series $\phi \in A[[X_1, \ldots, X_n]]$ satisfying:*
  (a) $\phi \equiv \phi_1 \bmod \deg 2$;
  (b) $f(\phi(X_1, \ldots, X_n)) = \phi(g(X_1), \ldots, g(X_n))$.

(we will denote by $(g, g, \ldots, g)$ the family $(g(X_1), \ldots, g(X_n))$, and by $\phi \circ g$ the formal series in n variables $\phi \circ (g, g, \ldots, g) := \phi(g(X_1), \ldots, g(X_n))$.)

Proof We construct ϕ by successive approximation: we will prove by induction on $r \geqslant 1$ that, for any $r \geqslant 1$, there exists a formal series $\phi^{(r)} \in A[[X_1, \ldots, X_n]]$, unique modulo $\deg(r + 1)$, satisfying (a), and also (b) modulo $\deg(r + 1)$. This yields, as a unique solution to the problem, the formal series ϕ defined by $\phi \equiv \phi^{(r)}$ modulo

$\deg(r+1)$ for any $r \geqslant 1$. In particular, note that $\phi^{(r+1)} - \phi^{(r)}$ should be congruent to 0 mod $\deg(r+1)$.

We start by taking $\phi^{(1)} = \phi_1$ (and it is unique modulo $\deg 2$ by a). Assume that $\phi^{(i)}$ has been constructed for $i \leqslant r$, and let $\phi_i = \phi^{(i)} - \phi^{(i-1)}$ (with the convention $\phi^{(0)} = 0$), so that we have $\phi_1 + \cdots + \phi_r = \phi^{(r)}$. Condition (b) yields $f \circ \phi^{(r)} = \phi^{(r)} \circ g \mod \deg(r+1)$, where we have, for simplicity, written g instead of $(g, g, \ldots, g)$. We are then looking for $\phi^{(r+1)}$ of the form $\phi^{(r+1)} = \phi^{(r)} + \phi_{r+1}$, and as we have seen ϕ_{r+1} is congruent to 0 modulo $\deg(r+1)$. We have

$$f \circ \phi^{(r)} \equiv \phi^{(r)} \circ g + E_{r+1} \quad \mod \deg(r+2)$$

with $E_{r+1} \equiv 0 \mod \deg(r+1)$, and we need to deal with the error term E_{r+1}. As the degree 1 term of f (its derivative at 0) is π and ϕ_{r+1} is congruent to 0 modulo $\deg(r+1)$, we obtain

$$f \circ \phi^{(r+1)} \equiv f \circ \phi^{(r)} + \pi \phi_{r+1} \quad \mod \deg(r+2)$$

and similarly

$$\phi^{(r)} \circ g + \phi_{r+1} \circ g \equiv \phi^{(r)} \circ g + \pi^{r+1} \phi_{r+1} \quad \mod \deg(r+2)$$

hence

$$\begin{aligned} f \circ \phi^{(r+1)} - \phi^{(r+1)} \circ g &\equiv f \circ \phi^{(r)} + \pi \phi_{r+1} - (\phi^{(r)} \circ g + \phi_{r+1} \circ g) \\ &\equiv E_{r+1} + (\pi - \pi^{r+1}) \phi_{r+1} \quad \mod \deg(r+2). \end{aligned}$$

We see that the unique solution is

$$\phi_{r+1} = -E_{r+1}/\pi(1 - \pi^r)$$

and we just need to check that ϕ_{r+1} has coefficients in A, or that E_{r+1} is divisible by π. As for $\phi \in \mathbf{F}_q[\![X]\!]$ we have $\phi(X^q) = (\phi(X))^q$ and $f(X) \equiv g(X) \equiv X^q \mod \pi$, we have:

$$f \circ \phi^{(r)} - \phi^{(r)} \circ g \equiv (\phi^{(r)}(X))^q - \phi^{(r)}(X^q) \equiv 0 \quad \mod \pi,$$

which concludes the proof of the proposition.   □

Proof of Theorem 11.3 We apply last proposition taking for $F_f(X, Y)$ the unique solution of $F_f(X, Y) \equiv X + Y \mod \deg 2$ and $f \circ F_f = F_f \circ (f, f)$. To check the formal group axioms, we use the uniqueness in the proposition: for example for (a), we note that $F_f(F_f(X, Y), Z)$ and $F_f(X, F_f(Y, Z))$ are both solutions of

$$\begin{cases} H(X, Y, Z) = X + Y + Z \text{ mod deg } 2 \\ H(f(X), f(Y), f(Z)) = f(H(X, Y, Z)). \end{cases} \qquad \square$$

Proposition 11.5 *Let $f \in \mathcal{F}_\pi$. Let F_f be the formal group law associated to f as in Theorem 11.3. Then for any a of $A = \mathcal{O}_K$, there exists a unique formal series $[a]_f \in A[\![X]\!]$ satisfying: $[a]_f \circ f = f \circ [a]_f$ and $[a]_f \equiv aX$ mod deg 2. Furthermore, $[a]_f$ is an endomorphism for the formal group law F_f. In particular, $[1]_f$ is the identity and $[\pi]_f = f$.*

Proof For f, g in $\mathcal{F}_\pi$, we define $[a]_{f,g}(T) \in A[\![T]\!]$ as the unique formal series (cf. Proposition 11.4) satisfying $[a]_{f,g}(T) \equiv aT$ mod deg 2 and $f([a]_{f,g}(T)) = [a]_{f,g}(g(T))$. Let $[a]_f = [a]_{f,f}$. Then the uniqueness in Proposition 11.4 yields the equality

$$F_f([a]_{f,g}(X), [a]_{f,g}(Y)) = [a]_{f,g}(F_g(X, Y)).$$

Indeed, each member of this equality is a formal series $H(X, Y) \in A[\![X, Y]\!]$, congruent to $aX + aY$ mod deg 2, and satisfying $H(g(X), g(Y)) = f(H(X, Y))$. This then means that $[a]_{f,g}$ is a homomorphism of formal groups from F_g to F_f, and in particular $[a]_f$ is an endomorphism of F_f. The two desired properties follow from definition of $[a]_{f,f}$. The uniqueness is clear, again by Proposition 11.4. $\square$

Proposition 11.6 *The map $a \mapsto [a]_f$ is an injective ring homomorphism from A to $\mathrm{End}(F_f)$.*

Proof As in the previous proposition, we show that

$$[a + b]_{f,g}(T) = F_f([a]_{f,g}(T), [b]_{f,g}(T))$$

and

$$[ab]_{f,h}(T) = ([a]_{f,g} \circ [b]_{g,h})(T). \qquad (11.1)$$

For example, the two members of the first equality are congruent to $aT + bT$ mod deg 2 and are solutions of the equation $f(H(T)) = H(g(T))$. By taking $f = g = h$ we obtain that $a \mapsto [a]_f$ is a ring homomorphism from A to $\mathrm{End}(F_f)$. It is injective since the term of degree 1 of $[a]_f$ is aX. $\square$

Remark 11.7 To define a ring homomorphism from A to $\mathrm{End}(F_f)$ as above, can be expressed by saying that we endow F_f with a structure of *formal A-module*. Thus, if L is an algebraic extension of K, then $F_f(\mathfrak{m}_L)$ (defined as the union of the $F_f(\mathfrak{m}_{L'})$ for every finite extension L' of K contained in L) is endowed with a structure of an A-module.

Proposition 11.8 *Let $f, g \in \mathcal{F}_\pi$. Then the group laws F_f and F_g associated to f and g, respectively, are isomorphic.*

Proof By definition of $[a]_f$, we have $[1]_f(T) = T$. We deduce, using (11.1), that if a is in A^*, then $[a]_{f,g}$ is invertible with inverse $[a^{-1}]_{g,f}$, and $[a]_{f,g}$ thus induces an isomorphism from F_g to F_f. $\square$

11.2 Change of the Uniformiser

We have seen in the last paragraph that a change of the element f in $\mathcal{F}_\pi$ produces isomorphic formal group laws. This is no longer true if we change the uniformiser π of $A := \mathcal{O}_K$. We will nevertheless see that we find isomorphic laws provided we rise to the ring of integers $\widehat{A}_{\mathrm{nr}} := \mathcal{O}_{\widehat{K}_{\mathrm{nr}}}$ of the completion $\widehat{K}_{\mathrm{nr}}$ of the maximal unramified extension K_{nr} of K. As usual we denote by $U_{\widehat{K}_{\mathrm{nr}}}$ the multiplicative group $\widehat{A}_{\mathrm{nr}}^*$ and $\overline{\kappa}$ the algebraic closure of κ. Thus $\overline{\kappa}$ is the residue field of the ring of integers of $\widehat{K}_{\mathrm{nr}}$. One denotes by σ the continuous extension of the Frobenius (cf. Example 7.15, b) to $\widehat{K}_{\mathrm{nr}}$.

If $x \in \widehat{K}_{\mathrm{nr}}$, we denote by σx the transform of x by σ, and similarly if $\theta \in \widehat{K}_{\mathrm{nr}}[\![X]\!]$ is a formal series, we denote $\sigma\theta$ the formal series obtained by applying σ to all the coefficients of θ. We also set $\tau x = \sigma(x)/x$ for any x in $\widehat{K}_{\mathrm{nr}}^*$.

Lemma 11.9 *Let $c \in \widehat{A}_{\mathrm{nr}}$ (resp. $c \in U_{\widehat{K}_{\mathrm{nr}}}$). Then the equation $\sigma x - x = c$ (resp. $\tau x = c$) has a solution in $\widehat{A}_{\mathrm{nr}}$ (resp. in $U_{\widehat{K}_{\mathrm{nr}}}$).*

Furthermore, if x is an element of $\widehat{A}_{\mathrm{nr}}$ satisfying $\sigma x = x$, then $x \in A$.

Proof Note that the uniformiser π remains of valuation 1 in $\widehat{K}_{\mathrm{nr}}$, hence we have (Theorem 7.18, a), valid without the finiteness assumption on the residue field) σ-equivariant isomorphisms of the multiplicative group $U_{\widehat{K}_{\mathrm{nr}}}/U_{\widehat{K}_{\mathrm{nr}}}^1$ with $\overline{\kappa}^*$, and also of $U_{\widehat{K}_{\mathrm{nr}}}^n/U_{\widehat{K}_{\mathrm{nr}}}^{n+1}$ with $\overline{\kappa}$ for $n \geqslant 1$. Let $c \in U_{\widehat{K}_{\mathrm{nr}}}$ with image $\overline{c}$ in $\overline{\kappa}^*$. The equation $\sigma\overline{x} = \overline{x} \cdot \overline{c}$ has a solution in $\overline{\kappa}^*$ because it is written $\overline{x}^q = \overline{x} \cdot \overline{c}$ (where q is the cardinality of κ) and $\overline{\kappa}$ is algebraically closed. This allows us to write $c = \tau x_1 \cdot a_1$ with $x_1 \in U_{\widehat{K}_{\mathrm{nr}}}$ and $a_1 \in U_{\widehat{K}_{\mathrm{nr}}}^1$. By induction we can write $c = \tau(x_1 \cdots x_n) \cdot a_n$ with $x_i \in U_{\widehat{K}_{\mathrm{nr}}}^{i-1}$ and $a_n \in U_{\widehat{K}_{\mathrm{nr}}}^n$; hence $c = \tau x$, where x is the infinite product of the x_i (which converges by completeness $U_{\widehat{K}_{\mathrm{nr}}}$, since its general term goes to 1). The case of the equation $\sigma x - x = c$ in $\widehat{A}_{\mathrm{nr}}$ is similar, via the isomorphisms $x \mapsto x/\pi^n$ (modulo $\mathfrak{m}_{\widehat{K}_{\mathrm{nr}}}$) of the additive group $\mathfrak{m}_{\widehat{K}_{\mathrm{nr}}}^n/\mathfrak{m}_{\widehat{K}_{\mathrm{nr}}}^{n+1}$ on $\overline{\kappa}$.

If now $x \in \widehat{A}_{\mathrm{nr}}$ satisfies $\sigma x = x$, we show by induction on $n > 0$ that x can be written as $x = x_n + \pi^n y_n$ with $x_n \in A$ and $y_n \in \widehat{A}_{\mathrm{nr}}$: for $n = 1$, the property $\sigma x = x$ shows that the image $\overline{x}$ of x in $\overline{\kappa}$ is in κ, hence x is written as $x = x_1 + \pi y_1$ with $x_1 \in A$ and $y_1 \in \widehat{A}_{\mathrm{nr}}$. It is then enough to apply the same reasoning to y_n instead of x. Then x is the limit of the x_n, hence remains in A since A is complete. □

Lemma 11.10 *Let π and ω be two uniformisers of K, let $\omega = u \cdot \pi$ with $u \in U_K$. Let $f \in \mathcal{F}_\pi$ and $g \in \mathcal{F}_\omega$. Let $H \in A[\![X]\!]$ be the unique formal series congruent to uX mod deg 2, and satisfying $f \circ H = H \circ f$. Then there exists $\varepsilon \in U_{\widehat{K}_{\mathrm{nr}}}$ and a formal series $\phi(X) \in \widehat{A}_{\mathrm{nr}}[\![X]\!]$, congruent to εX mod 2, such that:*

(a) $\sigma\phi = \phi \circ H$;
(b) $\sigma\phi \circ f = g \circ \phi$.

Proof The first step is to find a formal series $\alpha(X)$ satisfying the condition (a). We obtain it as a limit of the sequence of polynomials $\alpha_r(x) = \sum_{i=1}^r a_i X^i$ in $\widehat{A}_{\mathrm{nr}}[X]$, satisfying

$$^\sigma\alpha_r(X) - \alpha_r(H(X)) \equiv c_{r+1}X^{r+1} \quad \mathrm{mod}\ \deg(r+2),$$

with $c_{r+1} \in \widehat{A}_{\mathrm{nr}}$. For $r = 1$, this just means that $u = {}^\tau a_1$, which has a solution $a_1 = \varepsilon$ in $U_{\widehat{K}_{\mathrm{nr}}}$ by Lemma 11.9. Assume that α_r is constructed and let

$$\alpha_{r+1}(X) = \alpha_r(X) + a_{r+1}X^{r+1}$$

with $a_{r+1} = a \cdot \varepsilon^{r+1}$, where $a \in \widehat{A}_{\mathrm{nr}}$ is a solution of $^\sigma a - a = -c_{r+1}/(\varepsilon u)^{r+1}$ (such a solution exists by loc. cit.). As $^\sigma\varepsilon = \varepsilon u$, we obtain

$$^\sigma a_{r+1} - a_{r+1}u^{r+1} = (^\sigma a - a) \cdot (\varepsilon u)^{r+1} = -c_{r+1}$$

hence we deduce

$$^\sigma\alpha_{r+1}(X) - \alpha_{r+1}(H(X)) \equiv (c_{r+1} + (^\sigma a_{r+1} - a_{r+1}u^{r+1}))X^{r+1} \equiv 0$$

modulo $\deg(r+2)$, as desired.

Observe now that any formal series of type $\phi = \beta \circ \alpha$ with $\beta \in A[[X]]$ and $\beta \equiv X$ mod deg 2 again satisfies condition (a), since we then have

$$^\sigma\phi = {}^\sigma(\beta \circ \alpha) = {}^\sigma\beta \circ {}^\sigma\alpha = \beta \circ \alpha \circ H = \phi \circ H.$$

We now need to choose β such that ϕ also satisfies (b). To do this, let

$$h = {}^\sigma\alpha \circ f \circ \alpha^{-1} = \alpha \circ H \circ f \circ \alpha^{-1},$$

where α^{-1} is the formal series without constant term, inverse (for the composition of formal series) of α (well defined since $\alpha(X) \equiv \varepsilon X$ mod deg 2 since ε is invertible). We are looking for β such that $g \circ \beta = \beta \circ h$, which will imply

$$g \circ \phi = g \circ \beta \circ \alpha = \beta \circ h \circ \alpha = \beta \circ {}^\sigma\alpha \circ f = {}^\sigma\phi \circ f$$

as desired.

Recall that we have $^\sigma\alpha = \alpha \circ H$, and hence $\alpha^{-1} = H \circ {}^\sigma\alpha^{-1}$. First observe that

$$^\sigma h = {}^\sigma\alpha \circ H \circ f \circ {}^\sigma\alpha^{-1} = {}^\sigma\alpha \circ f \circ H \circ {}^\sigma\alpha^{-1} = (^\sigma\alpha \circ f) \circ \alpha^{-1} = h$$

hence $h \in A[[X]]$ by Lemma 11.9. On the other hand, $h(X) \equiv {}^\sigma\varepsilon\pi\varepsilon^{-1}X \equiv u\pi X \equiv \omega X$ mod deg 2, and

$$h(X) \equiv {}^\sigma\alpha(f(\alpha^{-1}(X))) \equiv {}^\sigma\alpha(\alpha^{-1}(X)^q) \equiv {}^\sigma\alpha(^\sigma\alpha^{-1}(X^q)) \equiv X^q \quad \mathrm{mod}\ \pi$$

(observe that modulo π, the Frobenius σ acts as raising to the qth power and that $^\sigma(\alpha^{-1}) = (^\sigma\alpha)^{-1}$). We then define $\beta = \sum_{i \geqslant 1} b_i X^i$ as the limit of polynomials β_r of degree $\leqslant r$, that we construct by induction to verify

$$g(\beta_r(X)) - \beta_r(h(X)) \equiv c_{r+1}X^{r+1} \mod X^{r+2},$$

with $c_{r+1} \in A$, which will give $g \circ \beta = \beta \circ h$. By what precedes, we can take $b_1 = 1$. One notes that c_{r+1} is zero $\mod \pi$ since

$$g(\beta_r(X)) \equiv \beta_r(X)^q \equiv \beta_r(X^q) \equiv \beta_r(h(X)) \mod \pi.$$

We then set $\beta_{r+1}(X) = \beta_r(X) + b_{r+1}X^{r+1}$ with

$$b_{r+1} = -c_{r+1}/(\omega - \omega^{r+1}),$$

which is in A. Then

$$g(\beta_{r+1}(X)) - \beta_{r+1}(h(X)) \equiv (c_{r+1} + (\omega - \omega^{r+1})b_{r+1})X^{r+1} \equiv 0 \mod \deg(r+2)$$

as desired. This finishes the induction. $\qquad\square$

Proposition 11.11 *Let π and $\omega = u \cdot \pi$ be two uniformisers of K. Let $f \in \mathcal{F}_\pi$ and $g \in \mathcal{F}_\omega$. Then there exists $\varepsilon \in U_{\widehat{K}_{nr}}$ and $\phi \in \widehat{A}_{nr}[\![X]\!]$ with $\phi(X) \equiv \varepsilon X \mod 2$, such that:*

(a) $^\sigma\phi = \phi \circ [u]_f$;
(b) $\phi \circ F_f = F_g \circ (\phi, \phi)$;
(c) $\phi \circ [a]_f = [a]_g \circ \phi$ for any $a \in \mathcal{O}_K$.
In particular, ϕ is an isomorphism of A-modules from F_f to F_g.

Proof We apply Lemma 11.10 to $H := [u]_f$ (see Proposition 11.5). We obtain $\phi \in \widehat{A}_{nr}[\![X]\!]$ with $\phi(X) \equiv \varepsilon X \mod 2$ and $\varepsilon \in U_{\widehat{K}_{nr}}$, already satisfying $^\sigma\phi = \phi \circ [u]_f$ and $^\sigma\phi \circ f = g \circ \phi$. Let us for example prove (b) (the proof of (c) is analogous). Let $\phi^{-1} \in \widehat{A}_{nr}[\![X]\!]$ be the formal series without constant term, inverse of ϕ (which exists since $\phi(0) = 0$ and $\phi'(0)$ is invertible). Let $G(X, Y) = \phi(F_f(\phi^{-1}(X), \phi^{-1}(Y))$. We need to show that $G = F_g$. But $G(X, Y) \equiv X + Y \mod \deg 2$, we also have $G \in A[\![X, Y]\!]$ since we have $G = \phi \circ F_f \circ \phi^{-1}$ hence

$$^\sigma G = {}^\sigma\phi \circ F_f \circ {}^\sigma\phi^{-1} = \phi \circ [u]_f \circ F_f \circ [u]_f^{-1} \circ \phi^{-1} = G$$

(Recall that $[u]_f$ commutes with F_f). It remains (using the uniqueness in Proposition 11.4) to check that $g \circ G = G \circ g$. But

$$g \circ G = g \circ \phi \circ F_f \circ \phi^{-1} = {}^\sigma\phi \circ f \circ F_f \circ \phi^{-1}$$
$$= {}^\sigma\phi \circ F_f \circ {}^\sigma\phi^{-1} \circ g = {}^\sigma G \circ g = G \circ g$$

(indeed, f commutes with F_f, and $G \in A[\![X, Y]\!]$ gives $^\sigma G = G$). $\qquad\square$

11.3   Fields Associated to Torsion Points

In this chapter, we fix a separable closure $\overline{K}$ of K. All algebraic separable extensions of K under consideration are assumed to be contained in $\overline{K}$. We let $A = \mathcal{O}_K$ be the ring of integers of K and we set $U_K = A^*$. For any $n > 0$, we have a multiplicative subgroup U_K^n of U_K, consisting of elements x such that $v(1 - x) \geqslant n$.

Let π be a uniformiser of K and $f \in \mathcal{F}_\pi$. We have previously defined the formal group (and even A-module) law F_f, and hence we can consider (cf. Remark 11.7) the A-module $F_f(\mathfrak{m}_{\overline{K}})$.

Let E_f^n be the kernel of the endomorphism $[\pi^n]_f$. It is thus a π^n-torsion submodule of $F_f(\mathfrak{m}_{\overline{K}})$ and $E_f := \bigcup_n E_f^n$ is the torsion submodule of $F_f(\mathfrak{m}_{\overline{K}})$. We can also view $F_f(\mathfrak{m}_{\overline{K}})$ and E_f as subsets of $\overline{K}$.

Let also $K_\pi^n = K(E_f^n)$ and $K_\pi = \bigcup_n K(E_f^n)$. Note that, by Proposition 11.8 these extensions do not depend on the choice f. Indeed we additionally have that an isomorphism between formal group laws acts on $\mathfrak{m}_{\overline{K}}$ by $x \mapsto F(x)$, where $F \in A[[X]]$; but for $x \in \mathfrak{m}_{\overline{K}}$, such an $F(x)$ remains in the field $K(x)$ generated by K and x ($F(x)$ is a limit of elements of $K(x)$, which is closed in $\overline{K}$ as it is a finite-dimensional vector space over K).

Each extension $K(E_f^n)$ is Galois over K since the conjugates over K of each element of E_f^n remain in E_f^n, by definition of E_f^n as π^n-torsion submodule of $F_f(\mathfrak{m}_{\overline{K}})$: indeed, π^n acts via the formal series $[\pi^n]_f$, which has coefficient in K. The Galois group $G_\pi := \mathrm{Gal}(K_\pi / K)$ is the projective limit of $G_{\pi,n} := \mathrm{Gal}(K(E_f^n)/K)$. We then obtain a homomorphism $G_\pi \to \mathrm{Aut}_A(E_f)$ by making G_π act on torsion points of $F_f(\mathfrak{m}_{\overline{K}})$.

Recall also that the composite of two abelian extensions is an abelian extension. This follows from the fact that if H and H' are two normal subgroups of a group G with G/H and G/H' abelian, then $G/(H \cap H')$ is abelian.

Lemma 11.12 *There exists an isomorphism of A-modules from E_f to K/A. In particular, the group $\mathrm{Aut}_A(E_f)$ is isomorphic to A^*.*

Proof By Proposition 11.8, we are free to take any f in $\mathcal{F}_\pi$. Let $f = \pi X + X^q$, where q is the cardinality of the residue field κ of $A = \mathcal{O}_K$. We have $[\pi]_f = f$ by Proposition 11.5, hence E_f^n is the set of roots of the nth iterated f_n of f. Note that for any $\alpha \in \mathfrak{m}_{\overline{K}}$, the equation $f(x) = \alpha$ has q distinct roots in $\overline{K}$ (the polynomial $f - \alpha$ is separable) and its solutions remain of > 0 valuation, hence these solutions belong to $\mathfrak{m}_{\overline{K}}$. Thus the A-module $F_f(\mathfrak{m}_{\overline{K}})$ is divisible (the multiplication by π is surjective), hence also its torsion submodule E_f. In particular, as A is a principal ideal domain and its only irreducible (up to a unit) element is π, the A-module E_f is isomorphic to a direct sum of copies of K/A ([8], Sect. 2, Exer. 3). But as E_f^1 (which is the π-torsion subgroup of the A-module E_f) has exactly q elements (the roots of f), the only possibility is that E_f is isomorphic to K/A. $\qquad\square$

Theorem 11.13 *(a) The composed morphism $\Phi : G_\pi \to \mathrm{Aut}_A(E_f) \simeq A^*$ is an isomorphism, which transforms the filtration of G_π by subgroups $\mathrm{Gal}(K_\pi/K_\pi^n)$ into a filtration of $A^* = U_K$ by subgroups U_K^n. In particular, the extension K_π/K is abelian.*

(b) For any $n > 0$, the uniformiser π is a norm of the extension K_π^n/K.

Proof (a) Let f and f_n be as in the proof of Lemma 11.12. Let $\tau \in G_\pi$. It induces an automorphism of the A-module E_f. This automorphism (by Lemma 11.12) is the multiplication by an element of $A^* = U_K$. We thus obtain a homomorphism $\Phi : G_\pi \to U_K$, which is injective by definition of K_π. It remains to check that Φ is surjective. Note that Φ induces an injective map $\Phi_n : \mathrm{Gal}(K_\pi^n/K) \to U_K/U_K^n$ since E_f^n corresponds to the π^n-torsion subgroup $\frac{1}{\pi^n}\mathcal{O}_K/\mathcal{O}_K$ of $K/A = K/\mathcal{O}_K$, on which U_K^n acts trivially. Let $\alpha \in E_f^n \setminus E_f^{n-1}$ and let $h = f_n/f_{n-1}$. As $f/X = X^{q-1} + \pi$, we have

$$h = f(f_{n-1})/f_{n-1} = (f_{n-1}(X))^{q-1} + \pi,$$

which shows that h is of degree $q^n - q^{n-1}$ and is irreducible since it is an Eisenstein polynomial (cf. Example 7.12). As α is a root of h, the degree $[K(\alpha) : K]$ is that of h (since h is the minimal polynomial of α). We thus obtain that the cardinality of $\mathrm{Gal}(K_\pi^n/K)$ is at least $q^n - q^{n-1}$, which is the cardinality of U_K/U_K^n (indeed, U_K/U_K^1 is isomorphic to κ^*, and the quotients U_K^i/U_K^{i+1} for $i > 0$ are isomorphic to κ, by Theorem 7.18). This implies that Φ_n is an isomorphism and $K(\alpha) = K_\pi^n$. By passing to the limit, we obtain that G_π is isomorphic to U_K via Φ.

(b) Let α be as above, we have $K_\pi^n = K(\alpha)$ and the polynomial h constructed in (a) is the minimal polynomial of α. Its constant term is π, which implies that π is the norm of $-\alpha$ for the extension K_π^n/K. $\qquad\square$

11.4 Computation of the Reciprocity Map

In the last paragraph, we have defined an abelian extension K_π of K associated to a uniformiser π of $A = \mathcal{O}_K$. It was constructed using the formal group of Lubin–Tate associated to π. The extension K_π depends on the choice of the uniformiser π. Nevertheless, we have:

Proposition 11.14 *Let π and ω be two uniformisers of K. Let $L_\pi = K_{\mathrm{nr}} \cdot K_\pi$ and $L_\omega = K_{\mathrm{nr}} \cdot K_\omega$. Then $L_\pi = L_\omega$.*

Note that this result is easier if we assume the existence theorem (cf. [40], where the existence theorem is proved directly in characteristic zero by the classical method of Kummer extensions, the method that we have seen in Sect. 9.2). The approach that we follow here is that of Serre ([9], Chap. VI) which has the advantage of providing a quick proof of the existence theorem in any characteristic using the Lubin–Tate theory.

Proof Let $f \in \mathcal{F}_\pi$ and $g \in \mathcal{F}_\omega$. The Proposition 11.11 says that the formal A-modules F_f and F_g are isomorphic over $\widehat{K}_{\mathrm{nr}}$. In particular, the fields generated by $\widehat{K}_{\mathrm{nr}}$ and torsion points of $F_f(\mathfrak{m}_{\overline{K}})$ and $F_g(\mathfrak{m}_{\overline{K}})$ respectively are the same, i.e., $\widehat{K}_{\mathrm{nr}} \cdot K_\pi = \widehat{K}_{\mathrm{nr}}.K_\omega$ A fortiori, the respective completions L_π and L_ω (seen as subfields of the completion $\widehat{\overline{K}}$ of $\overline{K}$) are equal. We conclude using the following lemma.

Lemma 11.15 *Let E be an algebraic extension (finite or infinite) of a local field, with completion $\widehat{E}$. Let $\alpha \in \widehat{E}$. If α is algebraic and separable over E, then $\alpha \in E$.*

Proof Let $\overline{E}$ be a separable closure of E. Let E' be the topological closure of E in $\overline{E}$. We can then view α as an element of E' (by definition of the completion). Observe that any element of $\mathrm{Gal}(\overline{E}/E)$ is a continuous map on $\overline{E}$. Indeed, $\overline{E}$ is a union of finite normal algebraic extensions of a local field F. It is thus enough to check that if L/F is a finite extension of local fields, then any automorphism of the F-vector space L is continuous, which follows from the fact that the topology defined on L by its absolute value is the same as its topology of a normed finite-dimensional F-vector space (one can also use Exercise 7.1).

We deduce that any element of $\mathrm{Gal}(\overline{E}/E)$ induces by continuity the identity on E', hence $E' = E$ by Galois theory, and thus $\alpha \in E$. $\qquad\square$

We now have the following link between the abelian extension K_π of K and the reciprocity map.

Proposition 11.16 *Let $L_\pi = K_{\mathrm{nr}} \cdot K_\pi$ be the abelian extension of K composite of K_{nr} and K_π. Let $\theta = \omega_{L_\pi/K} : K^* \to \mathrm{Gal}(L_\pi/K)$ be the reciprocity map, defined by passing to the limit over the reciprocity maps $\omega_{K'/K}$ for K' a finite extension of K contained in L_π. Then K_π is the subfield of L_π fixed by $\theta(\pi)$ and $\theta(\pi)$ induces the Frobenius on K_{nr}. Furthermore, the fields K_{nr} and K_π are linearly disjoint. In other words, the group $\mathrm{Gal}(L_\pi/K)$ is the direct product of the groups $\mathrm{Gal}(K_{\mathrm{nr}}/K)$ and G_π. In particular, all the intermediate finite extensions of K between K and K_π are totally ramified.*

Proof As $L_\pi \supset K_{\mathrm{nr}}$, we have an exact sequence

$$0 \longrightarrow H \longrightarrow \mathrm{Gal}(L_\pi/K) \longrightarrow \mathrm{Gal}(K_{\mathrm{nr}}/K) \longrightarrow 0,$$

where $H := \mathrm{Gal}(L_\pi/K_{\mathrm{nr}})$ and $\mathrm{Gal}(K_{\mathrm{nr}}/K)$ is topologically generated by the Frobenius σ. As $\theta(\pi) \in \mathrm{Gal}(L_\pi/K)$ is a lift of σ by Lemma 9.8 (passing to the limit over the finite extensions $K' \subset K_{\mathrm{nr}}$ of K), we obtain that $\mathrm{Gal}(L_\pi/K)$ is the direct product of H and of the closed subgroup I_π generated by $\theta(\pi)$. In other words, we have $L_\pi = K_{\mathrm{nr}} K'$ with K_{nr} and K' linearly disjoint over K, where K' is the fixed field of L_π by $\theta(\pi)$.

For any $n > 0$, the reciprocity map $K^* \to \mathrm{Gal}(K_\pi^n/K)$ is trivial on the image of the norm $(K_\pi^n)^* \to K^*$, which implies, by Theorem 11.13 (b), that $\omega_{K_\pi/K}(\pi)$ is trivial. By compatibility between $\omega_{L_\pi/K}$ and $\omega_{K_\pi/K}$ (Corollary 9.4), we obtain that the restriction of $\theta(\pi)$ to K_π is the identity, hence $K_\pi \subset K'$. As we also have $L_\pi = K_{\mathrm{nr}} \cdot K_\pi$, the only possibility is $K_\pi = K'$ since the subgroup $H' := \mathrm{Gal}(L_\pi/K_\pi)$ contains I_π and satisfies $H \cap H' = \{1\}$, hence is equal to I_π as $\mathrm{Gal}(L_\pi/K)$ is the direct product of H and I_π. $\qquad\square$

We can now prove the main theorem of this paragraph, which relates the construction of Lubin–Tate to the reciprocity map.

Theorem 11.17 *Let K be a local field. Let π be a uniformiser of K and $f \in \mathcal{F}_\pi$. Let $L_\pi = K_{nr}.K_\pi$. We define a homomorphism*

$$r_\pi : K^* \longrightarrow \mathrm{Gal}(L_\pi/K)$$

by the properties:

(a) $r_\pi(\pi)$ is the identity on K_π and coincides with the Frobenius σ on K_{nr};

(b) for any u of U_K, $r_\pi(u)$ is the identity on K_{nr} and acts by $[u^{-1}]_f$ on K_π.

Then r_π is independent of the choice of the uniformiser π and is equal to the reciprocity map $\omega_{L_\pi/K} : K^* \to \mathrm{Gal}(L_\pi/K)$.

Note that r_π is well defined since K^* is the direct product of subgroups U_K and $\pi^\mathbf{Z}$, and the extensions K_{nr} and K_π are linearly disjoint. Recall also that $[u^{-1}]_f$ can be seen as an A-endomorphism of E_f, hence induces an automorphism of K_π by Theorem 11.13 (a). We will see in the next paragraph that L_π is in fact the maximal abelian extension of K.

Proof We have already seen (Proposition 11.14) that the extension L_π does not depend on π. Let $\omega = u\pi$ be another uniformiser. We will prove that $r_\pi(\omega) = r_\omega(\omega)$, which will show that for any uniformiser ω, $r_\pi(\omega)$ is independent of π and hence that r_π is independent of π since K^* is generated by the uniformisers. We know already that $r_\pi(\omega)$ and $r_\omega(\omega)$ coincide on K_{nr} as $r_\pi(\omega) = r_\pi(u).r_\pi(\pi)$, which gives that $r_\pi(\omega)$ and $r_\omega(\omega)$ both induce the Frobenius on K_{nr}. It remains to show that $r_\pi(\omega)$ is the identity on K_ω.

We have by definition $K_\omega = K(E_g)$ with $g \in \mathcal{F}_\omega$. Let $\phi \in \widehat{A}_{nr}[\![X]\!]$ as in Proposition 11.11. Every element λ of E_g is written as $\lambda = \phi(\mu)$ with $\mu \in E_f$. Let $s = r_\pi(\omega)$. We have to check that $s(\lambda) = \lambda$, or in other words that $^s(\phi(\mu)) = \phi(\mu)$. Note that $s = r_\pi(\pi).r_\pi(u)$. As ϕ has coefficients in $\widehat{K}_{nr}$, we have

$$^s\phi = {}^\sigma\phi = \phi \circ [u]_f.$$

On the other hand

$$^s(\phi(\mu)) = {}^s\phi({}^s\mu) = {}^s\phi([u^{-1}]_f(\mu))$$

since $\mu \in K_\pi$ hence

$$^s(\phi(\mu)) = \phi \circ [u]_f([u^{-1}]_f(\mu)) = \phi(\mu),$$

as desired. This proves that $r_\pi = r_\omega$.

We know (Proposition 11.16, along with the equality $L_\pi = L_\omega$) that the reciprocity map $\theta : K^* \to \mathrm{Gal}(L_\pi/K)$ satisfies, for any uniformiser ω, the property that $\theta(\omega)$ induces the Frobenius σ on K_{nr} and identity on K_ω. By what we have just seen, $\theta(\omega) = r_\pi(\omega)$. As this holds for any uniformiser ω, we obtain that $r_\pi = r_\omega$ is the same map as θ. $\qquad\square$

To finish the description of the maximal abelian extension K^{ab} of K, it remains to prove that $K^{ab} = K_{nr} \cdot K_\pi$, which is the aim of the next paragraph.

11.5   The Existence Theorem (the General Case)

In this paragraph, we recover the existence theorem proved in Sects. 9.2 and 10.5 via a different method (and we extend it to local fields of characteristic $p > 0$).

Fix a uniformiser π of K and $f \in \mathcal{F}_\pi$. We have abelian extensions K_π^n (for $n > 0$) and K_π of K defined in the previous paragraphs. We set $L = K_{\mathrm{nr}} \cdot K_\pi$.

Definition 11.18 We call a subgroup M of the multiplicative group K^* a *group of norms* (or *norm group*) if there exists a finite abelian extension E of K such that $M = N_{E/K} E^*$.

In fact any subgroup of the form $N_{F/K} F^*$, where F is a finite separable extension (not necessarily abelian) of K is a group of norms by Proposition 9.7, (b) (see also Exercise 9.1). The existence theorem gives at once a characterisation of the groups of norms and an identification of the maximal abelian extension of K with $K_{\mathrm{nr}} \cdot K_\pi$. The main step (which will follow from the Lubin–Tate theory) is the next proposition:

Proposition 11.19 *Let K be a local field. Then every open finite index subgroup of K^* is of the form $N_{E/K} E^*$, where E is a finite extension of K contained in L.*

Proof Let M be an open finite index subgroup of K^*. The fact that M is open implies the existence of an $n > 0$ such that $M \supset U_K^n$, and the fact that it is of finite index implies the existence of an $m > 0$ such that $\pi^m \in M$. Thus M contains the subgroup $V_{n,m}$ generated by U_K^n and π^m. Let K_m be the unramified extension of K of degree m. Let $E = K_\pi^n \cdot K_m$: it is a subextension of L. Let $u \in U_K$ and $a \in \mathbf{Z}$. By Theorem 11.17, the image $(u \cdot \pi^a, E/K)$ of $u \cdot \pi^a$ under the reciprocity map $\omega_{E/K}$ is $[u^{-1}]_f$ on K_π^n and σ^a (where σ is the Frobenius) on K_m. It follows that the kernel of $\omega_{E/K}$ is exactly $V_{n,m}$ (recall that Theorem 11.13 (a) shows that $[u]_f$ acts trivially on K_π^n if and only if $u \in U_K^n$), hence $V_{n,m} = N_{E/K} E^*$. We conclude using Lemma 9.9 (a). $\qquad\square$

Theorem 11.20 (existence theorem) *Let K be a local field and let π be a uniformiser of K. Fix a separable closure $\overline{K}$ of K. Set $L = K_{\mathrm{nr}} \cdot K_\pi$. Then:*

(a) the map $E \mapsto N_{E/K} E^*$ is a bijection between the set of finite abelian extensions $E \subset \overline{K}$ of K and the set of open finite index subgroups of K^*.

(b) The extension L is the maximal abelian extension of K.

Proof Let F be a finite abelian extension of K. We know already (Lemma 9.9, c) that the subgroup $N_{F/K} F^*$ of K^* is open and of finite index. Part (a) follows: the injectivity follows from Lemma 9.9, (b) and the surjectivity from Proposition 11.19.

Let us prove (b). Proposition 11.19 says that we can write the open subgroup of finite index $N_{F/K} F^*$ of K^* in the form $N_{F/K} F^* = N_{E/K} E^*$, where E is a finite extension of K contained in L. Lemma 9.9, (b) then shows that $E = F$. In particular, $F \subset L$ and we have finally shown that L contains all the finite abelian extensions of K contained in $\overline{K}$, or that $L = K^{\mathrm{ab}}$ since L is an abelian extension of K. $\qquad\square$

We also recover a statement that we have obtained in the case of a p-adic field (Corollary 9.15 and Theorem 10.20).

Corollary 11.21 *The reciprocity map* $\omega_K : K^* \to \text{Gal}(K^{\text{ab}}/K)$ *induces an isomorphism of the completion of* K^* *(for the topology defined by open subgroups of finite index) with* $\text{Gal}(K^{\text{ab}}/K)$. *This last group is isomorphic to* $\widehat{\mathbf{Z}} \times U_K$.

Indeed, the existence theorem says that open subgroups of finite index of K^* are precisely the groups of norms and we know that the reciprocity map induces an isomorphism of $\varprojlim_L K^*/NL^*$ with $\text{Gal}(K^{\text{ab}}/K)$, the limit being taken over the finite abelian extensions of K.

Remark 11.22 If K is a p-adic field, Corollary 7.19 implies that every subgroup of finite index of K^* is open. On the other hand, if K is a local field of characteristic $p > 0$, the group K^* contains subgroups of finite index which are not closed (cf. Exercise 11.3) and we can only say that each subgroup of *prime to* p finite index of K^* is closed.

Example 11.23 *(The case of* $\mathbf{Q}_p$*)* For $K = \mathbf{Q}_p$, we can take $\pi = p$ and

$$f = (X + 1)^p - 1 = pX + C_p^2 X^2 + \cdots + X^p.$$

Then the formal group law F_f is simply

$$F_f(X, Y) = X + Y + XY,$$

and $F_f(\mathfrak{m}_{\overline{K}})$ is thus isomorphic to the group $1 + \mathcal{M}_{\overline{K}}$ endowed with the usual multiplication. Thus $E_f = U_{p^\infty}$ consists of roots of unity of order a power of p, and the maximal abelian extension $\mathbf{Q}_p^{\text{ab}}$ of $\mathbf{Q}_p$ is obtained as $\mathbf{Q}_p^{\text{ab}} = \mathbf{Q}_p^{\text{nr}}.\mathbf{Q}_{p^\infty}$, where $\mathbf{Q}_{p^\infty}$ is the field obtained by adjoining to $\mathbf{Q}_p$ the roots of unity of order a power of p. Finally, if $u \in \mathbf{Z}_p^*$, the associated element $[u] = [u]_f$ acts on U_{p^∞} via an identification of U_{p^∞} with the additive group $\mathbf{Q}_p/\mathbf{Z}_p$, i.e., by $x \mapsto x^u$. If we write $u = \sum_{n \geqslant 0} a_n p^n$ with a_0 not divisible by p, this gives $x^u = \prod_{n \geqslant 0} x^{a_n p^n}$, which makes sense when $x \in U_{p^\infty}$. Note also that $\mathbf{Q}_p^{\text{ab}}$ is simply the field obtained from $\mathbf{Q}_p$ by adding all the roots of unity.

From this example and from Theorem 11.17 we deduce the following condition, which will be useful for proving the global reciprocity law (Theorem 14.9).

Corollary 11.24 *Let* ζ *be a root of unity in* $\overline{\mathbf{Q}_p}$. *Let* $K = \mathbf{Q}_p(\zeta)$. *Let* $\theta_{K/\mathbf{Q}_p} : \mathbf{Q}_p^* \to \text{Gal}(K/\mathbf{Q}_p)$ *be the reciprocity map. Let* $x = p^m \cdot u$ *with* $m \in \mathbf{Z}$ *and* $u \in \mathbf{Z}_p^*$. *Then*

$$\theta_{K/\mathbf{Q}_p}(x) \cdot \zeta = \zeta^{p^m}$$

if the order of ζ *is prime to* p, *and*

$$\theta_{K/\mathbf{Q}_p}(x) \cdot \zeta = \zeta^{u^{-1}}$$

if the order ζ is a power of p.

Proof The condition that the order of ζ is prime to p means that $\zeta \in \mathbf{Q}_p^{\mathrm{nr}}$. To the contrary, ζ is of order power of p if and only if (with the notations of Theorem 11.17) $\zeta \in (\mathbf{Q}_p)_\pi$. In the first case, $\theta_{K/\mathbf{Q}_p}(u)$ acts trivially on ζ and $\theta_{K/\mathbf{Q}_p}(p^m)$ acts by the mth iterated Frobenius, i.e., by raising to the power p^m. In the second case it is $\theta_{K/\mathbf{Q}_p}(p^m)$ acting trivially on ζ and $\theta_{K/\mathbf{Q}_p}(u)$ acts by $[u^{-1}]$, i.e., sends ζ to $\zeta^{u^{-1}}$. $\square$

11.6   Exercises

Exercise 11.1 Let K be a local field. Let K_1 and K_2 be two finite abelian extensions of K. Denote by E the composite field $E = K_1 K_2$. Show that $N_{E/k} E^* = N_{K_1/K} K_1^* \cap N_{K_2/K} K_2^*$.

Exercise 11.2 Let K be a local field. Show that the *group of universal norms* $\bigcap_L N_{L/K} L^*$ (where L ranges through finite abelian extensions of K) is reduced to $\{1\}$. Deduce that the reciprocity map $\omega_K : K^* \to \mathrm{Gal}(K^{\mathrm{ab}}/K)$ is injective and induces an isomorphism between the unit group U_K and the abelian inertia group I_K^{ab} of K (cf. Corollaries 9.15 and 9.16 for the case where K is p-adic).

Exercise 11.3 Let K be a local field of characteristic $p > 0$. Recall (cf. Exercise 7.4) that the profinite group U_K^1 is then isomorphic to $G = \mathbf{Z}_p^{\mathbf{N}}$.

(a) Show that the subgroup $D = \mathbf{Z}_p^{(\mathbf{N})}$ (consisting of almost zero sequences of elements of $\mathbf{Z}_p$) is dense in G.

(b) Show that there exists a nonzero morphism of abelian groups $G \to \mathbf{Z}/p\mathbf{Z}$, which is zero on D.

(c) Deduce that K^* has a non-closed subgroup of finite index p.

Part III
Global Fields

This part is concerned with the global class field theory which is the analogue for number field (finite extensions of $\mathbf{Q}$) and function fields (of transcendence degree 1 over a finite field) of the local theory seen in Part II. The approach that we will follow in this book makes heavy use of the cohomological machinery from Part I (in particular, the Brauer group will play a very important role), as well as of the local results from Part II. The procedure is, by the way, very similar to the local case but technically much more complicated.

After a chapter where we recall basic facts of the theory of global fields (completions, idèles, class group, Dirichlet unit theorem), we present in Chap. 13 a detailed study of the cohomology of the idèle class group of a global field: the main results are the two main inequalities of the class field theory, leading to the *global class field axiom* (Theorem 13.23), for which we give a complete proof (including the case of a function field, assuming the knowledge of a little bit of algebraic geometry). In Chap. 14, we compute the Brauer group of a global field. An essential step is to show (using local calculations of Lubin–Tate) the *global reciprocity law* (Theorem 14.9). We finish the description of the abelian Galois group of a global field in Chap. 15, using an *existence theorem* similar to that in the local case. We also explain an older formulation of the results in terms of ideal and of ray class fields that we will revisit in the appendix devoted to analytic methods.

The class field theory has been developing in a complex manner, with different and complementary points of view. One can find good expositions of historical aspects[1] in articles by Hasse (exposé XI of [9]) and those by Iyanaga ([23, 24]) as well as in the book by Roquette [44]. Let us mention in particular an analytic approach, going back to Hey and Zorn, relying on the ζ functions of skew fields ([28, 55]). This approach remains relevant today with the development of the theory of automorphic forms and Langlands program which can be seen as a vast non-commutative generalisation of the class field theory.

---

[1]I am grateful to J.-B. Bost for having clarified these historical aspects to me and pointing me to the appropriate references.

Chapter 12
Basic Facts About Global Fields

In this chapter we will give a quick overview of the basic notions concerning global fields. For detailed proofs, we refer for example to the second article in [9].

12.1 Definitions, First Properties

Definition 12.1 A *Dedekind ring* is an integral domain A which is noetherian, integrally closed and of dimension $\leqslant 1$ (this last property means that every nonzero prime ideal is maximal).

Local Dedekind rings (other than fields) are exactly the discrete valuation rings. More generally, an integral noetherian domain A is Dedekind if and only if for every nonzero prime ideal $\mathfrak{p}$ of A, the localisation $A_\mathfrak{p}$ is a discrete valuation ring ([47], Chap. I, Prop. 4). For example, every principal ideal domain is Dedekind (and a Dedekind ring is a principal ideal domain if and only if it is a unique factorisation domain).

Definition 12.2 Let A be a Dedekind ring with field of fractions K. A *fractional ideal* I of A is a sub A-module of finite type of K (or equivalently, a sub A-module of K satisfying: there exists a nonzero d in A with $dI \subset A$). The *product* IJ of two fractional ideals I and J is the fractional ideal generated as an A-module by the products ij with $i \in I$ and $j \in J$. A fractional ideal is *principal* if it is of the form xA with $x \in K^*$.

The next theorem is the analogue for Dedekind rings of the decomposition into a product of irreducibles in unique factorisation domains. For a proof, see [47], Chap. I, Prop. 5 and 7.

Theorem 12.3 *Let A be a Dedekind ring with field of fractions K. Then each nonzero fractional ideal I in A is invertible (i.e., there exists a fractional ideal J of A such that $IJ = A$). Furthermore, such an I can be written uniquely as*

© Springer Nature Switzerland AG 2020
D. Harari, *Galois Cohomology and Class Field Theory*, Universitext,
https://doi.org/10.1007/978-3-030-43901-9_12

$$I = \prod_{\mathfrak{p} \in \text{Spec } A, \mathfrak{p} \neq \{0\}} \mathfrak{p}^{v_{\mathfrak{p}}(I)},$$

where Spec A denotes the set of prime ideals of A and $(v_{\mathfrak{p}}(I))$ is an almost zero family of integers. The $v_{\mathfrak{p}}(I)$ are all $\geqslant 0$ if and only if I is an ideal of A.

Corollary 12.4 *The set of nonzero fractional ideals of a Dedekind ring A forms an abelian group under multiplication. The quotient of this group by the subgroup of nonzero principal fractional ideals is called the* ideal class group *of A.*

The ideal class group will be denoted by $\text{Cl}(A)$, or $\text{Pic } A$ (it is the Picard group of Spec A in the language of algebraic geometry, cf. [21], Part. II, Chap. 6). Recall also ([5], Sect. 2, Prop. 1):

Proposition 12.5 *Let A be a Dedekind ring. If A has only finitely many prime ideals, it is a principal ideal domain (and hence its class group is trivial).*

We arrive at the definition of a global field that we will use in everything that follows:

Definition 12.6 A *global field* is either a finite extension of the field $\mathbf{Q}$ of rational numbers (called a *number field*), or a finite separable extension of $\mathbf{F}_q(t)$ (called a *function field* over $\mathbf{F}_q$). If k is a number field, its *ring of integers* $\mathcal{O}_k$ is the Dedekind ring defined as the integral closure of $\mathbf{Z}$ in k.

A *place* of a global field k is an equivalence class of non-trivial absolute values on k (two absolute values are equivalent if one is a power of the other). For any $\alpha \in k^*$, there are only finitely many places v for which the corresponding absolute value $|\alpha|_v$ of α is $\neq 1$ (cf. [9], Exp. II, Lem. p. 60). More precisely:

- In the case of a number field, there is a finite number of *archimedean* places v, which may be either real or complex (the completion k_v of k with respect to such a place is either $\mathbf{R}$ or $\mathbf{C}$, respectively). The other places (of which there are infinitely many) are called *non-archimedean* or *finite*: they correspond to nonzero prime ideals $\mathfrak{p}$ of the ring of integers $\mathcal{O}_k$ of k. For such a $\mathfrak{p}$, the intersection $\mathfrak{p} \cap \mathbf{Z}$ is a nonzero prime ideal of $\mathbf{Z}$, hence $\mathfrak{p} \cap \mathbf{Z} = p\mathbf{Z}$, where p is a prime number. The completion k_v at the place corresponding to $\mathfrak{p}$ is then a p-adic field with residue field $\mathcal{O}_k/\mathfrak{p}$ and the absolute values associated with v are of the form $|x| = a^{-v_{\mathfrak{p}}(x)}$ with $a > 1$, where $v_{\mathfrak{p}}$ is the valuation of k associated with the prime ideal $\mathfrak{p}$.
- In the case of a function field over $\mathbf{F}_q$, all places are non-archimedean and the completions k_v are isomorphic to local fields of characteristic $p = \text{Car } k$. If X is a smooth projective curve (here *smooth* means that the curve is non-singular in the usual sense) over $\mathbf{F}_q$ with function field k, the places of k correspond to closed points (in the sense of algebraic geometry) of the curve X. For example, for $k = \mathbf{F}_q(t)$, these closed points correspond to the point at infinity and to the irreducible monic polynomials in $\mathbf{F}_q[t]$ (note though that there is no natural analogue of the ring of integers of a number field). The absolute values associated with a closed

point x are of the form $|f| = a^{-v_x(f)}$ with $a > 1$, where v_x is the valuation of the discrete valuation ring $\mathcal{O}_{X,x}$ (the local ring of the curve X at x). Note also that if x_∞ is a closed point of a smooth projective curve X, the curve $Y := X - \{x_\infty\}$ is affine in the sense of algebraic geometry (it corresponds to the spectrum of a Dedekind ring, the ring of regular functions on Y), cf. [21], Part. IV, Exer. 1.3.

Definition 12.7 The *normalised absolute value* associated with a place v of a global field k is in the case where v is real, the usual absolute value of $\mathbf{R}$; in the case where v is complex, the *square* of the usual modulus; in the non-archimedean case, the absolute value given by $|x|_v = q^{-v(x)}$, where q is the cardinality of the residue field of the local field k_v, and the corresponding valuation is still denoted by v.

One of the reasons for adopting these conventions is ([9], Exp. II, p. 60):

Theorem 12.8 (product formula) *Let k be a global field. Let $\alpha \in k^*$. Then we have*

$$\prod_v |\alpha|_v = 1,$$

where v ranges through the set of places of k and $|\alpha|_v$ is the normalised absolute value associated to v.

Note that by the above, we have $|\alpha|_v = 1$ for almost all places v of k ($:=$ for all but finitely many places v of k).

Remark 12.9 Let k be a global field. Then for any $M > 0$, there is only a finite number of places of k whose residue field is of cardinality $\leqslant M$ (this follows immediately from cases $k = \mathbf{Q}$ and $k = \mathbf{F}_q(t)$, which are obvious).

Definition 12.10 If S is a set of places of k (containing archimedean places in the case of a number field), we denote by $\mathcal{O}_{k,S}$ the ring of S-*integers* of k, consisting of elements x of k which are integers outside S (i.e., such that $v(x) \geqslant 0$ if $v \notin S$).

Remark 12.11 If k is a number field, then $\mathcal{O}_{k,S}$ is a localisation of the ring of integers $\mathcal{O}_k$. If k is a function field of a curve X over $\mathbf{F}_q$, then for $S = \varnothing$ the ring $\mathcal{O}_{k,S}$ is a field, the algebraic closure of $\mathbf{F}_q$ in k. For $S \neq \varnothing$, the ring $\mathcal{O}_{k,S}$ is a localisation of the Dedekind ring of the regular functions on the affine curve $X \setminus \{v_0\}$, where $v_0 \in S$. In all cases $\mathcal{O}_{k,S}$ is a Dedekind ring.

For any place v of k and any finite separable extension K of k, we have a decomposition

$$k_v \otimes_k K = K_{w_1} \times \cdots \times K_{w_r}, \tag{12.1}$$

where $w_1, \ldots, w_r$ are the extensions of v to K ([9], Exp. II, Th. p. 57). We say that the places $w_1, \ldots, w_r$ *divide* v, or that they lie *above* v. In particular, $r \leqslant [K : k]$ and for a non-archimedean v, each local field K_{w_i} is a finite separable extension of the local field k_v. If $\mathfrak{p}$ is a prime ideal of $\mathcal{O}_k$ corresponding to v, the prime ideals of $\mathcal{O}_K$

corresponding to the w_i are those that appear in the decomposition of the ideal $\mathfrak{p}\mathcal{O}_K$ of $\mathcal{O}_K$. We will say that v is *unramified* in the extension K/k if all the extensions K_{w_i}/k_v are unramified, which is the case for almost all places v of k. Indeed, if we choose a finite non-empty set S of places of k (containing the archimedean places if k is a number field), and if v is a place of k not in S, then v is ramified in K/k if and only if the prime ideal $\mathfrak{p}$ corresponding to v does not divide the *discriminant* $\delta_{B/A}$ (which is a nonzero ideal in the Dedekind ring $A := \mathcal{O}_{k,S}$) of the extension B/A, where B is the integral closure of A in K ([47], Chap. III, Sect. 5, Cor. 1 and 2).

Note that there is a difference from the situation where k is a local field (or more generally a complete field for a discrete valuation, cf. Theorem 7.7): here the extension of v to an extension K is (generally) not unique. We will also see in the next paragraph that we can say a lot more if the extension K/k is Galois.

Example 12.12 Let $k = \mathbf{Q}$ and $K = \mathbf{Q}(\sqrt{-1})$. Let v be the place of $\mathbf{Q}$ corresponding to the prime 5. Then $K \otimes_{\mathbf{Q}} \mathbf{Q}_5 = \mathbf{Q}_5 \times \mathbf{Q}_5$ (indeed -1 is a square in $\mathbf{Q}_5$ by Hensel lemma), hence there are two places of K above v. On the other hand for $p = 2$ or $p = 3$, $K \otimes_{\mathbf{Q}} \mathbf{Q}_p$ remains a field and there is one place of K above p (ramified for $p = 2$, unramified for $p = 3$).

12.2 Galois Extensions of a Global Field

Let k be a global field. Let L be a finite Galois extension of k with group $G = \mathrm{Gal}(L/k)$. The group G acts on the set of places of L according to the formula $|a|_{\sigma w} := |\sigma^{-1}a|_w$ for any $a \in L$ and any $\sigma \in G$, which agrees with the natural action $\sigma \cdot \mathfrak{p} = \sigma(\mathfrak{p})$ of G on the nonzero prime ideals of $\mathcal{O}_k$ in the case of a number field (resp. closed points of a smooth projective curve associated with L in the case of a function field[1]). In particular, we have $(\sigma\tau)w = \sigma(\tau w)$ for all σ, τ in G. Each $\sigma \in G$ also induces a k_v-isomorphism between the completions L_w and $L_{\sigma w}$.

Definition 12.13 Let w be a place of L above a place v of k. The *decomposition group* G_w of w (with respect to the Galois extension L/k) is the subgroup of G consisting of σ such that $\sigma w = w$.

Observe that $G_{\sigma w} = \sigma G_w \sigma^{-1}$, so that up to conjugation, the decomposition group is determined by v (which allows for example to denote it G_v if G is abelian). For the proof of the next proposition, see for example the paper VII in [9], Prop. 1.2.

Proposition 12.14 *Let w be a place of L above a place v of k. The extension L_w/k_v is Galois and the injection $G_w \to \mathrm{Gal}(L_w/k_v)$ is an isomorphism. Furthermore, for each place v of k, the group G acts transitively on the places of L above v.*

---

[1]Stricto sensu, as J. Riou has pointed out to me, this action on the closed points is a right action as the functor which associates to a curve its function field is contravariant. By transforming this action into a left action, we recover the natural action at the level of ideals.

We deduce for example that the property that L_w is an unramified (resp. trivial) extension of k_v does not depend on the place w of L dividing v (if L_w/k_v is trivial, we say that v is *totally split* in the extension L/k). More generally, the ramification e and the residual degree f of L_w over k_v do not depend on v. Hence we have $[L : k] = n \cdot ef$, where n is the number of places of L above v ([47], Chap. I, Prop. 10). The totally split case corresponds to $n = [L : k]$, and $e = f = 1$.

In the case of a finite place v, unramified in the extension L/k, we obtain that, for any place w of L above v, the decomposition subgroup G_w of G is isomorphic to the Galois group of the residual extension $\mathbf{F}_w/\mathbf{F}_v$, where $\mathbf{F}_w$, $\mathbf{F}_v$ denote the residue fields of k_w, k_v respectively. In particular, the Frobenius $x \mapsto x^{N_v}$ (where N_v is the cardinality of the finite field $\mathbf{F}_v$) admits a unique lift F_w to G_w that we call the *Frobenius* associated with w. Up to conjugation, it depends on v only and we will denote by $F_{L/k}(v)$ its conjugacy class in G. As any element in this class is of order the residual degree f, we obtain that v decomposes in L into $[G : \langle\sigma\rangle]$ places, where $\sigma \in G$ is a representative of the Frobenius. In particular, it is totally split if and only if $F_{L/k}(v)$ is trivial. Note finally that for G abelian, the Frobenius at v is a well defined element of G for any place v of k unramified in L.

Example 12.15 Let L be a finite Galois extension of a number field k. Write $L = k(\alpha)$, with $\alpha \in \mathcal{O}_L$. The minimal polynomial g of α over k (which is of degree $[L : k]$) then has coefficients in $\mathcal{O}_k$. Furthermore $\mathcal{O}_k[\alpha]$ is a subgroup of finite index of $\mathcal{O}_L$ (note that there does not always exist an α such that $\mathcal{O}_L = \mathcal{O}_k[\alpha]$). Let $\mathfrak{p}$ be a prime ideal of $\mathcal{O}_k$. Assume that the discriminant of the polynomial g (it is the element of $\mathcal{O}_k$ given by the resultant of g and g') is not in $\mathfrak{p}$. In other words the reduction $\overline{g}$ of g modulo $\mathfrak{p}$ is a separable polynomial over the field $k(\mathfrak{p}) := (\mathcal{O}_k/\mathfrak{p})$. Then the extension L/k is unramified at $\mathfrak{p}$. Furthermore, if $\overline{g} = \overline{g}_1...\overline{g}_r$ is the decomposition of $\overline{g}$ as a product of irreducible polynomials pairwise distinct, each g_i has the residual degree of the extension L/k at $\mathfrak{p}$ and r is the number of prime ideals of L above $\mathfrak{p}$, cf. [13], Th. 5.1 and Exercise 5.6. A more general statement (where we do not assume $\overline{g}$ to be separable nor L/k to be Galois) relating the decomposition of $\mathfrak{p}\mathcal{O}_L$ to that of $\overline{g}$ is valid under the assumption that the prime number corresponding to $\mathfrak{p}$ does not divide the index $[\mathcal{O}_L : \mathcal{O}_k[\alpha]]$, see Th. I.8.3 of [41].

12.3 Idèles, Strong Approximation Theorem

Definition 12.16 Let k be a global field. Let Ω_k be the set of all places of k. An *idèle* of k is a family $(x_v)_{v\in\Omega_k}$ in $\prod_{v\in\Omega_k} k_v^*$, satisfying $v(x_v) = 0$ (i.e., $|x_v|_v = 1$) for almost all places v of k (In other words: $x_v \in \mathcal{O}_v^*$ for almost every v, where $\mathcal{O}_v$ is the ring of integers of the local field k_v).

Note that as there are only finitely many archimedean places (in the case of a number field), we can replace everywhere "almost every place" by "almost every finite place", which justifies talking of the valuation $v(x_v)$ or of the ring $\mathcal{O}_v$ in the above definitions.

The idèles form a multiplicative group that we denote I_k (or simply I when there is no ambiguity).

This group is thus the *restricted product* of the k_v^* relatively to the $\mathcal{O}_v^*$. Equipped with its restricted product topology (for which a basis of open neighbourhoods of 1 consists of subsets of type $\prod_{v \in S} U_v \times \prod_{v \notin S} \mathcal{O}_v^*$, where $S \subset \Omega_k$ is finite and U_v is an open subset of k_v^* for $v \in S$) is a locally compact group. This group is furthermore totally disconnected in the case of a function field (in the case of a number field, the group of *finite idèles*, consisting of idèles whose component at archimedean places is 1 and is totally disconnected). Note that the topology on I_k is not that induced by the product topology of $\prod_{v \in \Omega_k} k_v^*$.

Definition 12.17 A *principal idèle* is an idèle of the form $(x, x, \ldots, x, \ldots)$ with $x \in k^*$. The *idèle class group* of k is the quotient $C_k := I_k / k^*$, where we have still denoted by k^* the subgroup of principal idèles of k.

It is the group C_k that will in global class field theory play the role comparable to that of the multiplicative group in local class field theory.

The following theorem says that if S is a finite set of places (that we can assume to contain the archimedean places if k is a number field), we can approximate a family of elements (α_v) of $\prod_{v \in S} k_v$ by an element $\beta \in k$ ("weak approximation", which works in a much more general context, see [32], Th. XII.1.2.), imposing in addition (which is specific to global fields) that β is integral outside of S *and of a place v_0 fixed from the start*. We cannot expect better because of the product formula. More precisely, we have (for a proof, see [9], Exp. II, Sects. 14 and 15):

Theorem 12.18 (strong approximation) *Let k be a global field and let v_0 be a place of k. Fix a finite set S of places of k with $v_0 \notin S$, elements $\alpha_v \in k_v$ for $v \in S$, and $\varepsilon > 0$. Then there exists $\beta \in k$ with $|\beta - \alpha|_v \leqslant \varepsilon$ for $v \in S$ and $|\beta|_v \leqslant 1$ (i.e., $v(\beta) \geqslant 0$ if v is non-archimedean) for any $v \notin S \cup \{v_0\}$.*

The proof of the strong approximation theorem uses the following lemma (similar to Minkowski theorem in geometry of numbers) that we will use later on (see [9], Exp. II, Lem. p. 66 for a proof).

Lemma 12.19 *Let k be a global field. Then there exists a constant $C > 0$ (depending only on k) satisfying: for any idèle $(\alpha_v) \in I_k$ such that $\prod_{v \in \Omega_k} |\alpha_v|_v \geqslant C$, there exists $\beta \in k^*$ such that $|\beta|_v \leqslant |\alpha_v|_v$ for any place v of k.*

The next proposition implies in particular that the idèle class group C_k is Hausdorff (and hence locally compact as quotient of I_k).

Proposition 12.20 *The group k^* is discrete (and hence closed) in I_k.*

Proof Let S be a non-empty set of finite places of k, containing the set of archimedean places if k is a number field. Let U be an open neighbourhood of 1 in I_k defined by $(\alpha_v) \in U$ if and only if $|\alpha_v - 1|_v < 1$ for all $v \in S$ and $|\alpha_v| = 1$ for $v \notin S$. Then if $x \in k^*$ is not equal to 1, we have $\prod_{v \in \Omega_k} |x - 1|_v = 1$ by product formula, which excludes the case $x \in U$—else we would have $|x - 1|_v < 1$ for $v \in S$ and $|x - 1|_v \leqslant \max(|x|_v, 1)$ for $v \notin S$, which would imply $\prod_{v \in \Omega_k} |x - 1|_v < 1$. Thus 1 is an isolated point of the image of k^* in I_k. This shows that this subgroup is discrete. $\qquad \square$

Consider the continuous homomorphism

$$|.| : I_k \longrightarrow \mathbf{R}_+^*, \quad (\alpha_v) \longmapsto \prod_{v \in \Omega_k} |\alpha_v|_v$$

and let I_k^0 be its kernel. The product formula shows that we have $k^* \subset I_k^0$. We thus obtain a homomorphism $|.| : C_k \to \mathbf{R}_+^*$ with kernel $C_k^0 = I_k^0/k^*$. This group will play the role in the global class field theory, analogous to that of $\mathcal{O}_K^*$ when K is a local field (note for example that if k is a number field, we have $C_k \simeq C_k^0 \times \mathbf{R}_+^*$, just like we have $K^* = \mathcal{O}_K^* \times \mathbf{Z}$ for a local field K). In particular, we have:

Theorem 12.21 *The group C_k^0 is compact.*

To prove this theorem, we first recall the following:

Definition 12.22 An *adèle* of k is a family $(\alpha_v)_{v \in \Omega_k}$ with $\alpha_v \in k_v$ for every place v and $\alpha_v \in \mathcal{O}_v$ for almost all v. We denote by $\mathbf{A}_k$ the ring of adèles, endowed with the restricted product topology with respect to the $\mathcal{O}_v$.

We see immediately that the multiplicative group I_k of the idèles is simply the group of the invertible elements of $\mathbf{A}_k$. On the other hand the topology is different: the topology of I_k is not induced by that of $\mathbf{A}_k$. The strong approximation theorem is a statement about the density of k in the ring of "truncated adèles at v_0" (restricted product of the k_v for $v \neq v_0$) *for the $\mathbf{A}_k$-topology*

The proof of the Theorem 12.21 relies on the following lemma, which compares the two topologies on I_k^0.

Lemma 12.23 *The subset I_k^0 is closed in $\mathbf{A}_k$ for the $\mathbf{A}_k$-topology. Furthermore, the topologies induced by $\mathbf{A}_k$ and I_k on I_k^0 coincide.*

Proof Let $\alpha = (\alpha_v)$ be an adèle which is not in I_k^0. We need to find an open neighbourhood W of α in $\mathbf{A}_k$ that does not meet I_k^0. Note already that as almost all the $|\alpha_v|_v$ are $\leqslant 1$, the infinite product $C = \prod_{v \in \Omega_k} |\alpha_v|$ makes sense in $\mathbf{R}_+$, since the series with general term $\log(|\alpha_v|_v)$ has negative or zero terms after a certain rank. Let us consider two cases separately:

(a) Let us set $C < 1$. As α is an adèle, only a finite number of places v of k satisfy $|\alpha_v|_v \geqslant 1$. The property $C < 1$ then implies that we can find a finite set S of places of k such that S contains all the places v such that $|\alpha_v|_v \geqslant 1$, with furthermore $\prod_{v \in S} |\alpha_v|_v < 1$. It is enough to take W defined by $|\xi_v - \alpha_v|_v < \varepsilon$ for $v \in S$ (for ε small enough) and $|\xi_v|_v \leqslant 1$ for $v \notin S$.

(b) Suppose $C \geqslant 1$. Let us first show that α is necessarily an idèle, and hence $C > 1$ as by assumption $\alpha \notin I_k^0$. For v non-archimedean, the largest absolute value

< 1 in k_v is q^{-1}, where q is the cardinality of the residue field of $\mathcal{O}_v$. Remark 12.9 then says that there exists a finite set S of places of k such that, for v not in S, we have $|\alpha_v|_v = 1$, or $|\alpha_v|_v \leqslant 1/2$ (as α is an adèle, we can assume that $|\alpha_v|_v \leqslant 1$ for $v \notin S$). But as the infinite product $\prod |\alpha_v|_v$ converges to a > 0 real number, its general term tends to 1, which implies that only a finite number of $|\alpha_v|_v$ are $\leqslant 1/2$. Finally, after if necessary enlarging S, we can assume that $|\alpha_v|_v = 1$ for v not in S, and α is an idèle with $C > 1$.

After, if necessary, enlarging S, we can now also assume that for any v not in S, the property $|\xi_v|_v < 1$ implies $|\xi_v|_v < (2C)^{-1}$ (again, using Remark 12.9). Lastly, by definition of C, we can impose that

$$1 < \prod_{v \in S} |\alpha_v| < 2C.$$

Now, for $\varepsilon > 0$ small enough, the condition $|\xi_v - \alpha_v|_v < \varepsilon$ for $v \in S$ implies that

$$1 < \prod_{v \in S} |\xi_v|_v < 2C. \tag{12.2}$$

Then we can take W defined by $|\xi_v - \alpha_v|_v < \varepsilon$ for $v \in S$ and $|\xi_v|_v \leqslant 1$ for $v \notin S$. Indeed, an adèle $\xi = (\xi_v)$ of W cannot satisfy $|\xi| = 1$, in view of the inequality (12.2) and the fact that for $v \notin S$, $|\xi_v|$ is either equal to 1 or $< (2C)^{-1}$.

Finally, I_k^0 is closed in $\mathbf{A}_k$, and it remains to show that the topologies induced on I_k^0 by $\mathbf{A}_k$ and I_k are the same. Let $\alpha = (\alpha_v) \in I_k^0$. It is straightforward to see that an $\mathbf{A}_k$-neighbourhood of α contains a I_k-neighbourhood of α by definition of the restricted product topologies since $\mathcal{O}_v^* \subset \mathcal{O}_v$ for any finite place v of k. Conversely, let H be a I_k-neighbourhood of α. It contains an open I_k-neighbourhood of the type $|\xi_v - \alpha_v|_v < \varepsilon$ for $v \in S$ and $|\xi_v|_v = 1$ for $v \notin S$, where S contains the archimedean places (if k is a number field) and all places v such that $|\alpha_v|_v \neq 1$. In particular, $\prod_{v \in S} |\alpha_v| = 1$ as $\alpha \in I_k^0$. We can assume, as above, that for v not in S, the condition $|\xi_v|_v < 1$ implies $|\xi_v|_v < 1/2$. The property $\prod_{v \in S} |\alpha_v| = 1$ allows us to choose ε such that each element (ξ_v) of I_k^0 satisfying $|\xi_v - \alpha_v|_v < \varepsilon$ for $v \in S$ and $|\xi_v|_v \leqslant 1$ for $v \notin S$ satisfies $\prod_{v \in S} |\xi_v|_v < 2$, hence actually satisfies $|\xi_v|_v = 1$ for $v \notin S$ (otherwise we would have a place $v \notin S$ with $|\xi_v|_v < 1/2$, which would contradict $(\xi_v) \in I_k^0$). It follows that $H \cap I_k^0$ contains a $\mathbf{A}_k$-neighbourhood of α in I_k^0, as desired. $\square$

Proof of Theorem 12.21 By Lemma 12.23, it is enough to find a set which is $\mathbf{A}_k$-compact $W \subset \mathbf{A}_k$ such that the quotient map $W \cap I_k^0 \to I_k^0/k^*$ is surjective. Choose an idèle $\alpha = (\alpha_v)$ such that $|\alpha| := \prod_v \alpha_v$ is $> C$, where C is the constant provided by Lemma 12.19. We then choose for W the compact (as product of compacts) of $\mathbf{A}_k$ defined by $|\xi_v|_v \leqslant |\alpha_v|_v$ for every place v. Lemma 12.19 now says that if $\beta = (\beta_v)$ is in I_k^0, then we have an $\eta \in k^*$ such that $|\eta|_v \leqslant |\beta_v^{-1}\alpha_v|_v$ for any place v, i.e., $\eta\beta \in W$, as desired. $\square$

Theorem 12.21 allows to recover two fundamental results from the theory of global fields: the finiteness of the ideal class group and the Dirichlet unit theorem (we can also do the reverse and prove Theorem 12.21 using these). Let us start by saying a few words about the first of these results. To begin with, let us assume that k is a number field and denote by Ω_f its set of finite places. Let $\mathcal{I}_k$ be the *group of ideals* of k, i.e., the abelian group consisting of formal sums $\sum_{v \in \Omega_f} n_v v$ with $n_v \in \mathbf{Z}$ almost all equal to zero (in the language of algebraic geometry, it is the group of divisors $\mathrm{Div}(\mathrm{Spec}\, \mathcal{O}_k)$, cf. [21], Part. II, Chap. 6). The group $\mathcal{I}_k$ is isomorphic (by Theorem 12.3) to the multiplicative group of fractional ideals of $\mathcal{O}_k$, the ideals of $\mathcal{O}_k$ (in the usual sense) correspond to formal sums $\sum_v n_v \cdot v$ with $n_v \geqslant 0$ ("effective divisors "), almost all zero. The map

$$I_k \longrightarrow \mathcal{I}_k, \quad (\alpha_v) \longmapsto \sum_{v \in \Omega_f} v(\alpha_v) \cdot v$$

is continuous ($\mathcal{I}_k$ being endowed with discrete topology) by definition of the topology of I_k. Besides, the image of k^* is by definition the group of principal ideals $\mathcal{P}_k$, and the quotient $\mathcal{I}_k/\mathcal{P}_k$ is isomorphic to the group of ideal classes of $\mathcal{O}_k$. We obtain a continuous surjection (because there is at least one archimedean place in k) $I_k^0/k^* \to \mathcal{I}_k/\mathcal{P}_k$. Since a discrete and compact set is finite, Theorem 12.21 then gives

Theorem 12.24 *Let k be a number field. Then the ideal class group* $\mathrm{Pic}(\mathcal{O}_k) = \mathcal{I}_k/\mathcal{P}_k$ *is finite.*

The same proof shows that in the case of the function field of a smooth projective curve X over $\mathbf{F}_q$, the group $\mathcal{I}_k^0/\mathcal{P}_k = \mathrm{Pic}^0 X$ is finite, where $\mathcal{I}_k^0$ is the group of degree zero divisors on X. Here the group of divisors $\mathcal{I}_k$ consists of almost zero formal sums $\sum_{v \in \Omega_k} n_v v$ with $v \in \mathbf{Z}$. The degree of such a divisor is $\sum_v n_v d_v$, where d_v is the degree over $\mathbf{F}_q$ of the residue field of X at v. In particular, if we remove a closed point v_0 of X, the Dedekind ring of regular functions on the affine curve $Y = X \setminus \{v_0\}$ has a finite ideal class group (indeed, if we denote by d the degree of the closed point v_0, the divisors of Y whose degrees are in $d\mathbf{Z}$ form a subgroup of finite index of $\mathrm{Div}\, Y$, and this subgroup is isomorphic to $\mathrm{Div}^0 X$).

Let now k be a global field and let S be a finite set of places of k, such that (if k is a number field) S contains the set Ω_∞ of archimedean places. It is sometimes useful to work with the group $I_{k,S}$ of S-*idèles* of k, defined as

$$I_{k,S} = \prod_{v \in S} k_v^* \times \prod_{v \notin S} \mathcal{O}_v^*.$$

Note in particular that I_k is the union of the $I_{k,S}$ for finite S.

Proposition 12.25 *For S finite and large enough, we have $I_k = I_{k,S} \cdot k^*$ (and hence $C_k = I_{k,S} \cdot k^*/k^*$).*

Proof Let $\mathcal{O}_k$ be the ring of integers of k if k is a number field (resp. ring of regular functions on the affine curve $X \setminus \{v_0\}$, where X is a smooth projective curve over $\mathbf{F}_q$ with field of functions k, and v_0 is a closed point of X, in the case where k is a function field). The finiteness of the ideal class group of $\mathcal{O}_k$ allows to choose places $v_1, \ldots, v_r$ of k corresponding to prime ideals $\mathfrak{p}_1, \ldots, \mathfrak{p}_r$ of $\mathcal{O}_k$ such that the $\mathfrak{p}_i$ generate this ideal class group. Let S be a finite set of places of k containing archimedean places in the case of a number field (resp. v_0 in the case of a function field) and places $v_1, \ldots, v_r$.

Let $\alpha = (\alpha_v) \in I_k$. Denote by $\mathfrak{p}_v$ the prime ideal associated with a finite place v of k. The fractional ideal $\prod_{v \notin \Omega_\infty} \mathfrak{p}_v^{v(\alpha_v)}$ (resp. $\prod_{v \neq v_0} \mathfrak{p}_v^{v(\alpha_v)}$ if k is a function field) of $\mathcal{O}_k$ is written as $(x) \cdot I$, where I is a fractional ideal that belongs to the subgroup generated by the $\mathfrak{p}_i$. It follows that the idèle $\alpha \cdot x^{-1}$ is such that all of its components outside of S are invertible, and hence it is in $I_{k,S} \cdot k^*$. □

The following lemma is an easy consequence of Theorem 12.21 (cf. [9], Exp. II, Sect. 18) and will be useful to us later. Let S be a non-empty finite set of places of k, containing all the archimedean places if k is a number field. Let $E_{k,S} = k^* \cap I_{k,S}$ be the group of S-units of k. It is the group of invertible elements $\mathcal{O}_{k,S}^*$ of the ring $\mathcal{O}_{k,S}$ of S-integers of k.

Lemma 12.26 *With the above notations, let V be the **R**-vector space of maps from S to **R**. Let $\lambda : E_{k,S} \to V$ be the homomorphism defined by $\lambda(a) = f_a$, where $f_a(v) = \log |a|_v$ for any $v \in S$. Then λ has finite kernel, and its image is a lattice generating the **R**-vector space V^0 consisting of $f \in V$ such that $\sum_{v \in S} f(v) = 0$.*

Proof We first observe that if c and C are constants with $0 < c < C$, then the set of S-units η such that

$$\forall v \in S, \quad c \leqslant |\eta|_v \leqslant C$$

is finite since it is the intersection of a compact subset in I_k with the discrete subgroup (cf. Proposition 12.20) k^*. In particular, the elements x satisfying $|x|_v = 1$ for any place v are finite in number and form a multiplicative group. It is thus exactly the group of roots of units. This shows that $\ker \lambda$ is finite. To prove the statement about its image, observe that we can define λ by an analogous formula on S-idèles $I_{k,S}$, and the image of $I_S^0 := I_{k,S} \cap I_k^0$ generates[2] the **R**-vector space consisting of f satisfying $\sum_{v \in S} f(w) = 0$, which is of dimension $s - 1$. The group $\lambda(E_{k,S})$ is discrete (since there is a finite number of $\eta \in E_{k,S}$ with $1/2 \leqslant |\eta|_v \leqslant 2$ for every place v of S) and $\lambda(I_S^0)/\lambda(E_{k,S})$ is compact via the compactness of $I_S^0/E_{k,S}$ that follows from Theorem 12.21 (Note that $I_S^0/E_{k,S}$ is an open subgroup, hence closed in I_k^0/k^*). Finally $\lambda(E_{k,S})$ is a lattice generating the **R**-vector space V^0. □

As corollary, we obtain the Dirichlet unit theorem.

Theorem 12.27 *The group of S-units $E_{k,S}$ is isomorphic to a direct product of a finite group (roots of unity in k^*) and of $\mathbf{Z}^{s-1}$, where s is the cardinality of S.*

Let us finish this paragraph with a statement which will be useful for the proof of the Poitou–Tate duality (Proposition 17.10).

---

[2]And is even equal to it if k is a number field and S is the set of archimedean places.

Proposition 12.28 *Let k be a global field. Let S be a finite set of places of k (containing the archimedean places if k is a number field). Let n be a positive integer, prime to the characteristic of k if k is a function field. Let E be the set of elements $a \in k^*/k^{*n}$ such that the image a_v of a in k_v^*/k_v^{*n} is in $\mathcal{O}_v^*/\mathcal{O}_v^{*n}$ for all places $v \notin S$ such that n is invertible in $\mathcal{O}_v$. Then E is finite.*

Proof After possibly enlarging S, we can assume that n is invertible in all the $\mathcal{O}_v$ for $v \notin S$ (given the assumption made in the case of a function field), and also that the ring of S-integers $\mathcal{O}_{k,S}$ is principal (by finiteness of the ideal class group of $\mathcal{O}_{k,S}$). One observes then that E is the set of $a \in k^*/k^{*n}$ whose image in k_v^*/k_v^{*n} belongs to $\mathcal{O}_v^*/\mathcal{O}_v^{*n}$ for every place $v \notin S$. This means that the valuation $v(a)$ is a multiple of n for every $v \notin S$. As $\mathcal{O}_{k,S}$ is assumed to be principal, we can find an element $b \in k^*$ satisfying $v(b) = v(a)/n$ for every $v \notin S$. Then, the element $c := a/b^n$ satisfies $c \in \mathcal{O}_{k,S}^*$, which shows that E is contained in $\mathcal{O}_{k,S}^*/\mathcal{O}_{k,S}^{*n}$. This last group is finite by Dirichlet unit theorem. □

A finer version of the preceding statement is the following classical theorem (proved using the geometry of numbers, [41], Chap. III, Sect. 2).

Theorem 12.29 (Hermite–Minkowski) *Let k be a number field. We denote by $\delta_{k/\mathbf{Q}}$ the discriminant of the extension of Dedekind rings $\mathcal{O}_k/\mathbf{Z}$, that we can see as a positive integer. Then, for any integer $N > 0$, there is only a finite number of number fields $k \subset \overline{\mathbf{Q}}$ whose discriminant is at most N. Furthermore, the only number field with discriminant 1 is $\mathbf{Q}$.*

Corollary 12.30 *The only finite extension of $\mathbf{Q}$ unramified at every prime number is $\mathbf{Q}$.*

Proof If k is an extension of $\mathbf{Q}$ of degree > 1, its discriminant is also > 1, hence admits a prime divisor, which is then ramified in the extension $k/\mathbf{Q}$. □

Hermite–Minkowski theorem can be used to recover Proposition 12.28 (cf. [49], Part. II, Sect. 6, Lem. 6) by proving the *theorem of unramified extensions*, which says that if $S \supset \Omega_\infty$ is a finite set of places of k and d is an integer, there is only a finite number of extensions of k unramified outside S of degree $\leqslant d$ (see also Exercise 12.5 for a special case). There is a version of this theorem for the function fields of characteristic $p > 0$, provided we restrict ourselves to the extensions of degree prime to p.

12.4 Some Complements in the Case of a Function Field

In this paragraph, we assume that the reader has some familiarity with the language of algebraic geometry, in particular with the theory of algebraic curves. For more details about divisors and the Picard group, one may consult Chapter II.6 of [21]; one can consult [37] and [38] for facts about abelian varieties and Jacobians. Results presented here will be useful in the proof of the *class field axiom* (Corollary 13.29, see also Theorem 13.23) for the case of a global field of characteristic p, in particular in the case where the degree of the extension we consider is divisible by p.

We consider a smooth projective geometrically integral curve X over a perfect field κ (the case we are interested in is when κ is a finite field) with algebraic closure $\overline{\kappa}$. Thus κ is algebraically closed in the field of functions $k := \kappa(X)$ of the curve. Let us set $\overline{X} = X \times_\kappa \overline{\kappa}$. Denote by $\overline{\kappa}(X) = k \times_\kappa \overline{\kappa}$ the field of functions of $\overline{X}$ and by $\overline{X}^{(1)}$ the set of its closed points. Let $\Gamma_\kappa = \mathrm{Gal}(\overline{\kappa}/\kappa)$ be the absolute Galois group of κ. If $\mathrm{Div}\,\overline{X}$ stands for the group of divisors on $\overline{X}$ (i.e., the group $\bigoplus_{w \in \overline{X}^{(1)}} \mathbf{Z} \cdot w \simeq \bigoplus_{w \in \overline{X}^{(1)}} \mathbf{Z}$), we can associate to each function $f \in \overline{\kappa}(X)^*$ its divisor $\mathrm{Div}\,f \in \mathrm{Div}\,\overline{X}$, by taking the valuation of f at each $w \in \overline{X}^{(1)}$. The image of $\overline{\kappa}(X)^*$ by f is the subgroup of *principal divisors* $\mathrm{Div}_0\,\overline{X}$ and the kernel of f consists of constant functions (because $\overline{X}$ is projective). The group $\mathrm{Div}\,\overline{X}$ has a structure of Γ_κ-module, isomorphic to $\bigoplus_{v \in X^{(1)}} I_{\Gamma_{\kappa(v)}}^{\Gamma_\kappa}(\mathbf{Z})$, where $\kappa(v)$ designates the residue field of v. We thus obtain an exact sequence of Γ_κ-modules:

$$0 \longrightarrow \overline{\kappa}^* \longrightarrow \overline{\kappa}(X)^* \xrightarrow{\ \mathrm{Div}\ } \mathrm{Div}_0\,\overline{X} \longrightarrow 0. \tag{12.3}$$

On the other hand let us denote by $\mathrm{Pic}\,\overline{X}$ the quotient of $\mathrm{Div}\,\overline{X}$ by $\mathrm{Div}_0\,\overline{X}$ (it is the Picard group of $\overline{X}$, cf. [21], Part. II, Cor. 6.16). We have another exact sequence of Γ_κ-modules:

$$0 \longrightarrow \mathrm{Div}_0\,\overline{X} \longrightarrow \mathrm{Div}\,\overline{X} \longrightarrow \mathrm{Pic}\,\overline{X} \longrightarrow 0. \tag{12.4}$$

Proposition 12.31 *Assume that the field κ is of cohomological dimension $\leqslant 1$ (for example finite). Then the map $H^2(\kappa, \overline{\kappa}(X)^*) \to H^2(\kappa, \mathrm{Div}_0\,\overline{X})$ induced by Div is an isomorphism.*

Proof As $\mathrm{cd}(\kappa) \leqslant 1$, we have $\mathrm{Br}\,\kappa = H^2(\kappa, \overline{\kappa}^*) = 0$ by Theorem 6.17 and Remark 6.18. Furthermore, $H^3(\kappa, \overline{\kappa}^*) = 0$ since $\mathrm{scd}(\Gamma_\kappa) \leqslant 2$ by Proposition 5.8. The result follows from the long exact cohomology sequence associated to the sequence (12.3). $\qquad\qquad\square$

Besides, the structure of the group $\mathrm{Pic}\,\overline{X}$ is as follows: we have an exact sequence of Γ_κ-modules:

$$0 \longrightarrow \mathrm{Pic}^0\,\overline{X} \longrightarrow \mathrm{Pic}\,\overline{X} \xrightarrow{\ \deg\ } \mathbf{Z} \longrightarrow 0, \tag{12.5}$$

where deg is the degree map and the kernel $\mathrm{Pic}^0\,\overline{X}$ of deg is identified with the group $J(\overline{\kappa})$ of $\overline{\kappa}$-points of a κ- abelian variety J, called the *Jacobian* of X ([38], Th. 1.1). In particular, the group $\mathrm{Pic}^0\,\overline{X} = J(\overline{\kappa})$ is divisible ([37], Th. 8.2). Furthermore, the dimension of the Jacobian is the genus of the curve X.

We have the following important theorem, due to Lang (12.32):

Theorem 12.32 *Let A be a smooth algebraic group, assumed to be commutative and connected (for example an abelian variety) over a finite field κ. Then the group*

$H^1(\kappa, A) := H^1(\Gamma_\kappa, A(\overline{\kappa}))$ *is trivial. In particular, under the above assumptions and notations, we have* $H^1(\kappa, \mathrm{Pic}^0 \overline{X}) = 0$.

Remark 12.33 The theorem of Lang extends to non-abelian H^1 (that we do not treat in this book) of connected non-commutative algebraic groups, but not to abelian varieties over a field of arbitrary cohomological dimension $\leqslant 1$; it is for example not true in general over $\mathbf{C}((t))$.

Corollary 12.34 *Assume that the field* κ *is finite. Then* $H^1(\kappa, \mathrm{Pic}\,\overline{X}) = 0$. *Furthermore the map* $H^2(\kappa, \mathrm{Div}_0\,\overline{X}) \to H^2(\kappa, \mathrm{Div}\,\overline{X})$ *induced by the inclusion* $\mathrm{Div}_0\,\overline{X} \to \mathrm{Div}\,\overline{X}$ *is injective.*

Proof Using the long exact sequence associated with (12.4), we see that it is enough to show that $H^1(\kappa, \mathrm{Pic}\,\overline{X}) = 0$. As $H^1(\kappa, \mathbf{Z}) = 0$, the sequence (12.5) and Theorem 12.32 give the result. $\qquad\square$

12.5 Exercises

Exercise 12.1 Show that the Dedekind ring $\mathbf{Z}[\sqrt{-5}]$ is not a unique factorisation domain.

Exercise 12.2 Let k be a global field. Let K be a finite separable extension (not necessarily Galois) of k. We denote by F the Galois closure of K, and we set $G = \mathrm{Gal}(F/k)$. For any place λ of F, we denote by $G_\lambda \subset G$ its decomposition group.

(a) Let H be the subgroup $\mathrm{Gal}(F/K)$ of G. Show that

$$\bigcap_{g \in G} gHg^{-1} = \{1\}.$$

(b) Let v be a place of k and w be a place of K above v. We say that w is *split* in K/k if $K_w = k_v$, and that v is *totally split* in K/k if all the places of K above v are split (this is an extension to the non-Galois case of the classic definitions). Prove the equivalence of the three properties:
  (i) the place w is split in K/k.
  (ii) for every place λ of F above w, we have $G_\lambda \subset H$.
  (iii) there exists a place λ of F above w with $G_\lambda \subset H$.

(c) Show that a place v of k is totally split in K/k if and only if the decomposition group G_v of v in F/k is $\{1\}$ (this is not ambiguous, indeed the group G_v is defined up to conjugation in G). We will see in Chap. 18, as an application of the Čebotarev theorem, that there are infinitely many places v satisfying this condition.

(d) Let v be a place of k. Let D be the decomposition group of a place of F above v. Let $w_1, \ldots, w_r$ be the places of K above v. Show that the degrees $[K_{w_i} : k_v]$ for $i = 1, \ldots, r$ are the indices $[D : (D \cap gHg^{-1})]$ for $g \in G$ (or for $g \in G/H$).

Deduce from the above that there exists a place w of K above v, split in K/k if and only if there exists $g \in G$ such that $G_v \subset gHg^{-1}$. Show that if furthermore v is unramified in the extension F/k, the existence of such a place w is equivalent to $G_v \subset \bigcup_{g \in G} gHg^{-1}$ (we will see in Exercise 18.7 that there exist infinitely many places v of k which do not satisfy this condition).

(e) Show that there exists a unique place of K above v if and only if the image of G_v by the canonical surjection $G \to G/H$ is the whole of G/H (show first that this condition makes sense, even though G_v is only defined up to conjugation).

Exercise 12.3 Let k be a global field.

(a) Find a sequence of elements of I_k converging to 1 for the topology of $\mathbf{A}_k$, but not converging for the topology of I_k.

(b) Is the set I_k closed in $\mathbf{A}_k$ for the topology of $\mathbf{A}_k$?

Exercise 12.4 Let $k = \mathbf{Q}(\sqrt{2})$.

(a) Show that the subgroup H of I_k generated by $\{\pm 1\}$ and $1 + \sqrt{2}$ is not compact.

(b) Deduce that k^* is not closed in the restricted product of the k_v^* for v finite (relatively to $\mathcal{O}_v^*$), that is, in "finite idèles".

Exercise 12.5 Let $n > 0$ be an integer. Let k be a number field (whose algebraic closure $\bar{k}$ is fixed) containing the nth roots unity. Let S be a finite set of places of k (containing the archimedean places). Using Proposition 12.28, prove the following special case of the theorem of unramified extensions: the field k admits only finitely many cyclic extensions $k' \subset \bar{k}$ of degree dividing n and unramified outside S. Generalise the above to the case where we do not assume that k contains the nth roots unity, then to the abelian extensions (not necessarily cyclic).

Exercise 12.6 Recall (Corollary 12.30) that the only finite extension of the field $\mathbf{Q}$ that is, unramified at every prime number p is $\mathbf{Q}$. Let L be a finite abelian extension of $\mathbf{Q}$. The aim of this exercise is to show that L is a cyclotomic extension (Kronecker–Weber theorem) by purely local methods.

(a) Let p be a prime number ramified in $L/\mathbf{Q}$. Let L_p be the completion of L at a prime above p. Show that there exists an integer $n_p > 0$ such that $L_p \subset \mathbf{Q}_p(\mu_{n_p})$, where $\mathbf{Q}_p(\mu_{n_p})$ is the extension of $\mathbf{Q}_p$ obtained by adjoining the n_pth roots of unity (cf. Example 11.23).

In all that follows, we write $n_p = p^{e_p}.m_p$ with m_p prime to p, and we set $n = \prod_p p^{e_p}$, the product taken over all the prime numbers ramified in $L/\mathbf{Q}$ (whose set we denote by R). We also set $M = L(\mu_n)$.

(b) Show that if a prime number p is ramified in $M/\mathbf{Q}$, then it is ramified in $L/\mathbf{Q}$.

(c) Let (for $p \in R$) $I_p \subset \mathrm{Gal}(M_p/\mathbf{Q}_p)$ be the inertia subgroup. Show that I_p is isomorphic to $\mathrm{Gal}(\mathbf{Q}_p(\mu_{p^{e_p}})/\mathbf{Q}_p)$.

(d) Let I the subgroup of $\mathrm{Gal}(M/\mathbf{Q})$ generated by the I_p. Show that the fixed field M^I est $\mathbf{Q}$.

(e) Show that the cardinality of I is at most $[\mathbf{Q}(\mu_n) : \mathbf{Q}]$ and deduce that $L \subset \mathbf{Q}(\mu_n)$.

Exercise 12.7 (*by* [13], *Lemma 5.32*) Let E be a number field. Let $u \in \mathcal{O}_E$. We set $F = E(\sqrt{u})$. Let $\mathfrak{p}$ be a prime ideal of $\mathcal{O}_E$.

(a) Show that if $2u \notin \mathfrak{p}$, then $\mathfrak{p}$ is unramified in the extension F/E.

(b) Let $u = b^2 - 4c$ with $b, c \in \mathcal{O}_E$. Show that if $2 \in \mathfrak{p}$ and $u \notin \mathfrak{p}$, then $\mathfrak{p}$ is unramified in the extension F/E.

Chapter 13
Cohomology of the Idèles: The Class Field Axiom

In this chapter, we begin investigating cohomological properties of the groups I_k and C_k defined in the previous chapter for a global field k. The aim is to prove the global analogue of Proposition 8.2, where the idèle class group will replace the multiplicative group of a local field. It is this result that (like in local class field theory) is the first step in the computation of the Brauer group of a global field.

13.1 Cohomology of the Idèle Group

In this paragraph, we fix a global field k and a separable closure $\bar{k}$ of k. We will consider finite separable extensions K of k, that will always be implicitly assumed to be contained in $\bar{k}$. For any place v of k, we fix a separable closure $\bar{k}_v$ of k_v and a k-embedding $i_v : \bar{k} \to \bar{k}_v$, which fixes a place $\bar{v}$ of $\bar{k}$ above v: for $\bar{v}$ we take the restriction of the place of $\bar{k}_v$ induced by v. This allows for any separable algebraic extension $K \subset \bar{k}$ of k to have a distinguished place $v^\bullet$ of K above v. For simplicity we denote by $K_v := i_v(K)k_v$ the field Kk_v (for example $(\bar{k})_v = \bar{k}k_v$), which is the completion of K at $v^\bullet$ if $[K : k]$ is finite. We denote by $U_{K,v}$ the group of units of K_v^*, and $G_v(K/k)$ (or even G_v if there is no ambiguity) the decomposition group of $v^\bullet$ in $G = \mathrm{Gal}(K/k)$ if K is a Galois extension of k.

Let K be a finite Galois extension of k with group G. We can write the idèle group of K as the restricted product of the $I_K(v)$ for v a place of k with respect to the $U_K(v)$, where we have set

$$I_K(v) := \prod_{w|v} K_w^*; \quad U_K(v) := \prod_{w|v} U_{K,w}.$$

($U_{K,w}$ is the multiplicative group of the ring of integers of K_w for w a finite place of K). Each of the $I_K(v)$ and the $U_K(v)$ is a G-module via the k_v-isomorphism $K_w \simeq K_{\sigma w}$ induced by each $\sigma \in G$. We can also view $I_K(v)$ as the group of invertible elements of the ring $K \otimes_k k_v$, on which the group G acts k_v-linearly (via its natural action on K).

© Springer Nature Switzerland AG 2020
D. Harari, *Galois Cohomology and Class Field Theory*, Universitext,
https://doi.org/10.1007/978-3-030-43901-9_13

Proposition 12.14 easily implies that $I_K(v)$ (resp. $U_K(v)$) identifies with the induced module $I_G^{G_v}(K_v^*)$ of K_v^* (resp. to $I_G^{G_v}(U_{K,v})$).

Proposition 13.1 *With the above notations:*

(a) *we have* $I_k = H^0(G, I_K)$.
(b) *for any* $i \in \mathbf{Z}$, *we have*

$$\widehat{H}^i(G, I_K) = \bigoplus_{v \in \Omega_k} \widehat{H}^i(G_v, K_v^*).$$

Proof (a) We have a natural injection of I_k into I_K via the decomposition (12.1). Let $\alpha \in I_k$. For any $\sigma \in G$ and any place w of K above a place v of k, the component $(\sigma\alpha)_{\sigma w}$ of $\sigma\alpha$ at σw is $\sigma\alpha_w = \alpha_w = \alpha_{\sigma w}$, hence $\sigma\alpha = \alpha$. Conversely, if $\alpha = (\alpha_w)$ is in $H^0(G, I_K)$, then

$$(\sigma\alpha)_{\sigma w} = \sigma\alpha_w = \alpha_{\sigma w}$$

for any $\sigma \in G$. For $\sigma \in G_w$, this already gives $\alpha_w \in k_v^*$. The fact that G acts transitively on the places w above v (Proposition 12.14) then gives that the α_w for $w|v$ come from the same $\alpha_v \in k_v^*$ via embeddings $k_v^* \to K_w^*$, hence $\alpha \in I_k$.

(b) Shapiro's lemma gives

$$\widehat{H}^i(G, I_K(v)) = \widehat{H}^i(G_v, K_v^*)$$

and besides if v is unramified in the extension K/k, we have

$$\widehat{H}^i(G, U_K(v)) = \widehat{H}^i(G_v, U_{K,v}) = 0$$

since $U_{K,v}$ is a cohomologically trivial G_v-module by Proposition 8.3. Consider the finite sets S of places of k, containing places ramified in K/k and the archimedean places. Let us set

$$I_{K,S} = \prod_{v \in S} I_K(v) \times \prod_{v \notin S} U_K(v).$$

Then I_K is the inductive limit of the $I_{K,S}$. We deduce (with the help of the analogue for modified groups of Proposition 1.25, cf. also the comment after Corollary 2.7):

$$\widehat{H}^i(G, I_K) = \varinjlim_S \widehat{H}^i(G, I_{K,S})$$

$$= \varinjlim_S \left(\widehat{H}^i\left(G, \prod_{v \in S} I_K(v)\right) \times \widehat{H}^i\left(G, \prod_{v \notin S} U_K(v)\right)\right)$$

$$= \varinjlim_S \left(\prod_{v \in S} \widehat{H}^i(G, I_K(v)) \times \prod_{v \notin S} \widehat{H}^i(G, U_K(v)) \right)$$

$$= \varinjlim_S \prod_{v \in S} \widehat{H}^i(G_v, K_v^*) = \bigoplus_{v \in \Omega_k} \widehat{H}^i(G_v, K_v^*). \qquad \Box$$

Corollary 13.2 *We have* $H^1(G, I_K) = H^3(G, I_K) = 0$.

Proof By the previous proposition, we need to see that for every place v of k, we have $H^1(G_v, K_v^*) = H^3(G_v, K_v^*) = 0$. The first assertion follows from the Hilbert 90 theorem. The second follows from it for real v by Theorem 2.16. For v finite, Tate–Nakayama theorem (Theorem 3.14) gives an isomorphism between $H^1(G_v, \mathbf{Z}) = 0$ and $H^3(G_v, K_v^*)$. $\qquad \Box$

Note also the following lemma about the behaviour of the idèle groups with respect to the norm.

Lemma 13.3 *Let K be a finite Galois extension of k. An idèle $\alpha = (\alpha_v) \in I_k$ is in $N_{K/k} I_K$ if and only if for every place v of k, its local component $\alpha_v \in k_v^*$ is a norm for the extension K_v/k_v (recall that K_v is the completion of K for a place w above v).*

Proof This follows immediately from Proposition 13.1(b) applied to $i = 0$. We can also simply observe that the norm $N_{K/k} : I_K \to I_k$ is obtained by taking the map $(\prod_{w|v} N_{K_w/k_v}) : I_K(v) \to k_v^*$ on each $I_K(v)$, where v ranges through the set of places of k. $\qquad \Box$

We would like to "pass to the limit" in some of the above assertions. Denote by $I := \varinjlim_K I_K$ the *idèle group* of $\overline{k}$, the limit being taken over the finite separable extensions K of k (warning: note that it is not the restricted product of the $(\overline{k})_v^*$). We also call $C := I/\overline{k}^*$ the *idèle class group* of $\overline{k}$, which is the inductive limit of the C_K. Proposition 13.1(a) and Hilbert 90 theorem yield:

Proposition 13.4 *If K is a finite Galois extension of k with group G, we have $H^0(G, C_K) = C_k$. Similarly, we have $H^0(k, I) = I_k$ and $H^0(k, C) = C_k$.*

By passing to the limit in Proposition 13.1(b), we obtain

$$H^i(k, I) = \bigoplus_{v \in \Omega_k} H^i(G_v(\overline{k}/k), (\overline{k})_v^*)$$

for any $i \geqslant 1$. In this statement we would like to replace $(\overline{k})_v$ by the separable closure $\overline{k_v}$ of k_v. We effectively have that $(\overline{k})_v = \overline{k} k_v$ is also $\overline{k_v}$, but this statement is non-trivial; it uses the following lemma.

Lemma 13.5 (Krasner) *Let F be a complete field for an ultrametric absolute value (not necessarily associated to a discrete valuation), with separable closure $\overline{F}$. Let $\alpha \in \overline{F}$, we denote by $\alpha_1 = \alpha, \ldots, \alpha_n$ its conjugates over F. Let $\beta \in \overline{F}$ satisfying*

$$|\alpha - \beta| < |\alpha - \alpha_i|$$

for $i = 2, \ldots, n$, where $|.|$ is the unique extension of the absolute value of F to $\overline{F}$. Then we have the inclusion of fields $F(\alpha) \subset F(\beta)$.

Proof Let L be the Galois closure over $F(\beta)$ of the field extension $F(\alpha, \beta)/F(\beta)$. Let $\sigma \in G := \mathrm{Gal}(L/F(\beta))$. Then $\sigma(\beta - \alpha) = \beta - \sigma(\alpha)$. We also have that σ preserves $|.|$ by uniqueness of the extension of the absolute value in the complete case, which follows from the equivalence of the norms on finite dimensional vector spaces over a complete field (the existence for a finite extension of a complete field can for example be obtained using the formula (7.1) of Theorem 7.7). Thus we obtain

$$|\beta - \sigma(\alpha)| = |\beta - \alpha| < |\alpha_i - \alpha|$$

for $i = 2, \ldots, n$. We deduce that

$$|\alpha - \sigma(\alpha)| < |\alpha - \alpha_i|$$

for $i = 2, \ldots, n$ since $|\cdot|$ is ultrametric. Finally $\sigma(\alpha) = \alpha$, that is: $\alpha \in F(\beta)$, as desired. $\square$

We deduce the announced result.

Proposition 13.6 *Let k be a global field. Let v be a place of k. Then $(\overline{k})_v$ (with the conventions explained at the beginning of the paragraph) identifies with the separable closure $\overline{k_v}$ of the completion k_v of k at v.*

Proof We have a natural inclusion $(\overline{k})_v \to \overline{k_v}$. We can assume that v is a finite place (the archimedean case is trivial). Let $\alpha \in \overline{k_v}$ with minimal polynomial $f \in k_v[X]$. As k is dense in k_v, we can find a separable polynomial $g \in k[X]$ very close to f, which implies that $|g(\alpha)|$ can be made as small as we wish. Write $g(X) = \prod(X - \beta_j)$ with $\beta_j \in \overline{k} \subset (\overline{k})_v$. Then g has a root β which can be made as close as we wish to α, hence in particular made to satisfy $|\beta - \alpha| < |\alpha_i - \alpha|$ for all the conjugates $\alpha_i \in \overline{k_v}$ of α other than α. Krasner's lemma then implies $\alpha \in k_v(\beta) = k(\beta)_v \subset (\overline{k})_v$. $\square$

Corollary 13.7 *For any $i \geqslant 1$, we have $H^i(k, I) = \bigoplus_{v \in \Omega_k} H^i(k_v, \overline{k_v}^*)$, where $\overline{k_v}$ is the separable closure of k_v. In particular, $H^1(k, I) = 0$.*

Note that by what we have just seen, the notation $\overline{k_v}$ is not ambiguous.

13.2 The Second Inequality

The aim of this paragraph is to show:

Theorem 13.8 *Let k be a global field. Let K be a Galois extension of k with Galois group G assumed to be cyclic of order n. Let $h(G, C_K)$ be the Herbrand quotient of the G-module C_K, where $C_K = I_K / K^*$ is the idèle class group of K. Then $h(G, C_K) = n$.*

We deduce:

Corollary 13.9 ("the second inequality") *Under the assumptions of the preceding theorem, the cardinality of the group $C_k / N_{K/k} C_K = I_k / k^* N_{K/k} I_K$ is at least n.*

Proof Applying Hilbert 90 and Proposition 13.1(a), we have $H^0(G, C_K) = C_k$, hence $\widehat{H}^0(G, C_K) = C_k / N_{K/k} C_K = I_k / k^* N_{K/k} I_K$. The corollary then follows from the definition of the Herbrand quotient. □

The aim of this chapter is to show that we in fact have $\#\widehat{H}^0(G, C_K) = n$ in the above corollary. To do this, we will need the *first inequality*. Even though it can be proved using Corollary 13.9 (see next paragraphs), historically it was proved earlier using analytic methods (cf. [9], Exp. VIII or Appendix B of this book, Corollary B.23). This explains the terminology we adopt here (some authors use the inverse convention, it is for example the case of [9], Exp. VII). Corollary 13.9 is mainly used to show that $K = k$ when one can verify that $I_k = k^* N_{K/k} I_K$.

The Proof of Theorem 13.8 uses the following lemma.

Lemma 13.10 *Let G be a finite group.*

(a) *Let M and M' be two modules over the ring $\mathbf{Q}[G]$, of finite dimension over $\mathbf{Q}$, and such that the $\mathbf{R}[G]$-modules $M_\mathbf{R} := M \otimes_\mathbf{Q} \mathbf{R}$ and $M'_\mathbf{R}$ are isomorphic. Then the $\mathbf{Q}[G]$-modules M and M' are isomorphic.*

(b) *Assume that G is cyclic. Let E be a finite dimensional $\mathbf{R}$-vector space endowed with an action of G. Let L and L' be two lattices in E, stables under the action of G, and generating the $\mathbf{R}$-vector space E. Then if one of the Herbrand quotients $h(L)$, $h(L')$ is defined, the other is too and we have $h(L) = h(L')$.*

Proof (a) Each $\mathbf{Q}[G]$-homomorphism $\varphi : M \to M'$ induces a $\mathbf{R}[G]$-homomorphism $\varphi \otimes 1 : M_\mathbf{R} \to M'_\mathbf{R}$ and the map $\varphi \to \varphi \otimes 1$ induces an isomorphism of $\mathbf{R}$-vector spaces:

$$(\mathrm{Hom}_{\mathbf{Q}[G]}(M, M')) \otimes_\mathbf{Q} \mathbf{R} \longrightarrow \mathrm{Hom}_{\mathbf{R}[G]}(M_\mathbf{R}, M'_\mathbf{R}).$$

Fix bases of the $\mathbf{Q}$-vector spaces M and M' (which have the same dimension). The determinant of an element of $\mathrm{Hom}_{\mathbf{Q}[G]}(M, M')$ or of $\mathrm{Hom}_{\mathbf{R}[G]}(M_\mathbf{R}, M'_\mathbf{R})$ in these bases is then well defined. Let then (ξ_i) be a basis of the $\mathbf{Q}$-vector space $\mathrm{Hom}_{\mathbf{Q}[G]}(M, M')$. The above isomorphism shows that it is a basis of the $\mathbf{R}$-vector space $\mathrm{Hom}_{\mathbf{R}[G]}(M_{\mathbf{R}}, M'_{\mathbf{R}})$. The assumption that $M_{\mathbf{R}}$ and $M'_{\mathbf{R}}$ are isomorphic implies that the polynomial

$$F(t_1, \ldots, t_n) := \det(\sum_i t_i \xi_i) \in \mathbf{Q}[t_1, \ldots, t_m]$$

is not identically zero on the whole of $\mathbf{R}^m$, hence it is nonzero on $\mathbf{Q}^m$ since $\mathbf{Q}$ is infinite. Thus we have a $\mathbf{Q}[G]$-isomorphism between M and M'.

(b) Let us set $M = L \otimes \mathbf{Q}$ and $M' = L' \otimes \mathbf{Q}$. Then $M_{\mathbf{R}}$ and $M'_{\mathbf{R}}$ are both $\mathbf{R}[G]$-isomorphic to E. By a), there exists a $\mathbf{Q}[G]$-isomorphism $\varphi : M \to M'$. Then φ induces an injective homomorphism $\varphi : L \to (1/N)L'$ for a certain integer $N > 0$. This implies that $f := N\varphi$ is an injective homomorphism from L to L'. As L and L' are lattices of the same rank, the cokernel of f is finite and we conclude by applying Theorem 2.20, (c).

$\square$

Proof of Theorem 13.8 By Proposition 12.25, we can find a finite non-empty set S of places of k (containing the archimedean places and the places ramified in K/k) such that $I_K = K^* I_{K,S}$, where

$$I_{K,S} = \prod_{v \in S} I_K(v) \times \prod_{v \notin S} U_K(v) = \prod_{v \in S}(\prod_{w|v} K^*_w) \times \prod_{v \notin S}(\prod_{w|v} U^*_{K,w}).$$

Let T be the (finite) set of places of K which are above a place of S. We then have

$$C_K = I_K/K^* = I_{K,S}/E_{K,T},$$

where we recall that $E_{K,T} := K^* \cap I_{K,S}$ is the group of T-units of K. By Theorem 2.20, we have, shortening $h(G, \cdot)$ to $h(\cdot)$,

$$h(C_K) = h(I_{K,S})/h(E_{K,T})$$

whenever the right hand side member is defined (the advantage to introduce S and T is precisely that it allows this condition to be satisfied, while $h(K^*)$ is not defined since $\widehat{H}^0(G, K^*) = K^*/NL^*$ is in general infinite).

We first compute $h(I_{K,S})$ purely "locally". For $v \notin S$, the place v is unramified in K/k, which implies as we have already seen (consequence of Shapiro's lemma and of Proposition 8.3) that the G-module $U_K(v)$ is cohomologically trivial. Let now $v \in S$, denote by $n_v := [K_v : k_v]$ the local degree of K/k at v. Shapiro's lemma implies that for v in S, we have:

$$h(I_K(v)) = h(G_v, K^*_v) = n_v$$

by local class field axiom (Proposition 8.2) hence finally

$$h(I_{K,S}) = \prod_{v \in S} n_v.$$

We now compute $h(E_{K,T})$, which is the "global" part of the proof. We need to show that $n \cdot h(E_{K,T}) = \prod_{v \in S} n_v$. To do this, we will use Lemma 13.10. Let V be the vector space of maps from T to $\mathbf{R}$, which is isomorphic to $\mathbf{R}^t$ with $t = \#T$. The group G acts on V by the formula $(\sigma f)(w) = f(\sigma^{-1} w)$ for all $f \in V, \sigma \in G$, $w \in T$. Let N be the lattice consisting of $f \in V$ whose image is contained in $\mathbf{Z}$. It is clear that N generates the $\mathbf{R}$-vector space V and is stable for the action of G. The G-module N is isomorphic to $\prod_{v \in S}(\prod_{w|v} \mathbf{Z}_w)$, with $\mathbf{Z}_w = \mathbf{Z}$. More precisely, if we set (for a fixed v in S) $N_v := \prod_{w|v} \mathbf{Z}_w$, then N_v is a sub-G-module of N and G acts on N_v by permutation of the $\mathbf{Z}_w$. Shapiro's lemma then yields, for any $q \in \mathbf{Z}$:

$$\widehat{H}^q(G, N) = \prod_{v \in S} \widehat{H}^q(G_v, \mathbf{Z})$$

which implies immediately

$$h(N) = \prod_{v \in S} n_v.$$

For $a \in E_{K,T}$, denote by $f_a : T \to \mathbf{R}$ the map defined by $f_a(w) = \log |a|_w$ and apply Lemma 12.26. It implies that the map

$$\lambda : E_{K,T} \longrightarrow V, \quad a \longmapsto f_a$$

has a finite kernel, and that the image of f is a lattice M^0 which generates the $\mathbf{R}$-vector space $V^0 \subset V$ consisting of f such that $\sum_{w \in T} f(w) = 0$. We have $h(E_{K,T}) = h(M^0)$ by Theorem 2.20, (c). Let $g \in V$ be defined by $g(w) = 1$ for any $w \in T$. Let us set $M = M^0 + \mathbf{Z}g$. Then the lattice M generates the $\mathbf{R}$-vector space $V = V^0 + \mathbf{R}g$. As M^0 and $\mathbf{Z} \cdot g$ are both stable by G, we can write

$$h(M) = h(M^0) \cdot h(\mathbf{Z}) = nh(M^0) = nh(E_{K,T}).$$

As M and N generate the same $\mathbf{R}$-vector space, Lemma 13.10 implies that $h(N) = h(M)$. Finally $n \cdot h(E_{K,T}) = h(N)$, or

$$n \cdot h(E_{K,T}) = \prod_{v \in S} n_v,$$

as desired. $\square$

We deduce:

Proposition 13.11 *Let K be a finite abelian extension of a global field k. We assume that there exists a subgroup D of I_k satisfying: $D \subset N_{K/k}I_K$ and k^*D is dense in I_k. Then $K = k$.*

Proof If F is a cyclic extension of k contained in K, we have $D \subset N_{K/k} I_K \subset N_{F/k} I_F$ by transitivity of the norms. Thus, if we can deal with the case of a cyclic extension, we obtain that every subextension F of K with F/k cyclic is trivial, which implies that K/k is trivial by Galois theory as K/k is assumed to be abelian. We can thus assume that K is a cyclic extension of k.

Now, Theorem 11.20, (a) tells us that each $N_{K_v/k_v} K_v^*$ is an open subgroup of K_v^*. Furthermore $N_{K_v/k_v} K_v^*$ contains $U_v := \mathcal{O}_v^*$ for almost all places v since v unramified implies $\widehat{H}^0(G_v, U_{K,v}) = 0$ with $G_v = \mathrm{Gal}(K_v/k_v)$ (Proposition 8.3). We deduce, using Lemma 13.3, that $N_{K/k} I_K$ is an open subgroup of I_k, hence also $k^* N_{K/k} I_K$ (union of open sets). Thus $k^* N_{K/k} I_K$ is a closed subgroup (as it is open) and dense (as it contains $k^* D$) of I_k, hence $k^* N_{K/k} I_K = I_k$. Corollary 13.9 then implies $[K : k] = 1$, i.e., $K = k$. $\qquad\square$

Proposition 13.12 *Let S be a finite set of places of k, containing all archimedean places if k is a number field. Let K be a finite abelian extension of k, unramified outside S. Then the Galois group $G = \mathrm{Gal}(K/k)$ is generated by the Frobenii $F_{K/k}(v)$ for $v \notin S$.*

Proof Note that since G is abelian, the Frobenius $F_v := F_{K/k}(v)$ is well defined as an element of G (and not just up to conjugation). Let G' be the subgroup of G generated by the F_v for $v \notin S$ and let E be its fixed field. Then the image of F_v in $\mathrm{Gal}(E/k) = G/G'$ is trivial for $v \notin S$, which gives $E_v = k_v$, and hence each element of k_v is a norm of E_v/k_v for $v \notin S$. Denote then by D the subgroup of I_k consisting of the idèles (α_v) such that $\alpha_v = 1$ for all $v \in S$. Then by Lemma 13.3 and what we just saw, we have $D \subset N_{E/k} I_E$. On the other hand, strong approximation Theorem 12.18 (or even its "weak" form, which is the one without a condition outside S) yields that $k^* D$ is dense in I_k. Proposition 13.11 then says that $E = k$, or that $G' = G$. $\qquad\square$

Corollary 13.13 *Let K be a non-trivial abelian extension of k. Then there are infinitely many places v of k which are not totally split in K. If furthermore K/k is cyclic of degree ℓ^m with ℓ prime, there are infinitely many places of k which are inert in K/k (that is, unramified and such that the associated decomposition group is the whole of $G := \mathrm{Gal}(K/k)$).*

Note that an inert place v corresponds to a Frobenius $F_{K/k}(v)$ that generates $\mathrm{Gal}(K/k)$, or to a place that is, unramified and there is only one place w of K above v.

Proof The first assertion follows from Proposition 13.12 and the fact that places v totally split in K correspond to $F_{K/k}(v) = 1$. The second assertion is obtained by applying the first to the intermediate extension E/k of degree ℓ, that corresponds to the unique subgroup of order ℓ^{m-1} of $\mathrm{Gal}(K/k)$: we obtain infinitely many places v for which the Frobenius $F_{K/k}(v)$ has a non-trivial image in the quotient of order ℓ of G, hence generates G. $\qquad\square$

For example, the above corollary implies that an element of $\mathbf{Z}$ that is, not a square can not become a square modulo all primes except finitely many. There is in fact a more general and precise statement than Corollary 13.13, the *Čebotarev theorem* (cf. Chap. 18).

13.3 Kummer Extensions

In this paragraph, we establish general results on Kummer extensions, that will be useful in the next paragraph and also in the proof of the global existence theorem.

Fix a field k and an integer $n > 0$ not divisible by the characteristic of k, such that k contains a primitive nth root of unity ζ.

Definition 13.14 A *Kummer extension* of k is a field extension K of k of the form $K = k(\sqrt[n]{\Delta})$, where Δ is a subgroup of k^*, containing k^{*n}, and such that Δ/k^{*n} is finite.[1]

Note that as $\zeta \in k$, the notation $k(\sqrt[n]{\Delta})$ is not ambiguous, and the choice of a primitive nth root of unity in k identifies the Galois modules μ_n and $\mathbf{Z}/n$. As for any $a \in \Delta$, the extension $k(\sqrt[n]{a})$ is cyclic of degree dividing n (it corresponds to the element of $H^1(k, \mathbf{Z}/n) \simeq H^1(k, \mu_n) = k^*/k^{*n}$ given by the class of a), a Kummer extension is a finite abelian extension of k whose Galois group is of exponent dividing n (isomorphic to a subgroup of $(\mathbf{Z}/n)^r$, where r is the cardinality of Δ/k^{*n}). Conversely, a cyclic extension of k of degree dividing n corresponds to an element of $H^1(k, \mathbf{Z}/n) \simeq k^*/k^{*n}$, hence is of the form $k(\sqrt[n]{a})$. More generally, we have:

Proposition 13.15 *Let K be a finite abelian extension of k whose Galois G is of exponent n. Then $K = k(\sqrt[n]{\Delta})$ with $\Delta = K^{*n} \cap k^*$. Furthermore, we have an isomorphism*

$$\overline{u} : \Delta/k^{*n} \longrightarrow \operatorname{Hom}(G, \mathbf{Z}/n).$$

Conversely, if Δ' is a subgroup of k^* containing k^{*n} and if we set $K = k(\sqrt[n]{\Delta'})$, then we have $\Delta' = K^{*n} \cap k^*$.

Thus, for an abelian extension K of k with Galois group of exponent n, we have one and only one subgroup Δ of k^* containing k^{*n} such that $K = k(\sqrt[n]{\Delta})$.

Proof We clearly have $k(\sqrt[n]{\Delta}) \subset K$. Conversely, the extension K/k is the composite of all its cyclic subextensions E/k as it is abelian (as a finite abelian group is a direct product of cyclic groups). Such an extension E is of degree dividing n, hence is of the form $E = k(\sqrt[n]{a})$ with $a \in k^* \cap K^{*n}$, hence $E \subset k(\sqrt[n]{\Delta})$ and finally $K \subset k(\sqrt[n]{\Delta})$. We define a morphism $u : \Delta \to \operatorname{Hom}(G, \mathbf{Z}/n)$ by $u(a) = \chi_a$, where

$$\chi_a(\sigma) = \sigma(\sqrt[n]{a})/\sqrt[n]{a} \tag{13.1}$$

is defined using the identification of $\mathbf{Z}/n$ with the subgroup of nth roots of unity of k^*. The kernel of u is k^{*n} since χ_a is trivial if and only if $\sqrt[n]{a} \in k^*$. It remains to show that u is surjective. Let $\chi \in \operatorname{Hom}(G, \mathbf{Z}/n) = \operatorname{Hom}(G, \mu_n)$, then χ becomes a 1-coboundary (by Hilbert 90) when viewed as a map from G to K^*. This means that we have $b \in K^*$ such that $\chi(\sigma) = \sigma(b)/b$ for any $\sigma \in G$. Then

---

[1] Some authors do not require this condition, but we will not need the infinite case.

$$\sigma(b^n) = (\sigma b)^n = \chi(\sigma)^n b^n = b^n$$

hence $a := b^n \in k^* \cap K^{*n} = \Delta$, and finally $\chi = \chi_a$.

Lastly, let Δ' be a subgroup of k^* containing k^{*n} and $K := k(\sqrt[n]{\Delta'})$. Let us set $\Delta = K^{*^n} \cap k^*$ and let us show that $\Delta = \Delta'$. The inclusion $\Delta' \subset \Delta$ is obvious. Denote by H the subgroup of $G = \mathrm{Gal}(K/k)$ consisting of σ such that $\chi_a(\sigma) = 1$ for every $a \in \Delta'$. That means that H is the orthogonal of the family $(\chi_a)_{a \in \Delta'}$ for the duality between G and $\mathrm{Hom}(G, \mathbf{Z}/n)$. By biduality, the orthogonal of H is the subgroup of $\mathrm{Hom}(G, \mathbf{Z}/n)$ consisting of the χ_a for $a \in \Delta'$, which means that the image of Δ'/k^{*n} by the isomorphism $\bar{u} : \Delta/k^{*n} \to \mathrm{Hom}(G, \mathbf{Z}/n)$ is exactly $\mathrm{Hom}(G/H, \mathbf{Z}/n)$. Besides, the formula (13.1) shows that H fixes each element of $\sqrt[n]{\Delta'}$, hence fixes the whole of K (which is the extension of k generated by $\sqrt[n]{\Delta'}$). Galois theory then tells us that H is the trivial group. As $\bar{u}$ is an isomorphism, this implies that $\Delta/k^{*n} = \Delta'/k^{*n}$ and finally $\Delta = \Delta'$, as desired. □

Remark 13.16 The link between the last proposition and cohomological Kummer theory seen in Sect. 6.2 (Corollary 6.6) is the following: the group $H^1(G, \mathbf{Q}/\mathbf{Z}) = H^1(\mathrm{Gal}(K/k), \mathbf{Z}/n)$ is (by restriction-inflation sequence) given by

$$H^1(G, \mathbf{Q}/\mathbf{Z}) = \mathrm{Ker}[H^1(k, \mathbf{Z}/n) \longrightarrow H^1(K, \mathbf{Z}/n)].$$

Identifying (via a choice of ζ) the Galois modules $\mathbf{Z}/n$ and μ_n, we obtain an isomorphism between $H^1(G, \mathbf{Q}/\mathbf{Z})$ and the kernel of the map $k^*/k^{*n} \to K^*/K^{*n}$, that is, with Δ/k^{*n}.

We will need the following "local" result.

Lemma 13.17 *Let K be a local field. Let $n > 0$ be an integer not divisible by the characteristic of the residue field κ of K, Assume that K contains primitive nth roots of unity (and hence n divides $q - 1$, where $q = \#\kappa$). Then, for $x \in K^*$, the extension $L = K(\sqrt[n]{x})/K$ is unramified if and only if $x \in U_K K^{*n}$.*

Proof Assume that $x = u \cdot y^n$ with $u \in U_K$ and $y \in K^*$. We want to show that $K(\sqrt[n]{u})$ is unramified over K. By Hensel lemma, the polynomial $X^n - u$ factors as a product of linear factors over the unramified extension K' of K whose residue field is the decomposition field of the reduction $X^n - \bar{u}$ over κ. Thus $\sqrt[n]{u} \in K'$ and we have that $K(\sqrt[n]{u})$ is unramified over K. Conversely, if $K(\sqrt[n]{x})/K$ is unramified, let us write $x = u \cdot \pi^r$ with $u \in U_K$ and π a uniformiser of K. Then the valuation $v_L(\sqrt[n]{u \cdot \pi^r})$ is $\frac{1}{n} v_L(\pi^r) = \frac{1}{n} v_K(\pi^r)$, which shows that n divides r. □

Lastly, Kummer extensions are related to another "local" result, that we will use in the next paragraph.

Proposition 13.18 *Let k be a global field. Let v be a place of k. Let $n > 0$ be an integer not divisible by the characteristic of k (which is automatic if k is a number field), and such that $\mu_n \subset k_v$. Then the cardinality of k_v^*/k_v^{*n} is $n^2/|n|_v$ and if v is finite, the cardinality of $\mathcal{O}_v^*/\mathcal{O}_v^{*n}$ is $n/|n|_v$.*

Proof The case where v is archimedean is straightforward, observing that for real v we necessarily have $n = 1$ or $n = 2$, and for a complex v $|n|_v = n^2$ by convention. Let us now assume that v is finite. As $k_v^* \simeq \mathbf{Z} \times \mathcal{O}_v^*$, it is enough to compute the cardinality of $\mathcal{O}_v^*/\mathcal{O}_v^{*n}$. To do this, make the group $\mathbf{Z}/n$ acts trivially on $\mathcal{O}_v^*$, and consider the corresponding Herbrand quotient $h_n(\mathcal{O}_v^*)$. As $H^1(\mathbf{Z}/n, \mathcal{O}_v^*) = \mathrm{Hom}(\mathbf{Z}/n, \mathcal{O}_v^*)$ is of cardinality n (because $k_v \supset \mu_n$), we have

$$h_n(\mathcal{O}_v^*) = \frac{\#\mathcal{O}_v^*/\mathcal{O}_v^{*n}}{n}.$$

We are reduced to showing that $h_n(\mathcal{O}_v^*) = |n|_v^{-1}$. If k is a function field of characteristic p, the assumption made on n implies that $|n|_v = 1$. Besides, the subgroup of finite index U_v^1 of $U_v = \mathcal{O}_v^*$ is a pro-p-group (Theorem 7.18, a), hence satisfies $h_n(U_v^1) = 1$ as p does not divide n. We conclude using Theorem 2.20. If k is a number field, the group U_v^1 has a subgroup of finite index isomorphic to the additive group $\mathcal{O}_v$ (Theorem 7.18, b). As $H^1(\mathbf{Z}/n, \mathcal{O}_v) = \mathcal{O}_v[n] = 0$, Theorem 2.20 implies that

$$h_n(U_v) = h_n(U_v^1) = h_n(\mathcal{O}_v) = \#(\mathcal{O}_v/n\mathcal{O}_v) = |n|_v^{-1},$$

as desired. $\qquad\square$

13.4 First Inequality and the Axiom of Class Field Theory

We start with a linear algebra lemma:

Lemma 13.19 *Let ℓ be a prime number and let m be a non negative integer. We set $n = \ell^m$. Let A be the ring $\mathbf{Z}/n\mathbf{Z}$. Let M be a free finite type A-module and M_1 a sub-A-module of M such that M/M_1 is free. Then M_1 is free of finite type.*

Proof As $M_2 := M/M_1$ is free, we have $M \simeq M_1 \oplus M_2$. As an abelian group, M is isomorphic to $(\mathbf{Z}/n\mathbf{Z})^r$ with $r \geqslant 0$. The theorem on the structure of finite abelian groups implies that the abelian n-torsion group M_1 is isomorphic to a direct sum of groups of the form $(\mathbf{Z}/n'\mathbf{Z})$, where n' is a power of ℓ. But the uniqueness part of this same theorem imposes that $n' = n$ (we could also have used a more advanced theorem which says that every projective finite type module over a local ring is free cf. [36], Part. I, Th. 2.9). $\qquad\square$

We are going back to global fields. In the remaining part of this paragraph, k is a global field; in all the statements before Theorem 13.23, we denote by n a power of a prime number ℓ different from the characteristic of k, such that k contains a primitive nth root of unity ζ. The first aim will be to compute (by a method due to Chevalley) the norm group $N_{K/k}C_K$ for a Kummer extension K/k (in the sense of Definition 13.14) with Galois group $G = (\mathbf{Z}/n)^r$. We start by fixing a finite set S of places of k, containing all the archimedean places (if k is a number field), all places above ℓ, all places ramified in K/k, and such that $I_k = I_{k,S}k^*$ (cf. Proposition 12.25). We denote by s the cardinality of S.

Proposition 13.20 *With the above notation, we have $s \geqslant r$. Furthermore, there exists a set T of places of k, disjoint from S and of cardinality $s - r$, such that $K = k(\sqrt[n]{\Delta})$, where Δ is the kernel of the diagonal map*

$$E_{k,S} \longrightarrow \prod_{v \in T} k_v^* / k_v^{*n},$$

with $\Delta = K^{*n} \cap E_{k,S}$.

(Recall that $I_{k,S}$ is the group of S-idèles and $E_{k,S} = k^* \cap I_{k,S}$ is the group of S-units).

Proof We start by showing that $K = k(\sqrt[n]{\Delta})$ with $\Delta = K^{*n} \cap E_{k,S}$. Proposition 13.15 gives $K = k(\sqrt[n]{D})$ with $D = K^{*n} \cap k^*$. For $x \in D$, the choice of S yields that $k(\sqrt[n]{x})/k$ is unramified for $v \notin S$, hence $x = u_v y_v^n$ with $u_v \in U_v := \mathcal{O}_v^*$ and $y_v \in k_v^*$ (Lemma 13.17). Let us set $y_v = 1$ for $v \in S$. We obtain an idèle $y = (y_v)$ (indeed, x is of valuation zero outside a finite number of places), that we can write (again, by the choice of S) $y = \alpha \cdot z$ with $\alpha \in I_{k,S}$ and $z \in k^*$. Then $x/z^n \in I_{k,S} \cap k^* = E_{k,S}$ hence $x/z^n \in \Delta$. Thus $D \subset \Delta \cdot k^{*n}$, and the reverse inclusion is obvious, hence $K = k(\sqrt[n]{\Delta})$.

Let us set $N = k(\sqrt[n]{E_{k,S}})$, then $N \supset K$ as $E_{k,S} \supset \Delta$. As $E_{k,S}k^{*n}/k^{*n} = E_{k,S}/E_{k,S}^n$ (observe that $k^{*n} \cap E_{k,S} = E_{k,S}^n$, where $E_{k,S}^n$ is the subgroup of nth powers in $E_{k,S}$). Proposition 13.15 yields

$$\mathrm{Gal}(N/k) \simeq \mathrm{Hom}(E_{k,S}/E_{k,S}^n, \mathbf{Z}/n).$$

On the other hand $E_{k,S}$ contains the nth roots of unity and hence is an abelian group isomorphic (by the Dirichlet unit theorem) to $\mathbf{Z}^{s-1} \times \mu_q$, where q is an integer such that n divides q. This implies that $E_{k,S}/E_{k,S}^n$ is a free $\mathbf{Z}/n$-module of rank s, hence it is also the case for $\mathrm{Gal}(N/k)$. As its quotient $G = \mathrm{Gal}(K/k)$ is by assumption a free $\mathbf{Z}/n$-module of rank r, we already obtain that $r \leqslant s$ and $\mathrm{Gal}(N/K)$ is a free $\mathbf{Z}/n$-module by Lemma 13.19.

Choose a $\mathbf{Z}/n$-basis $\sigma_1, \ldots, \sigma_{s-r}$ of $\mathrm{Gal}(N/K)$, and call N_i the fixed field of σ_i for $i = 1, \ldots, s - r$. Then $K = \bigcap_{1 \leqslant i \leqslant s-r} N_i$. By Corollary 13.13, we can find for each i a prime ideal $\mathfrak{p}_i$ of N_i, above a finite place v_i of k, such that: the places v_i are pairwise distinct, are not in S, and each $\mathfrak{p}_i$ is inert in the extension N/N_i (Recall that the N/N_i are cyclic of order n, with n a power of a prime number ℓ). This means that the Frobenius associated to $\mathfrak{p}_i$ generates $\mathrm{Gal}(N/N_i)$, or that there is only one prime $\mathfrak{p}_i' = \mathfrak{p}_i \mathcal{O}_N$ of N above N_i. We will show that we can take $T = \{v_1, \ldots, v_{s-r}\}$.

Let us first show that $\mathrm{Gal}(N/N_i)$ is the decomposition group D_i of v_i in N/k. The place v_i is unramified in the extension N/k by Lemma 13.17. In particular, D_i is cyclic, generated by the Frobenius $F_{N/k}(v_i)$. On the other hand $D_i \supset \mathrm{Gal}(N/N_i)$ as every element of $\mathrm{Gal}(N/k)$ which induces the identity on N_i fixes $\mathfrak{p}_i$, hence $\mathfrak{p}_i'$.

As $\mathrm{Gal}(N/N_i)$ is of order n and D_i is cyclic of exponent $\leqslant n$, the only possibility is $D_i = \mathrm{Gal}(N/N_i)$, as desired.

As $\mathrm{Gal}(N/K)$ is the direct product of the $\mathrm{Gal}(N/N_i)$, we obtain that K/k is the maximal subextension of N/k in which all the places v_i are totally split: indeed, this property comes down to saying that K is the maximal subextension of N/k such that all the $F_{N/k}(v_i)$ are in $\mathrm{Gal}(N/K)$. Let then $x \in E_{k,S}$. We obtain:

$$x \in \Delta \iff k(\sqrt[n]{x}) \subset K \iff k_{v_i}(\sqrt[n]{x}) = k_{v_i}, \; \forall i = 1, \ldots, s - r,$$

which means exactly that Δ is the kernel of $E_{k,S} \to \prod_{i=1}^{s-r} k_{v_i}^* / k_{v_i}^{*n}$. $\qquad\square$

We are coming to the most difficult result of this paragraph, which includes in particular the "first inequality" $[C_k : N_{K/k}C_K] \leqslant [K : k]$ in the case of a Kummer extension.

Theorem 13.21 *Under the assumptions and notation of Proposition 13.20, we set (for any non-archimedean place v of k) $U_v := \mathcal{O}_v^*$ and*

$$I_k(S, T) := \prod_{v \in S} k_v^{*n} \times \prod_{v \in T} k_v^* \times \prod_{v \notin S \cup T} U_v.$$

Let $C_k(S, T) = I_k(S, T) \cdot k^*/k^*$. Then we have

$$N_{K/k}C_K \supset C_k(S, T)$$

and $[C_k : C_k(S, T)] = [K : k]$. If furthermore the extension K/k is cyclic, we have $C_k(S, T) = N_{K/k}C_K$ (and hence $[C_k : N_{K/k}C_K] = [K : k]$).

We start with a lemma.

Lemma 13.22 *We have $I_k(S, T) \cap k^* = E_{k, S \cup T}^n$.*

Proof It is straightforward to see that we have the inclusion $E_{k,S \cup T}^n \subset I_k(S, T) \cap k^*$. Let conversely $y \in I_k(S, T) \cap k^*$. Let us set $M = k(\sqrt[n]{y})$. To show that $M = k$, it is enough by Corollary 13.9 to see that $C_k = N_{M/k}C_M$. Let us first show that the diagonal map $f : E_{k,S} \to \prod_{v \in T} U_v/U_v^n$ is surjective, where U_v^n is the subgroup of nth powers in U_v. Its kernel is Δ by Proposition 13.20, and we have

$$\#(E_{k,S}/\Delta) = \frac{\#E_{k,S}/E_{k,S}^n}{\#\Delta/E_{k,S}^n}.$$

As we have seen in the Proof of Proposition 13.20 (consequence of the Dirichlet unit theorem), the cardinality of $E_{k,S}/E_{k,S}^n$ is n^s. On the other hand, the cardinality of $\Delta/E_{k,S}^n = \Delta k^{*n}/k^{*n}$ is that of $G = \mathrm{Gal}(K/k)$ by Proposition 13.15, it is thus n^r. Finally the cardinality of $E_{k,S}/\Delta$ is n^{s-r}, and to obtain the surjectivity it is enough to see that it is also the cardinality of $\prod_{v \in T} U_v/U_v^n$. This follows from Proposition 13.18, since v does not divide ℓ if $v \in T$ and n is a power of ℓ.

Let now $\alpha = (\alpha_v) \in I_{k,S}$, with image $[\alpha] \in C_k = I_{k,S}k^*/k^*$. We want to show that $[\alpha] \in N_{M/k}C_M$. By surjectivity of f we find an $x \in E_{k,S}$ such that for every place v in T, we have $\alpha_v = x \cdot u_v^n$ with $u_v \in U_v$. Let us set $\alpha' = \alpha/x$, it is enough to see that α' is in $N_{M/k}I_M$, which can be checked component by component by Lemma 13.3. For $v \in S$, we have $y \in k_v^{*n}$ hence $M_v = k_v$, which obviously implies that α_v' is a norm of the extension M_v/k_v. For $v \in T$, it also works because $\alpha_v' = u_v^n$ is an nth power. Lastly, for $v \notin S \cup T$, we have that M_v/k_v is unramified and $\alpha_v' \in U_v$, which is enough to guarantee that α_v' is a local norm, by Proposition 8.3. Finally we have $M = k$, and therefore $y \in k^{*n} \cap I_k(S,T) \subset E_{k,SUT}^n$, as desired. □

Proof of Theorem 13.21 We use the exact sequence:

$$1 \longrightarrow I_{k,SUT} \cap k^*/I_k(S,T) \cap k^* \longrightarrow I_{k,SUT}/I_k(S,T)$$
$$\longrightarrow I_{k,SUT} \cdot k^*/I_k(S,T)k^* \longrightarrow 1$$

and we calculate the cardinalities of various terms. As $I_k = I_{k,SUT} \cdot k^*$, the order of the group on the right is $[C_k : C_k(S,T)]$. By Lemma 13.22, the order of the group on the left is $[E_{k,SUT} : E_{k,SUT}^n] = n^{2s-r}$ as we have already seen (as the cardinality of $S \cup T$ is $2s - r$). Lastly, the cardinality of the middle group is that of $\prod_{v \in S}[k_v^* : k_v^{*n}]$, which is $\prod_{v \in S} n^2/|n|_v$ by Proposition 13.18. But as S contains all the archimedean places and those dividing ℓ, by the product formula we have (recall that n is a power of ℓ):

$$1 = \prod_{v \in \Omega_k} |n|_v = \prod_{v \in S} |n|_v$$

hence $\prod_{v \in S} n^2/|n|_v = n^{2s}$. We obtain

$$[C_k : C_k(S,T)] = n^r = [K : k].$$

as desired.

Let us now show that $C_k(S,T) \subset N_{K/k}C_K$. Let $\alpha \in I_k(S,T)$, it is enough to check (Lemma 13.3) that each component α_v of α is a local norm. For $v \in S$, we have $\alpha_v \in k_v^{*n}$, which is a local norm via local reciprocity isomorphism $k_v^*/NK_v^* \simeq \mathrm{Gal}(K_v/k_v)$ and the fact that $\mathrm{Gal}(K/k)$ is of exponent n. For $v \in T$, we have $K_v = k_v$ since $\Delta \subset k_v^{*n}$, hence the required condition is automatically satisfied. Lastly, for $v \notin S \cup T$, α_v is a unit and K_v/k_v is unramified by Lemma 13.17 and hence it still works (Proposition 8.3).

If now K/k is cyclic, we have $r = 1$ and the Corollary 13.9 implies

$$[K : k] \leqslant [C_k : N_{K/k}C_K] \leqslant [C_k : C_k(S,T)] = [K : k],$$

which proves the equality $N_{K/k}C_K = C_k(S,T)$. □

At last we deduce:

Theorem 13.23 (The axiom of the global class field theory) *Let k be a global field. Let K be a finite Galois extension of k with cyclic Galois group G. Then $\widehat{H}^0(G, C_K)$ is of cardinality $[K : k]$, and $H^1(G, C_K) = 0$.*

We will see later on that the triviality of $H^1(G, C_K)$ remains true if G is an arbitrary finite group. The assertion on $\widehat{H}^0(G, C_K)$ is also true (in a more precise form) when G is finite and abelian. This will be a consequence of results that we will see in the next two chapters. We will deduce that in Theorem 13.21, the equality $C_k(S, T) = N_{K/k}C_K$ remains true if we only assume that the extension K/k is abelian.

Proof in the case where the characteristic of k does not divide $[K : k]$ We postpone to the last paragraph the case where k is a field of functions of characteristic $p > 0$ dividing $[K : k]$, which needs to be treated separately. We can do this using the method of [1], Chap. 6, but we thought it would be more insightful to use a geometric approach, based on the results of Sect. 12.4. For now, we assume that k is a number field, or a function field whose characteristic is prime to $[K : k]$. As we know that the Herbrand quotient $h(G, C_K)$ is $n := [K : k]$ by Theorem 13.8, it is enough to see that $\widehat{H}^{-1}(G, C_K)$ (which is also $H^1(G, C_K)$ by Theorem 2.16) is of cardinality 1. We proceed by induction on n. Consider a subextension M/k of K/k of prime degree ℓ. If $\ell < n$, the induction assumption and the restriction-inflation exact sequence imply $H^1(G, C_K) = 0$. Let us then assume that $n = \ell$ is prime. Denote by μ_ℓ the group of ℓth roots of unity (in the algebraic closure of k) and let us set $k' = k(\mu_\ell)$, $K' = K(\mu_\ell)$. Then K'/k' is a cyclic Kummer extension and Theorem 13.21 implies that the group $\widehat{H}^0(\mathrm{Gal}(K'/k'), C_{K'})$ is of cardinality $[K' : k']$, hence $H^1(\mathrm{Gal}(K'/k'), C_{K'}) = 0$ (still by Theorem 13.8). As $[k' : k] \leqslant \ell - 1$, the induction assumption implies also that $H^1(\mathrm{Gal}(k'/k), C_{k'}) = 0$ hence $H^1(\mathrm{Gal}(K'/k), C_{K'}) = 0$ by the restriction-inflation sequence, and a fortiori $H^1(G, C_K) = 0$ by the same sequence.   □

Corollary 13.24 (Hasse norm principle) *Let k be a global field and let K be a finite cyclic extension of k. Then an element $x \in k^*$ is a norm of the extension K/k if and only if its image x_v in k_v^* is a norm in K_v/k_v for every place v of k.*

Note that this result is not true for an arbitrary abelian extension (see for example Exercise 5 of [9]).

Proof Let $G = \mathrm{Gal}(K/k)$. As $\widehat{H}^{-1}(G, C_K) = 0$, the exact modified cohomology sequence yields an injection $\widehat{H}^0(G, K^*) \to \widehat{H}^0(G, I_K)$, which allows us to conclude using Proposition 13.1, (b).   □

Theorem 13.25 *Let K/k be a finite Galois extension of global fields with group G. Then $H^1(G, C_K) = 0$.*

Proof For G cyclic, the result is part of the class field axiom (Theorem 13.23). On the other hand, recall that (for any prime number p) the abelianisation of a non-trivial p-group is a non-trivial p-group. Hence every non-trivial p-group has a quotient of order p. We deduce that the theorem is true when G is a p-group by induction on the cardinality of G, taking a Galois degree p subextension E/k and using the restriction-inflation exact sequence

$$0 \longrightarrow H^1(\mathrm{Gal}(E/k), C_E) \longrightarrow H^1(G, C_K) \longrightarrow H^1(\mathrm{Gal}(K/E), C_K).$$

Finally, for an arbitrary G, we consider, for every p dividing $\#G$, a p-Sylow G_p of G and its fixed fixed field K_p. Then, as $H^1(G, C_K)$ is the direct sum of the $H^1(G, C_K)\{p\}$, the map

$$H^1(G, C_K) \longrightarrow \prod_p H^1(\mathrm{Gal}(K/K_p), C_K)$$

obtained via the restrictions is injective by Lemma 3.2, which yields the result by the case where G is a p-group. $\qquad\square$

Corollary 13.26 Let $C = \varinjlim_K C_K = I_{\overline{k}}/\overline{k}^*$. Then $H^1(k, C) = 0$.

13.5   Proof of the Class Field Axiom for a Function Field

In this paragraph, we consider a global field k of characteristic p, that we view as the field of functions $k = \kappa(X)$ of a smooth projective geometrically integral curve X over a finite field κ. We will follow the method of [39] (Part. I, App. A) to prove the axiom of the global class field theory (Theorem 13.23) in this situation. In a certain sense, class field theory for a function field is easier than for a number field, once one knows some geometric results from Sect. 12.4. Indeed, this approach will for example yield the axiom of class field theory (as well as Theorem 13.25 and Corollary 13.26) much easier than with the method of Kummer extensions that we have followed in the last paragraph. Furthermore it has the advantage to work in the case where p divides the degree of the extension as well. The geometric approach also admits vast generalisations to higher dimensions whose very complete exposition can be found in [48].

The notations of this paragraph are analogous to those of Sect. 12.4. For any finite extension $\kappa' \subset \overline{\kappa}$ of κ, set $X_{\kappa'} = X \times_\kappa \kappa'$ and denote by $\kappa'(X) = \kappa(X) \otimes_\kappa \kappa'$ the field of functions of the curve $X_{\kappa'}$. Thus $\kappa'(X)$ is a finite extension of the global field k. Let $I_{\overline{\kappa}(X)}$ be the inductive limit over these κ' of the idèle groups $I_{\kappa'(X)}$. We have a natural inclusion i of $\overline{\kappa}(X)^*$ in $I_{\overline{\kappa}(X)}$.

Proposition 13.27 The map $H^2(\kappa, \overline{\kappa}(X)^*) \to H^2(\kappa, I_{\overline{\kappa}(X)})$ induced by i is injective.

Proof A closed point w of $X_{\kappa'}$ defines a discrete valuation val_w on the completion $\kappa'(X)_w$, hence by definition of $I_{\kappa'(X)}$ an induced map

$$\bigoplus_{w \in X_{\kappa'}^{(1)}} \mathrm{val}_w : I_{\kappa'(X)} \longrightarrow \bigoplus_{w \in X_{\kappa'}^{(1)}} \mathbf{Z} \cdot w = \mathrm{Div} X_{\kappa'}.$$

We thus have, by definition of the map Div (cf. Sect. 12.4), a commutative diagram:

$$
\begin{array}{ccc}
\kappa'(X)^* & \longrightarrow & I_{\kappa'(X)} \\
\mathrm{Div} \downarrow & & \downarrow \\
\mathrm{Div}_0\, X_{\kappa'} & \longrightarrow & \mathrm{Div} X_{\kappa'}.
\end{array}
$$

By passing to the limit on the κ', we obtain a commutative diagram:

$$
\begin{array}{ccc}
\overline{\kappa}(X)^* & \overset{i}{\longrightarrow} & I_{\overline{\kappa}(X)} \\
\mathrm{Div} \downarrow & & \downarrow \\
\mathrm{Div}_0\, \overline{X} & \longrightarrow & \mathrm{Div}\overline{X}.
\end{array}
$$

Proposition 12.31 and Corollary 12.34 tell us that the maps $H^2(\kappa, \overline{\kappa}(X)^*) \to H^2(\kappa, \mathrm{Div}_0\, \overline{X})$ and $H^2(\kappa, \mathrm{Div}_0\, \overline{X}) \to H^2(\kappa, \mathrm{Div}\overline{X})$ induced by this diagram are injective. The result follows. $\qquad\square$

Let us fix a separable closure $\overline{k}$ of $k = \kappa(X)$ containing $\overline{\kappa}$ and consider a finite Galois extension $k' \subset \overline{k}$ of k. Let κ' be the algebraic closure of κ in k'. Denote by $I_{k'}$ the group of idèles of the global field k', and similarly by $I_{\kappa'(X)}$ that of $\kappa'(X)$. We have a commutative diagram:

$$
\begin{array}{ccc}
k'^* & \longrightarrow & I_{k'} \\
\uparrow & & \uparrow \\
\kappa'(X)^* & \longrightarrow & I_{\kappa'(X)}
\end{array}
$$

hence, by passing to the limit over the k', we obtain the commutative diagram:

$$
\begin{array}{ccc}
\overline{k}^* & \longrightarrow & I_{\overline{k}} \\
\uparrow & & \uparrow \\
\overline{\kappa}(X)^* & \longrightarrow & I_{\overline{\kappa}(X)}
\end{array}
\qquad\qquad (13.2)
$$

Here $I := I_{\overline{k}}$ is the idèle group of $\overline{k}$ as defined in Sect. 13.1. We also have the idèle class group $C = I/\overline{k}^*$ of $\overline{k}$.

Theorem 13.28 *The group* Br k *is isomorphic to* $H^2(\kappa, \overline{\kappa}(X)^*)$. *The map* Br $k = H^2(k, \overline{k}^*) \to H^2(k, I_{\overline{k}})$ *induced by the diagram (13.2) is injective.*

Proof Consider the algebraic extensions (of infinite degree) of the field $k = \kappa(X) \subset \overline{\kappa}(X) \subset \overline{k}$. The Galois group of $\overline{\kappa}(X)/\kappa(X)$ identifies with $\Gamma_\kappa = \mathrm{Gal}(\overline{\kappa}/\kappa)$. As $H^1(\overline{\kappa}(X), \overline{k}^*) = 0$ by Hilbert 90, we have by Corollary 1.45 (in its profinite version) an exact sequence

$$0 \longrightarrow H^2(\kappa, \overline{\kappa}(X)^*) \xrightarrow{\text{Inf}} H^2(k, \overline{k}^*) = \mathrm{Br}\, k \longrightarrow H^2(\overline{\kappa}(X), \overline{k}^*) = \mathrm{Br}(\overline{\kappa}(X)).$$

But the field $\overline{\kappa}(X)$ is C_1 by Tsen theorem (Example 6.20, b), hence has a trivial Brauer group (Theorem 6.22), which yields an isomorphism

$$H^2(\kappa, \overline{\kappa}(X)^*) \simeq \mathrm{Br}\, k.$$

The first statement follows.

Let us now show that the map

$$\text{Inf} : H^2(\kappa, I_{\overline{\kappa}(X)}) \longrightarrow H^2(k, I_{\overline{k}})$$

deduced from the diagram (13.2) is injective. Let $k' \subset \overline{k}$ be a finite Galois extension of k and κ' be the algebraic closure of κ in k'. It is enough to show that $\text{Inf} : H^2(\mathrm{Gal}(\kappa'(X)/k), I_{\kappa'(X)}) \to H^2(\mathrm{Gal}(k'/k), I_{k'})$ is injective (we then pass to the limit on k'). Let us set $K = \kappa'(X)$. We apply Proposition 13.1(b) to finite extensions of the global fields K/k and k'/k, which reduces to showing that if v is a place of k then the map

$$\text{Inf} : H^2(\mathrm{Gal}(K_v/k_v), K_v^*) \longrightarrow H^2(\mathrm{Gal}(k'_v/k_v), k_v'^*)$$

is injective. This is the case by Hilbert 90 and Corollary 1.45.

We deduce from diagram (13.2) a commutative diagram:

$$
\begin{array}{ccc}
\mathrm{Br}\, k & \longrightarrow & H^2(k, I_{\overline{k}}) \\
\text{Inf} \uparrow & & \text{Inf} \uparrow \\
H^2(\kappa, \overline{\kappa}(X)^*) & \longrightarrow & H^2(\kappa, I_{\overline{\kappa}(X)}).
\end{array}
$$

We have seen that the vertical left hand side map is an isomorphism and the vertical right hand side map is injective. Proposition 13.27 implies that the horizontal bottom map is also injective. The result follows. □

Corollary 13.29 (Class field axiom) *We have $H^1(k, C) = 0$. If K is a finite Galois extension of k with Galois group G, then denoting by C_K the idèle class group of K, we have $H^1(G, C_K) = 0$. If furthermore G is cyclic, the group $\widehat{H}^0(G, C_K)$ is of cardinality $[K : k]$.*

Proof The long cohomology exact sequence associated to

$$0 \longrightarrow \overline{k}^* \longrightarrow I \longrightarrow C \longrightarrow 0$$

and Theorem 13.28 along with the fact that $H^1(k, I) = 0$ (Corollary 13.7) yield the equality $H^1(k, C) = 0$. The restriction-inflation exact sequence then implies (using Proposition 13.4) that $H^1(G, C_K) = 0$, and we then conclude that for G cyclic, the cardinality of $\widehat{H}^0(G, C_K)$ is $[K : k]$ by Theorem 13.8.   □

Note that we have proved the triviality of $H^1(k, C)$ and that of $H^1(G, C_K)$ directly without having to deal with the case the case where G is cyclic like we did in the case of a number field.

Remark 13.30 The exact sequence (12.3) shows that $\mathrm{Br}\, k$ is isomorphic to $H^2(\kappa, \mathrm{Div}_0\, \overline{X})$ since we just saw that it is isomorphic to $H^2(\kappa, \overline{\kappa}(X)^*)$, while the groups $\mathrm{Br}\,\kappa$ and $H^3(\kappa, \overline{\kappa}^*)$ are trivial since $\mathrm{cd}(\kappa) \leqslant 1$. As $H^1(\kappa, \mathrm{Pic}\,\overline{X}) = 0$ (Corollary 12.34) and $\mathrm{scd}(\kappa) \leqslant 2$, the exact sequence (12.4) then yields the exact sequence

$$0 \longrightarrow \mathrm{Br}\, k \longrightarrow H^2(\kappa, \mathrm{Div}\overline{X}) \longrightarrow H^2(\kappa, \mathrm{Pic}\,\overline{X}) \longrightarrow 0.$$

By Shapiro's lemma, we have

$$H^2(\kappa, \mathrm{Div}\overline{X}) \simeq \bigoplus_{v \in X^{(1)}} H^2(\kappa(v), \mathbf{Z}) \simeq \bigoplus_{v \in X^{(1)}} H^1(\kappa(v), \mathbf{Q}/\mathbf{Z}),$$

and this last group is also isomorphic to $\bigoplus_{v \in X^{(1)}} \mathrm{Br}\, k_v$ by Theorem 8.9. On the other hand we have $H^i(\kappa, \mathrm{Pic}^0\, \overline{X}) = 0$ for $i \geqslant 2$ since as $\mathrm{Pic}^0\, \overline{X}$ is divisible, we have for any $n > 0$ a surjection from $H^i(\kappa, (\mathrm{Pic}^0\, \overline{X})[n])$ onto $H^i(\kappa, \mathrm{Pic}^0\, \overline{X})[n]$. But $H^i(\kappa, (\mathrm{Pic}^0\, \overline{X})[n]) = 0$ as $\mathrm{cd}(\kappa) \leqslant 1$. Using the exact sequence (12.5), we deduce that $H^2(\kappa, \mathrm{Pic}\,\overline{X}) \simeq H^2(\kappa, \mathbf{Z}) \simeq \mathbf{Q}/\mathbf{Z}$. Finally, we obtain an exact sequence

$$0 \longrightarrow \mathrm{Br}\, k \longrightarrow \bigoplus_{v \in X^{(1)}} \mathrm{Br}\, k_v \longrightarrow \mathbf{Q}/\mathbf{Z} \longrightarrow 0,$$

which is the Brauer–Hasse–Noether sequence that we will encounter a bit later (Theorem 14.11). Here again, we see that this result can be obtained directly in the case of a function field.

Nevertheless, the identification of the maps obtained by this method with those of Theorem 14.11 is not completely obvious.

13.6   Exercises

Exercise 13.1 Let K be a local field. Let n be an integer prime to the characteristic of K, and such that K contains the nth roots of unity. We set $L = K(\sqrt[n]{K^*})$. Show that L is a finite abelian extension of K and that the group of norms $N_{L/K}L^*$ is exactly K^{*n}.

Exercise 13.2 Let k be a global field. Let K be a finite Galois extension of k. Show that we have an injection

$$\mathrm{Br}(K/k) \longrightarrow \bigoplus_{v \in \Omega_k} \mathrm{Br}(K_v/k_v).$$

Exercise 13.3 Let k be a global field. Let K be a finite non-trivial Galois extension of k (not necessarily abelian). Show that there are infinitely many places v of k such that the extension K/k is not totally split at v. Using Exercise 12.2, show that this result (that will be refined in Exercise 18.7) still holds true for K/k which is not Galois.

Exercise 13.4 Let k be a global field. Let F be a finite non-trivial Galois extension of k with group G. For any place v of k, we denote by F_v the completion of F at a place above v. Show that we have an exact sequence

$$k^*/N_{F/k}F^* \longrightarrow \bigoplus_{v \in \Omega_k} k_v^*/N_{F_v/k_v}F_v^* \longrightarrow \widehat{H}^0(G, C_F).$$

Using Exercise 13.3, show that in this exact sequence, the middle term is infinite (for the sequel, see Exercise 15.1).

Chapter 14
Reciprocity Law and the Brauer–Hasse–Noether Theorem

In this chapter, we compute the Brauer group of a global field, using a method which is quite similar to the local case, except that the role played by the unramified extensions will be played by *cyclic* extensions of a special type, called *cyclotomic* (i.e., generated by the roots of unity). A very important step will be to prove the *global reciprocity law* associated to the norm residue symbol (a very special case of which is the classical quadratic reciprocity law). To do this, we will use local calculations coming from the Lubin–Tate construction for $\mathbf{Q}_p$.

In this chapter, we denote by k a global field, by I_k the idèle group of k, and by $C_k = I_k/k^*$ its idèle class group. We will also use the groups

$$ I = I_{\bar{k}} := \varinjlim_K I_K \quad \text{and} \quad C = C_{\bar{k}} := \varinjlim_K C_K = I/\bar{k}^*, $$

the limit taken over all the finite Galois extensions K of k. We denote by Ω_k (resp. Ω_f, $\Omega_{\mathbf{R}}$) the set of places (resp. of finite places, of real places) of k.

14.1 Existence of a Neutralising Cyclic Extension

In this paragraph, we show that for any element α of the Brauer group $\operatorname{Br} k$ of a global field, there exists a finite cyclic extension K/k of k such that the restriction of α to $\operatorname{Br} K$ is zero.

Proposition 14.1 *Let p be a prime number. Let L/k be an infinite Galois extension of k. If $p = 2$ and k is a number field, assume that L is totally imaginary (i.e., does not have any real embeddings). We make the assumption that for every finite place v of k, the extension L_v/k_v is of degree divisible by p^∞ (as defined in Definition 4.5). Then we have*

© Springer Nature Switzerland AG 2020

D. Harari, *Galois Cohomology and Class Field Theory*, Universitext,
https://doi.org/10.1007/978-3-030-43901-9_14

$$H^2(L, I)\{p\} = 0, \quad H^2(\mathrm{Gal}(L/k), I_L)\{p\} \simeq H^2(k, I)\{p\},$$

where $I_L = I^{\mathrm{Gal}(\overline{k}/L)}$ is the inductive limit of the I_K for the finite Galois subextensions K/k of L/k.

Recall that L_v is the field Lk_v, which is also the union of the K_v for K finite extension of k contained in L (with the conventions recalled at the beginning of the Sect. 13.1).

Proof As $H^1(K, I) = 0$ (Corollary 13.2) for any finite extension K of k contained in L, we have $H^1(L, I) = 0$ by passing to the limit (Proposition 4.18). We thus have a restriction–inflation exact sequence

$$0 \to H^2(\mathrm{Gal}(L/k), I_L) \to H^2(k, I) \to H^2(L, I)$$

which reduces us to show that $H^2(L, I)\{p\} = 0$. Corollary 13.7 implies (again by passing to the limit)

$$H^2(L, I)\{p\} = \bigoplus_{v \in \Omega_k} H^2(L_v, \overline{L}_v^*)\{p\}.$$

Suppose first that v is a finite place of k. Then the assumption that p^∞ divides the degree of L_v over k_v gives $H^2(L_v, \overline{L}_v^*)\{p\} = (\mathrm{Br}\, L_v)\{p\} = 0$ by Theorem 8.11, (a). If v is an archimedean place, we again have $H^2(L_v, \overline{L}_v^*)\{p\} = 0$: it is straightforward if $p \neq 2$, and follows from the assumption for $p = 2$. $\qquad\square$

Let $k(\mu)$ be the extension of k obtained by adjoining all the roots of unity to k. If k is the field of functions of a curve over $\mathbf{F}_q$, with $\mathbf{F}_q$ algebraically closed in k (recall that we can always reduce to this situation, by replacing $\mathbf{F}_q$ by its algebraic closure in k), the field $k(\mu)$ is simply obtained by taking the field of functions $\overline{\mathbf{F}_q}(C) = k\overline{\mathbf{F}_q}$. We then set $\widetilde{k} = k(\mu)$, and observe that $\mathrm{Gal}(\widetilde{k}/k)$ is isomorphic to $\widehat{\mathbf{Z}}$. In the case of a number field, denote by (for any p prime) $k(\mu_{p^\infty})$ the field obtained by adjoining to k all roots of unity of order a power of p. The group $G_p := \mathrm{Gal}(k(\mu_{p^\infty})/k)$ is an open subgroup of $\mathrm{Gal}(\mathbf{Q}(\mu_{p^\infty})/\mathbf{Q}) \simeq \mathbf{Z}_p^*$, hence $\Gamma_p := G_p/(G_p)_{\mathrm{tors}}$ is isomorphic (by Theorem 7.18) to the additive group $\mathbf{Z}_p$. Let $\widetilde{k}(p)$ be the subfield of $k(\mu_{p^\infty})$ fixed by $(G_p)_{\mathrm{tors}}$. It is an extension of k with Galois group isomorphic to $\mathbf{Z}_p$.

Definition 14.2 We call $\widetilde{k}(p)$ the $\mathbf{Z}_p$-*cyclotomic extension* of k. The composite $\widetilde{k}$ of all extensions $\widetilde{k}(p)$ for p prime is called the $\widehat{\mathbf{Z}}$-*cyclotomic extension*[1] of k.

We thus have $\mathrm{Gal}(\widetilde{k}/k) \simeq \widehat{\mathbf{Z}}$, which implies that each finite extension K of k contained in $\widetilde{k}$ is cyclic (as finite quotients of $\widehat{\mathbf{Z}}$ are cyclic). In the case of a number field, all archimedean places of k are totally split in $\widetilde{k}$, since $\widehat{\mathbf{Z}}$ is torsion-free.

We also have:

---

[1] Thanks to JoëRiou for pointing out a lack of precision in the initial definition and in the first version of Lemma 14.3.

Lemma 14.3 *Let v be a finite place of k. Then the decomposition group $D_v := G_v(\widetilde{k}/k)$ is isomorphic to $\widehat{\mathbf{Z}}$.*

Proof The group D_v is a closed subgroup of $\widehat{\mathbf{Z}}$. As every abelian profinite group is the direct product of its pro-p-Sylow (cf. Proposition 4.10, d), we obtain that D_v is isomorphic to a group of the form $\prod_p H_p$, where H_p is a closed subgroup of $\mathbf{Z}_p$ and the product is taken over all the primes p. It is then enough to show that each H_p is isomorphic to $\mathbf{Z}_p$, or that each H_p is nonzero. But, by construction of $\widetilde{k}(p)$, there exists a finite extension of $\widetilde{k}(p)$ (and hence also a finite extension $L(p)$ of $\widetilde{k}$) which contains all the roots of unity of order a power of p. If one of the H_p were zero, a completion of $L(p)$ at a place above v would be a finite extension of k_v (as $\widetilde{k}/k$ would be totally split at v), and hence a finite extension of $\mathbf{Q}_\ell$ (where ℓ is the prime number that divides v) would contain all the roots of unity of order a power of p, which is impossible (e.g. by Theorem 7.18). $\qquad\square$

Proposition 14.4 *The group $H^2(k, I)$ is the union of the $H^2(\mathrm{Gal}(K/k), I_K)$ where $K \subset \widetilde{k}$ ranges through the finite cyclic extensions of k. We can furthermore restrict to subextensions K of $\widetilde{k}$ if k has no real places (resp. of $\widetilde{k}_1$, where k_1 is any quadratic totally imaginary extension of k, if k has at least one real place).*

In particular, taking $k_1 = k(\sqrt{-1})$, we see that we can always choose for K a cyclic cyclotomic extension (i.e., generated by the roots of unity) of k. The case of a function field over $\mathbf{F}_q$ is in a certain way more pleasant, since K can then be chosen (like in the local case) everywhere unramified, which simplifies for example the proof of the reciprocity law (Theorem 14.9).

Proof Let K be a finite Galois extension of k with group G. As $H^1(K, I) = 0$ (Corollary 13.2), we have an exact sequence

$$0 \to H^2(G, I_K) \to H^2(k, I) \to H^2(K, I),$$

which allows us to identify $H^2(G, I_K)$ with the subgroup of $H^2(k, I)$ consisting of elements whose restriction to $H^2(K, I)$ is zero. Let $x \in H^2(k, I)$. By Corollary 13.7, we can decompose x as $x = x_a + x_f$ with

$$x_a \in \bigoplus_{v \in \Omega_\mathbf{R}} H^2(k_v, \overline{k}_v^*) \quad \text{and} \quad x_f \in \bigoplus_{v \in \Omega_f} H^2(k_v, \overline{k}_v^*).$$

Proposition 14.1 (or rather its proof) along with the last lemma implies that the restriction of x_f to $H^2(\widetilde{k}, I)$ is zero. Hence we obtain a finite extension $K \subset \widetilde{k}$ such that the restriction of x_f to $H^2(K, I)$ is zero. This finishes the proof in the case where k has no real places since the extension K/k is then automatically cyclic.

In the case where k is a number field with at least one real place and k_1 is a totally imaginary quadratic extension of k, denote by K_1 the cyclic extension of degree $2[K : k]$ of k contained in $\widetilde{k}$. In particular K_1 is a quadratic extension of K, and it is disjoint from k_1 over k (as it is totally split at real places of k while k_1 has no real places). We thus have $K_1 = K(\sqrt{a})$ and $k_1 = k(\sqrt{b})$ with a, b in K. We also have $\mathrm{Gal}(K_1/k) \simeq \mathbf{Z}/2n$ (where $n := [K : k]$) and $\mathrm{Gal}(k_1 K_1/k)$ is isomorphic to $\mathbf{Z}/2n \times \mathbf{Z}/2$. If we set $L = K(\sqrt{ab})$, the extension L of k is cyclic: indeed, its Galois group is isomorphic to the quotient of $\mathbf{Z}/2n \times \mathbf{Z}/2$ by the subgroup of order 2 generated by $(n, 1)$. The number field L is totally imaginary because for any real place v of K, we have $a_v > 0$ and $b_v < 0$ since K_1 is totally split at real places of k and k_1 is totally imaginary. This implies that the restriction of x to $H^2(L, I)$ is zero. $\square$

We deduce the first step in the computation of $\mathrm{Br}\, k$:

Proposition 14.5 *The Brauer group* $\mathrm{Br}\, k$ *is the union of the* $\mathrm{Br}(K/k)$ *for finite cyclic extensions* K/k *(we can even restrict ourselves to K as in Proposition 14.4).*

Proof Let K be a finite Galois extension of k with group G. As the groups $H^1(K, C)$, $H^1(k, C)$, $H^1(G, C_K)$ are zero and so are $H^1(K, \bar{k}^*)$ and $H^1(K, I)$, we have a commutative diagram with exact rows and columns:

$$
\begin{array}{ccc}
0 \longrightarrow \mathrm{Br}\, K \longrightarrow & H^2(K, I) \\
\uparrow & \uparrow \\
0 \longrightarrow \mathrm{Br}\, k \longrightarrow & H^2(k, I) \\
\uparrow & \uparrow \\
0 \rightarrow H^2(G, K^*) \rightarrow & H^2(G, I_K) \\
\uparrow & \uparrow \\
0 & 0
\end{array}
$$

We conclude with Proposition 14.4. $\square$

14.2   The Global Invariant and the Norm Residue Symbol

In this section, we will construct analogues of the local invariant and of the local reciprocity map in the global case.

Definition 14.6 Let K/k be a finite Galois extension with group G. For any place v of k, we denote by $G_v = \mathrm{Gal}(K_v/k_v)$ the decomposition subgroup associated to v. We define

$$
\mathrm{inv}_{K/k} : H^2(G, I_K) \rightarrow \frac{1}{[K : k]}\mathbf{Z}/\mathbf{Z} \subset \mathbf{Q}/\mathbf{Z}
$$

by the formula:

$$
\mathrm{inv}_{K/k}(c) = \sum_{v \in \Omega_k} \mathrm{inv}_{K_v/k_v}(c_v),
$$

where c_v is the component at v of $c \in H^2(G, I_K) = \bigoplus_{v \in \Omega_k} H^2(G_v, K_v^*)$.

Here inv_{K_v/k_v} is the local invariant induced by $\mathrm{inv}_v : \mathrm{Br}\, k_v \rightarrow \mathbf{Q}/\mathbf{Z}$ (for a real v it is simply the isomorphism between $\mathrm{Br}\, \mathbf{R}$ and $\mathbf{Z}/2$, and for a complex v it is the zero map). On the other hand for any place v of k, we have the reciprocity map $(., K_v/k_v) : k_v^* \to G_v^{ab}$ and its composite (still denoted by $(., K_v/k_v)$) with the natural map $G_v^{ab} \to G^{ab}$ (if v is archimedean we take for $(\cdot, K_v/k_v)$ the map induced by the surjective homomorphism from k_v^*/k_v^{*2} to G_v^{ab}). We then define the *norm residue symbol* associated to the extension K/k by

$$(., K/k) : I_k \to G^{ab}, \qquad (\alpha, K/k) = \prod_{v \in \Omega_k} (\alpha_v, K_v/k_v). \qquad (14.1)$$

The product is well defined since if v is unramified in K/k and $\alpha_v \in U_v = \mathcal{O}_v^*$, then $(\alpha_v, K_v/k_v) = 1$ (indeed, all the elements of U_v are norms of K_v/k_v by Proposition 8.3).

Proposition 9.3 easily implies the following:

Proposition 14.7 *Let K/k be a finite Galois extension with group G. Then for any $\chi \in H^1(G, \mathbf{Q}/\mathbf{Z})$ and for any $\alpha \in I_k$, we have*

$$\chi((\alpha, K/k)) = \mathrm{inv}_{K/k}(\alpha \cup \chi),$$

where, for the purposes of taking the cup-product, α is seen as an element of $H^0(G, I_K)$ and χ as an element of $H^2(G, \mathbf{Z})$.

We deduce that if M is a finite Galois extension (that we can assume to be abelian) of k containing K, there is a compatibility between $(., K/k)$ and $(., M/k)$ analogous to that of the local case (Corollary 9.4).

The properties of the local invariant (Theorem 8.9) also imply:

Proposition 14.8 *Let L be a finite Galois extension of k. Let K be a subextension. Then we have the commutative diagrams:*

$$
\begin{array}{ccc}
H^2(\mathrm{Gal}(L/K), I_L) & \xrightarrow{\ \mathrm{inv}_{L/K}\ } & \frac{1}{[L:K]}\mathbf{Z}/\mathbf{Z} \\
\mathrm{Res} \uparrow & & \uparrow \cdot[K:k] \\
H^2(\mathrm{Gal}(L/k), I_L) & \xrightarrow{\ \mathrm{inv}_{L/k}\ } & \frac{1}{[L:k]}\mathbf{Z}/\mathbf{Z}
\end{array}
$$

and

$$
\begin{array}{ccc}
H^2(\mathrm{Gal}(L/K), I_L) & \xrightarrow{\ \mathrm{inv}_{L/K}\ } & \frac{1}{[L:K]}\mathbf{Z}/\mathbf{Z} \\
\mathrm{Cores} \downarrow & & \downarrow j \\
H^2(\mathrm{Gal}(L/k), I_L) & \xrightarrow{\ \mathrm{inv}_{L/k}\ } & \frac{1}{[L:k]}\mathbf{Z}/\mathbf{Z}
\end{array}
$$

where the map

$$j : \frac{1}{[L:K]}\mathbf{Z}/\mathbf{Z} \to \frac{1}{[L:k]}\mathbf{Z}/\mathbf{Z}$$

is the inclusion. If furthermore K/k is Galois, then $\mathrm{inv}_{L/k}$ *extends* $\mathrm{inv}_{K/k}$ *via the inclusion* $H^2(\mathrm{Gal}(K/k), I_K) \to H^2(\mathrm{Gal}(L/k), I_L)$.

We also have, by passing to the limit over the finite Galois extensions K of k, a map

$$\mathrm{inv} : H^2(k, I) \to \mathbf{Q}/\mathbf{Z}.$$

An essential property for all this theory is the following theorem:

Theorem 14.9 (global reciprocity law) *Let K be a finite extension of k with abelian Galois group G. Then:*

(a) *the map* $\mathrm{inv}_{K/k} : H^2(G, I_K) \to \mathbf{Q}/\mathbf{Z}$ *is zero on the image of $H^2(G, K^*)$ in $H^2(G, I_K)$.*

(b) *Let a be a principal idèle. Then we have:* $(a, K/k) = 1$.

Proof Note first that (a) implies (b) by Proposition 14.7 and the fact that an element of the abelian group G is trivial if and only if its image by every character of G is trivial. To prove (a), it is enough to show that the map $\mathrm{inv} : H^2(k, I) \to \mathbf{Q}/\mathbf{Z}$ is trivial on the image of $\mathrm{Br}\, k = H^2(k, \overline{k}^*)$ (by the last assertion of Proposition 14.8).

Let $\alpha \in \mathrm{Br}\, k$. By Proposition 14.5, in the case of a function field, we can assume that $\alpha \in \mathrm{Br}(K/k)$ with $K = k(\zeta_n)$, where n is prime to the characteristic of k. In the case of a number field, we first reduce ourselves to $k = \mathbf{Q}$ by observing that if $\alpha \in \mathrm{Br}(L/k)$ with L finite and Galois over k, then $\mathrm{inv}(\alpha) = \mathrm{inv}_{L/\mathbf{Q}}(\mathrm{Cores}\,\alpha)$, where Cores is the corestriction from $\mathrm{Br}\, k$ to $\mathrm{Br}\, \mathbf{Q}$ (this follows from Proposition 14.8). Proposition 14.5 allows to further assume that $\alpha \in \mathrm{Br}(K/\mathbf{Q})$, where K is a cyclic subextension of $\mathbf{Q}(\zeta_n)/\mathbf{Q}$ for a certain n.

Let then $G = \mathrm{Gal}(K/k)$ and let χ be a generator of the group $H^1(G, \mathbf{Q}/\mathbf{Z})$. The cup-product with $\delta_\chi \in H^2(G, \mathbf{Z})$ is an isomorphism from $\widehat{H}^0(G, K^*)$ to $H^2(G, K^*)$ (Remark 2.29). Thus we can write every element of $H^2(G, K^*)$ in the form $\overline{a} \cup \delta_\chi$ with $a \in k^*$. By Proposition 14.7, we are reduced to prove the assertion (b) of the theorem in two special cases:

(i) k function field over $\mathbf{F}_q$ and $K = k(\zeta_n)$ with n prime to the characteristic of k;

(ii) $k = \mathbf{Q}$ and $K = k(\zeta)$, where ζ is a root of unity (recall that if E is a subextension of $k(\zeta)$, then $(a, E/k)$ is the image of $(a, k(\zeta)/k)$ in the quotient $\mathrm{Gal}(E/k)$ of $\mathrm{Gal}(k(\zeta)/k)$).

The case (i) is easier: indeed, for any place v of k, the extension K_v/k_v is unramified and Lemma 9.8(a) applies. We obtain that $(a, K_v/k_v)$ is $F(v)^{v(a)}$, where $F(v)$ is the Frobenius at v (which acts on ζ_n by raising to the power $n(v)$, where $n(v)$ is the cardinality of the residue field at v). We thus have

$$(a, K_v/k_v) \cdot \zeta_n = \zeta_n^{n(v)^{v(a)}}$$

hence

$$(a, K/k) \cdot \zeta_n = \zeta_n^{\prod_{v \in \Omega_k} n(v)^{v(a)}} = \zeta_n$$

by the product formula since $n(v)^{v(a)} = |a|_v^{-1}$, which concludes the proof in this case.

In the case (ii), we will use explicit local computations coming from the Lubin–Tate theory. Let $\theta_p : \mathbf{Q}_p^* \to \mathrm{Gal}(K_p/\mathbf{Q}_p)$ be the local reciprocity map defined for p prime or $p = \infty$ (with the convention that $\mathbf{Q}_\infty = \mathbf{R}$). We already have that

$$\theta_\infty(x) \cdot \zeta = \zeta^{\varepsilon(x)},$$

where $\varepsilon(x)$ is the sign of x. For p prime, let $x = p^m u$ with $u \in \mathbf{Z}_p^*$ and $m \in \mathbf{Z}$. Corollary 11.24 implies:

$$\theta_p(x) \cdot \zeta = \zeta^{p^m}, \quad \forall \zeta \in U_p'$$

$$\theta_p(x) \cdot \zeta = \zeta^{u^{-1}}, \quad \forall \zeta \in U_{p^\infty},$$

where U_p' (resp. U_{p^∞}) is the set of roots of unity of order prime to p (resp. of order a power of p). To prove the formula $\prod_p \theta_p(a) = 1$ ($a \in \mathbf{Q}^*$) in the group $\mathrm{Gal}(K/k)$ (where the product corresponds to the composition of automorphisms), it is enough to deal with the case $a = -1$ and the case $a = q$ with q prime, and look at the action of $\prod_p \theta_p(a) = 1$ on a root of unity $\zeta \in U_{\ell^\infty}$ with ℓ prime. This follows from the formulas (for p prime or $p = \infty$):

$$\theta_\infty(-1) \cdot \zeta = \zeta^{-1}; \quad \theta_\ell(-1) \cdot \zeta = \zeta^{-1}; \quad \theta_p(-1) \cdot \zeta = \zeta, \ \forall p \neq \ell, \infty.$$

For $q \neq \ell$:

$$\theta_p(q) \cdot \zeta = \zeta, \ \forall p \neq \ell, q; \quad \theta_\ell(q) \cdot \zeta = \zeta^{q^{-1}}; \quad \theta_q(q) \cdot \zeta = \zeta^q.$$

and lastly:

$$\theta_p(\ell) \cdot \zeta = \zeta; \quad \forall p. \qquad \qquad \Box$$

Corollary 14.10 *Let K be a finite cyclic extension of k with group G. Then we have the exact sequence*

$$0 \to H^2(G, K^*) \to H^2(G, I_K) \xrightarrow{\mathrm{inv}_{K/k}} \frac{1}{[K:k]}\mathbf{Z}/\mathbf{Z} \to 0. \qquad (14.2)$$

Proof Let us first show the surjectivity of $\mathrm{inv}_{K/k}$ in the case where $[K:k] = p^m$ with p prime. By Corollary 13.13, there exists a finite place v of k which is inert in K. As we then have $[K_v : k_v] = [K:k]$, the fact that the local invariant inv_{K_v/k_v} is an isomorphism from $H^2(G_v, K_v^*)$ onto $\frac{1}{[K_v:k_v]}\mathbf{Z}/\mathbf{Z}$ implies that $\mathrm{inv}_{K/k}$ is surjective onto $\frac{1}{[K:k]}\mathbf{Z}/\mathbf{Z}$. We deduce immediately the same surjectivity for any cyclic G by Proposition 14.8.

Now, the equalities $H^1(G, C_K) = 0$ and $H^3(G, K^*) = H^1(G, K^*) = 0$ (recall that G is cyclic) yield the exact sequence

$$0 \to H^2(G, K^*) \to H^2(G, I_K) \to H^2(G, C_K) \to 0.$$

Thus the cokernel of the map $H^2(G, K^*) \to H^2(G, I_K)$ has the same cardinality as $H^2(G, C_K) = \widehat{H}^0(G, C_K)$, which is $[K : k]$ by the class field axiom. On the other hand, Theorem 14.9 implies that the sequence (14.2) is a complex and we have furthermore seen that its last arrow is surjective. We finally deduce that this sequence is exact.   □

We deduce the main theorem of this chapter:

Theorem 14.11 (Brauer–Hasse–Noether) *Let k be a global field. Then we have the exact sequence*

$$0 \to \mathrm{Br}\, k \to \bigoplus_{v \in \Omega_k} \mathrm{Br}\, k_v \xrightarrow{\ \mathrm{inv}_k\ } \mathbf{Q}/\mathbf{Z} \to 0,$$

where the map inv_k is defined as the sum of local invariants $\mathrm{inv}_v : \mathrm{Br}\, k_v \to \mathbf{Q}/\mathbf{Z}$.

Proof We take the limit over the cyclic extensions K of k in Corollary 14.10. Propositions 14.4 and 14.5 tell us that we obtain an exact sequence

$$0 \to \mathrm{Br}\, k \to H^2(k, I) \to \mathbf{Q}/\mathbf{Z} \to 0.$$

The map $H^2(k, I) \to \mathbf{Q}/\mathbf{Z}$ is obtained by identifying $H^2(k, I)$ with $\bigoplus_{v \in \Omega_k} \mathrm{Br}\, k_v$ (cf. Proposition 13.1), and then taking the sum of local invariants (by Definition 14.6, which constructs $\mathrm{inv}_{K/k}$ when K is a finite Galois extension of k). The result follows.   □

Remark 14.12 Let $n > 0$. If k contains a primitive nth root of unity, then for $a, b \in k^*$, the sum (over all the places v of k) of the local symbols (cf. Definition 9.10) $(a, b)_v \in \mathrm{Br}\, k_v \hookrightarrow \mathbf{Q}/\mathbf{Z}$ is zero.

Let us take $n = 2$ and $k = \mathbf{Q}$. If p and q are two distinct odd primes, we can compute local Hilbert symbols $(p, q)_v$ for any place v of $\mathbf{Q}$ by the formulas of the Example 9.11. For any odd integer x, let $\varepsilon(x)$ be the class of $(x - 1)/2$ modulo 2. We obtain, using multiplicative notation (i.e., considering local Hilbert symbols $(p, q)_v$ as taking values in $\{\pm 1\}$):

$$(p, q)_2 = (-1)^{\varepsilon(p)\varepsilon(q)}; \quad (p, q)_\infty = 1; \quad (p, q)_p = \left(\frac{q}{p}\right); \quad (p, q)_q = \left(\frac{p}{q}\right),$$

and $(p, q)_\ell = 1$ for primes numbers $\ell \notin \{p, q\}$. The global reciprocity law then yields the classical *quadratic reciprocity law*:

$$\left(\frac{p}{q}\right) = \left(\frac{q}{p}\right) \cdot (-1)^{\varepsilon(p)\varepsilon(q)} = \left(\frac{q}{p}\right) \cdot (-1)^{(p-1)(q-1)/4}.$$

Remark 14.13 There exists a geometric interpretation of the Brauer group of a field k: it classifies the *Severi–Brauer varieties*, that is, algebraic k-varieties that become isomorphic to the projective space over the separable closure $\bar{k}$. In particular, over a field of characteristic different from 2, the Hilbert symbol (a, b) corresponds to the projective conic with equation $x^2 - ay^2 - bz^2 = 0$. Saying that this Hilbert symbol is trivial is equivalent to saying that the conic has a k-point (or that it is isomorphic over k to the projective line).

When k is a number field, the injectivity of the map $\mathrm{Br}\, k \rightarrow \bigoplus_{v \in \Omega_k} \mathrm{Br}\, k_v$ from Brauer–Hasse–Noether theorem means that a projective conic which has a k_v-point for every place v has in fact a k-point. This result (originally due to Legendre) extends to a projective quadric of arbitrary dimension: it's the *Hasse–Minkowski theorem*, whose complete proof over $\mathbf{Q}$ can be found in [46], Chap. IV.

Corollary 14.14 *Let p be a prime number. Let L be an algebraic separable field extension of k, assumed totally imaginary if k is a number field and $p = 2$. Assume that p^∞ divides $[L_v : k_v]$ for every finite place v of k. Then $\mathrm{cd}_p(L) \leq 1$.*

Recall that $L_v := Lk_v$. Of course, this corollary does not apply to finite extensions of k.

Proof It is enough to check that $(\mathrm{Br}\, L')\{p\} = 0$ for any separable algebraic extension L' of L by Theorem 6.17. As L' satisfies the same assumptions as L, we are reduced to proving that $(\mathrm{Br}\, L)\{p\} = 0$. But $\mathrm{Br}\, L$ injects into the direct sum (for v a place of k) of the $\mathrm{Br}\, L_v$ (pass to the limit in Brauer–Hasse–Noether theorem). We conclude using Theorem 8.11. $\qquad\square$

14.3 Exercises

Exercise 14.1 Let k be a number field. Let p be a prime number. Assume that $p \neq 2$ or that k is totally imaginary. Show that the cohomological dimension $\mathrm{cd}_p(k)$ is 2 (use Corollary 14.14 for a well chosen extension of k). State and prove an analogue for global fields of positive characteristic.

Exercise 14.2 Let k be a global field. Let S be a set of places of k. Give a necessary and sufficient condition on S for the diagonal map

$$\mathrm{Br}\, k \rightarrow \bigoplus_{v \in S} \mathrm{Br}\, k_v$$

to be surjective. Same question replacing "surjective" by "injective".

Chapter 15
The Abelianised Absolute Galois Group of a Global Field

We keep the notations of the previous chapter. In particular k is a global field whose idèle group is denoted by I_k and idèle class group is denoted by C_k. We denote by $G_k = \mathrm{Gal}(\bar{k}/k)$ the absolute Galois group of k. We also denote by $I = I_{\bar{k}}$ the inductive limit of the I_K for K finite Galois extensions of k, and $C = I/\bar{k}^*$ that of the C_K.

15.1 Reciprocity Map and the Idèle Class Group

In the last chapter we have defined a map $\mathrm{inv}_k : H^2(k, I) \to \mathbf{Q}/\mathbf{Z}$ by passing to the limit over the maps $\mathrm{inv}_{K/k} : H^2(\mathrm{Gal}(K/k), I_K) \to \mathbf{Q}/\mathbf{Z}$. We would like to deduce from it analogous maps on $H^2(k, C)$. One difficulty is that in general the group $H^2(\mathrm{Gal}(K/k), I_K)$ does not surject onto $H^2(\mathrm{Gal}(K/k), C_K)$, but we will see that this problem disappears when we pass to the limit.

Lemma 15.1 *Let K be a finite Galois extension of k with group G. Then the cardinality of $H^2(G, C_K)$ divides $[K : k]$.*

Proof This is very analogous to the corresponding local statement (Lemma 8.7). The case of a cyclic extension is part of the class field axiom (Theorem 13.23). We prove the statement in the case where G is a p-group by induction on $[K : k]$, using the exact sequence (which is valid by Corollary 1.45, because $H^1(\mathrm{Gal}(K/E), C_K) = 0$ for any Galois subextension E of k):

$$0 \longrightarrow H^2(\mathrm{Gal}(E/k), C_E) \longrightarrow H^2(G, C_K) \longrightarrow H^2(\mathrm{Gal}(K/E), C_K).$$

The general case is dealt with by considering (for p dividing $\#G$) a p-Sylow $G_p = \mathrm{Gal}(K/K_p)$ of G and using the injectivity of the restriction map

$$H^2(G, C_K) \longrightarrow \bigoplus_p H^2(\mathrm{Gal}(K/K_p), C_K).$$

$\square$

© Springer Nature Switzerland AG 2020
D. Harari, *Galois Cohomology and Class Field Theory*, Universitext,
https://doi.org/10.1007/978-3-030-43901-9_15

Let now $\widetilde{k}$ be the $\widehat{\mathbf{Z}}$-cyclotomic extension of k (Definition 14.2). Let us set $\widetilde{G} = \mathrm{Gal}(\widetilde{k}/k)$. As scd $\widehat{\mathbf{Z}} = 2$ (Example 5.12), we have $H^3(\widetilde{G}, \widetilde{k}^*) = 0$, hence we have an exact sequence

$$0 \longrightarrow H^2(\widetilde{G}, \widetilde{k}^*) \longrightarrow H^2(\widetilde{G}, I_{\widetilde{k}}) \longrightarrow H^2(\widetilde{G}, C_{\widetilde{k}}) \longrightarrow 0.$$

But on the other hand by passing to the limit in the Corollary 14.10, we also have an exact sequence

$$0 \longrightarrow H^2(\widetilde{G}, \widetilde{k}^*) \longrightarrow H^2(\widetilde{G}, I_{\widetilde{k}}) \longrightarrow \mathbf{Q}/\mathbf{Z} \longrightarrow 0,$$

where the last map is induced by inv_k. Recall that if $K \subset \widetilde{k}$ is a finite extension of k, the map $H^2(\mathrm{Gal}(K/k), C_K) \to H^2(k, C)$ is injective by Corollary 1.45 as $H^1(K, C) = 0$. We obtain an isomorphism

$$\mathrm{inv}_{\widetilde{k}/k} : H^2(\widetilde{G}, C_{\widetilde{k}}) \simeq \mathbf{Q}/\mathbf{Z}.$$

The next proposition is a global analogue of the local Theorem 8.6.

Proposition 15.2 *The sequence*

$$0 \longrightarrow \mathrm{Br}\, k \longrightarrow H^2(k, I) \longrightarrow H^2(k, C) \longrightarrow 0$$

is exact.

Proof Let K be a finite Galois extension of k of degree n, with Galois group $G := \mathrm{Gal}(K/k)$. Let $\widetilde{K}$ be the $\widehat{\mathbf{Z}}$-cyclotomic extension of K. Let k_n be the unique subextension of $\widetilde{k}$ of degree n over k. We will prove exactly as in Lemma 8.8 that the subgroups $H^2(G, C_K)$ and $H^2(\mathrm{Gal}(k_n/k), C_{k_n})$ of $H^2(k, C)$ coincide. The class field axiom implies that the cardinality of $H^2(\mathrm{Gal}(k_n/k), C_{k_n})$ (which is also this of $\widehat{H}^0(\mathrm{Gal}(k_n/k), C_{k_n})$ as k_n/k is cyclic) is n, and Lemma 15.1 then says that the cardinality of $H^2(G, C_K)$ divides $n = \#H^2(\mathrm{Gal}(k_n/k), C_{k_n})$. We observe that $H^2(\widetilde{G}, C_{\widetilde{k}})$ is a subgroup of $H^2(\mathrm{Gal}(\widetilde{K}/k), C_{\widetilde{K}})$ by Theorem 13.25 and Corollary 1.45. Similarly, we have inclusions

$$H^2(\mathrm{Gal}(k_n/k), C_{k_n}) \hookrightarrow H^2(\widetilde{G}, C_{\widetilde{k}}) \hookrightarrow H^2(k, C).$$

Using Proposition 14.8, we obtain a commutative diagram whose first line is exact:

$$0 \longrightarrow H^2(G, C_K) \longrightarrow H^2(k, C) \xrightarrow{\text{Res}} H^2(K, C)$$

$$H^2(\text{Gal}(k_n/k), C_{k_n}) \longrightarrow H^2(\widetilde{G}, C_{\widetilde{k}}) \xrightarrow{\text{Res}} H^2(\text{Gal}(\widetilde{K}/K, C_{\widetilde{K}})$$

$$\text{inv}_{\widetilde{K}/k} \Big\downarrow \qquad\qquad \Big\downarrow \text{inv}_{\widetilde{K}/K}$$

$$\mathbf{Q}/\mathbf{Z} \xrightarrow{\quad \cdot n \quad} \mathbf{Q}/\mathbf{Z}.$$

As $H^2(\text{Gal}(k_n/k), C_{k_n})$ is of cardinality n and $\text{inv}_{\widetilde{K}/K}$ is an isomorphism, the second row of the diagram is a complex. We deduce that $H^2(\text{Gal}(k_n/k), C_{k_n})$ is included in $H^2(G, C_K)$ and as we already know that the cardinality of $H^2(G, C_K)$ divides that of $H^2(\text{Gal}(k_n/k), C_{k_n})$, we finally obtain $H^2(G, C_K) = H^2(\text{Gal}(k_n/k), C_{k_n})$, as desired.

This implies that $H^2(k, C)$ is the union of the $H^2(\text{Gal}(k_n/k), C_{k_n})$ and, in particular, the union of the $H^2(\text{Gal}(K/k), C_K)$ for K a cyclic extension of k. Besides, we have an analogous statement for K^* and I_K (Propositions 14.4 and 14.5). We obtain the desired result by taking the limit of the exact sequence

$$0 \longrightarrow \text{Br}(K/k) \longrightarrow H^2(G, I_K) \longrightarrow H^2(G, C_K) \longrightarrow 0$$

which is valid for any cyclic extension K of k. □

Corollary 15.3 *The map* $\text{inv} : H^2(k, I) \to \mathbf{Q}/\mathbf{Z}$ *induces an isomorphism* $\text{inv}_k : H^2(k, C) \to \mathbf{Q}/\mathbf{Z}$.

Proof This follows immediately from Brauer–Hasse–Noether theorem, combined with the last proposition. □

Let now K/k be a finite Galois extension with group G. As the restriction $H^2(k, C) \to H^2(K, C)$ corresponds to multiplication by $[K : k]$ in $\mathbf{Q}/\mathbf{Z}$ by Proposition 14.8, we obtain an isomorphism

$$\text{inv}_{K/k} : H^2(G, C_K) \longrightarrow \frac{1}{[K : k]} \mathbf{Z}/\mathbf{Z}.$$

As in the local case, we call the *fundamental class* of $H^2(G, C_K)$ the element which is sent to $1/[K : k]$ by $\text{inv}_{K/k}$.

The above properties along with the fact that $H^1(G, C_K) = 0$, will in Chap. 16 imply that the isomorphisms $\text{inv}_K : H^2(G_K, C) \to \mathbf{Q}/\mathbf{Z}$ form a *class formation* on the G_k-module C. The next theorem is by the way a special case of a general result on class formation (Proposition 16.6) and it is analogous to the one we saw in the local case (Theorem 9.2).

Theorem 15.4 *Let K be a finite Galois extension of k with group G. The cup-product with the fundamental class $u_{K/k}$ of $H^2(G, C_K)$ induces an isomorphism $G^{\text{ab}} \to C_k/N_{K/k} C_K$. The reciprocal isomorphism*

$$\omega_{K/k} = (., K/k) : C_k/N_{K/k}C_K \longrightarrow G^{ab}$$

is obtained by passing to the quotient from the norm residue symbol $(., K/k) : I_k \to$
G^{ab}. *We call it the* reciprocity isomorphism.

Proof The fact that the cup-product wit $u_{K/k}$ induces an isomorphism between
$G^{ab} \simeq H^{-2}(G, \mathbf{Z})$ and $C_k/N_{K/k}C_K \simeq \widehat{H}^0(G, C_K)$ is an immediate consequence of
the Tate–Nakayama theorem (Theorem 3.14). The second assertion follows from the
definition of the local reciprocity maps and that of the norm residue symbol given
by the formula (14.1).   □

Remark 15.5 Let K be a finite abelian extension of k. Let S be a finite set of places
of k, containing the archimedean places and the ramified places in the extension K/k.
Let $\mathcal{I}_k$ be the group of fractional ideals of $\mathcal{O}_k$ and $\mathcal{I}_k^S$ the subgroup of $\mathcal{I}_k$ generated by
the prime ideals of $\mathcal{O}_k$ which are not in S. In the traditional formulation of the class
field theory in terms of ideals (cf. for example [26], Sect. IV.8 of [40], or Sect. B.5
of the Appendix B of this book), the *Artin map* is defined by

$$\psi_{K/k} : \mathcal{I}_k^S \longrightarrow \mathrm{Gal}(K/k), \quad \prod_{i=1}^m \wp_i^{e_i} \mapsto \prod_{i=1}^m \sigma_{\wp_i}^{e_i},$$

where for each prime ideal $\wp_i$ not in S, we denote by $\sigma_{\wp_i} \in \mathrm{Gal}(K/k)$ the Frobenius
at $\wp_i$ (which is well defined as the extension K/k is abelian and unramified at $\wp_i$).
The disadvantage of this definition is that it does not extend to the whole of $\mathcal{I}_k$,
and therefore one has to work hard to show that $\psi_{K/k}$ is surjective (this can be
deduced easily from the analytic statement of the Appendix B, Proposition B.20)
and determine its kernel (which is more difficult). This is one of the reasons why
the cohomological definition of Theorem 15.4 and the idelic approach (rather than
in terms of ideals) is more convenient. Note that by formula (14.1) and Lemma 9.8,
(a), the map $\psi_{K/k}$ allows to recover the reciprocity map $\omega_{K/k}$ on the idèles whose
component at each place of S is 1, by sending such an idèle (x_v) to $\prod_{v \notin S} v^{\mathrm{val}(x_v)} \in$
$\mathcal{I}_k^S$, where $\mathrm{val}(x_v)$ designates the valuation of $x_v \in k_v^*$ at the place v (this place
corresponds to a prime ideal of $\mathcal{O}_k$ not in S).

The compatibility of the norm residue symbols (which follows from Proposition 14.7) allows, by passing to the limit to define a *reciprocity homomorphism*

$$\mathrm{rec}_k = \mathrm{rec} : C_k \longrightarrow G_k^{ab}$$

which is continuous. Its image is dense and its kernel $\bigcap_K N_{K/k}C_K$ (the intersection
taken over all the finite Galois extensions or all finite abelian extensions; we will
see in Lemma 15.8(c) that this is also the intersection of the $N_{K/k}C_K$ over finite
separable extensions K) is the *universal norm group*. Theorem 15.4 immediately
implies:

Corollary 15.6 *For any finite abelian extension of k with group G, the cardinality
of* $\widehat{H}^0(G, C_K)$ *is* $[K : k]$.

Thus the statement of the axiom of global class field theory generalises to each abelian extension (not necessarily cyclic). We also deduce Lemma 15.7, which will be useful in the next paragraph.

Lemma 15.7 *With the assumptions and notation of Theorem 13.21, the equality*

$$C_k(S, T) = N_{K/k}C_K$$

is valid without the assumption that K/k is cyclic.

Indeed, the theorem says that $[C_k : C_k(S, T)] = [K : k]$ and $C_k(S, T) \subset N_{K/k}C_K$, but we know that $[C_k : N_{K/k}C_K] = [K : k]$ as soon as K is a finite abelian extension of k.

15.2 The Global Existence Theorem

The aim of this paragraph is to state the global version of the local Theorem 11.20, and to prove it in the case of a number field (in the case of characteristic $p > 0$, we will see that there is a difficulty related to subgroups of C_k of index divisible by p). If k is a global field, we define a *group of norms* (or *norm group*) in C_k as a subgroup of C_k of the form $N_{K/k}C_K$, where K is a finite abelian extension of k. We start with a lemma which collects the results analogous to the local case. These results will be extended later to class formations (Propositions 16.31 and 16.29).

Lemma 15.8 *(a) Each subgroup of C_k containing a norm group is a norm group.*
(b) The intersection of two norm groups is a norm group.
(c) If K is a finite separable extension of k (not necessarily Galois), then $N_{K/k}C_K$ is a norm group.

Proof (a) The argument is exactly the same as in the local case (Lemma 9.9, a).
(b) (cf. also Exercise 11.1 for the local case). If K and L are two finite abelian extensions of k and $E = KL$ is the composite field, then we have by transitivity of the norms:

$$N_{K/k}C_K \cap N_{L/k}C_L \supset N_{E/k}C_E,$$

which allows to conclude using (a). The above inclusion is in fact an equality: indeed, the two members correspond to the kernel of the reciprocity map $\omega_{E/k}$ by compatibility of the reciprocity maps (which comes from Proposition 14.7) and the triviality of the kernel of $\mathrm{Gal}(E/k) \to \mathrm{Gal}(K/k) \times \mathrm{Gal}(L/k)$.
(c) Once we know Theorem 15.4, the argument is identical to the one in the local case, which is given in Proposition 9.7, (a) and (b). □

Recall also that we denote by $C_k^0 = I_k^0/k^*$ the kernel of the homomorphism

$$| \cdot | : C_k \longrightarrow \mathbf{R}_+^*, \quad (\alpha_v) \longmapsto \prod_{v \in \Omega_k} |\alpha_v|_v.$$

It is a compact subgroup of C_k (Theorem 12.21).

Theorem 15.9 (global existence theorem) *Let k be a global field. Then the open subgroups of finite index of C_k are exactly the norm subgroups, i.e., subgroups of the form $N_{K/k} C_K$, where K is a finite abelian extension of k. Furthermore, every norm subgroup N is associated to a unique finite abelian extension $K \subset \overline{k}$ of k, that we call the* class field *of N.*

Proof Let K be a finite abelian extension of k. Then $N_{K/k} C_K$ is of finite index in C_k by Theorem 15.4. Let us show that $N_{K/k} C_K$ is a closed subgroup of C_k. The map $N_{K/k}$ is continuous on C_K, which means that the image of the compact subgroup C_K^0 is compact. Besides, if k is a number field, the topological group C_k is isomorphic to $C_k^0 \times \mathbf{R}_+^*$: more precisely we obtain a subgroup Γ of representatives of C_k / C_k^0 isomorphic to $\mathbf{R}_+^*$ by considering an archimedean place v_0 of k, then by defining Γ as the image in C_k of the idèles of the form $(x, 1, 1, \ldots, 1, \ldots)$, with x in $\mathbf{R}_+^*$, where the first component is that at v_0. The image of Γ in C_K is then also a subgroup of representatives of C_K / C_K^0 and we have

$$N_{K/k} C_K = N_{K/k} C_K^0 \times N_{K/k} \Gamma = N_{K/k} C_K^0 \times \Gamma^n = N_{K/k} C_K^0 \times \Gamma$$

with $n := [K : k]$, since Γ is divisible. As $N_{K/k} C_K^0$ is compact, $N_{K/k} C_K^0 \times \Gamma$ is closed in $C_k = C_k^0 \times \Gamma$. The argument is analogous in the case where k is the function field of a smooth projective curve X over a finite field, by choosing a closed point v_0 of X and replacing Γ by the subgroup $d\mathbf{Z}$ of $\mathbf{Z}$, where d is the degree of v_0.

On the other hand if N is a norm group, the uniqueness of the extension K such that $N = N_{K/k} C_K$ is proved exactly as in the local case (cf. Lemma 9.9, (b), which will be generalised to class formations in Proposition 16.31). Let N be an open finite index subgroup of C_k. We wish to show that N is a subgroup of norms. We distinguish two cases.

(a) The case of a number field, or of a function field when the index $[C_k : N]$ is not divisible by the characteristic of k.

It remains, by Lemma 15.8 (a), to show that N contains a norm group. We first reduce to the case where n is a power of a prime number ℓ (different from Car k), by factoring n as $n = \prod_i p_i^{m_i}$ (with p_i prime), and then by observing that if $N_i \subset C_k$ is the subgroup of index $p_i^{n_i}$ containing N, then N is the intersections of a (finite) family of the N_i (and each N_i contains N, hence is open). Hence by Lemma 15.8(b), it is enough to know that each N_i is a norm group.

Let $J \subset I_k$ be the preimage of N. It is an open subgroup of I_k. This implies that J contains a subgroup of the form

$$U_k^S := \prod_{v \in S} \{1\} \times \prod_{v \notin S} U_v,$$

where S is a finite set of places of k, containing the archimedean places. We can furthermore assume that S contains the places dividing ℓ and (by Proposition 12.25) that $I_k = I_{k,S}k^*$. Besides, as J is of index n in I_k, it also contains the group $\prod_{v \in S} k_v^{*n} \times \prod_{v \notin S}\{1\}$, and hence J contains the group

$$I_k(S) = \prod_{v \in S} k_v^{*n} \times \prod_{v \notin S} U_v.$$

It is thus enough to show that $C_k(S) := I_k(S).k^*/k^*$ contains a norm group.

We start with the case where k contains the nth roots of 1. Let us set $K = k(\sqrt[n]{E_{k,S}})$. It is a Kummer extension (cf. Chap. 13, Sect. 13.3) which, by Proposition 13.15, satisfies

$$\mathrm{Gal}(K/k) \simeq \mathrm{Hom}((E_{k,S}/E_{k,S}^n), \mathbf{Z}/n).$$

In particular, the extension K/k is of degree $[E_{k,S} : E_{k,S}^n] = n^s$, where s is the cardinality of S (cf. Proposition 13.20). Theorem 13.21 and its consequence (Lemma 15.7) apply with $r = s$, which means $T = \emptyset$. We thus obtain $C_k(S) = N_{K/k}C_K$.

In the general case, we denote by k' the cyclotomic extension of k obtained by adjoining to k the nth roots of 1. By the case where $\mu_n \subset k$, we can suppose (enlarging S if necessary) that $I_{k'} = I_{k',S'}k'^*$ and $C_{k'}(S') = N_{K'/k'}C_{K'}$, where S' is the set of places of k' above a place of S and $K' := k'(\sqrt[n]{k'_{S'}})$. The formula (easy to check), valid for $\beta \in I_{k'}$:

$$(N_{k'/k}(\beta))_v = \prod_{w|v} N_{k'_w/k_v}\beta_w$$

gives $N_{k'/k}(I_{k'}(S')) \subset I_k(S)$ hence

$$N_{K'/k}(C_{K'}) \subset N_{k'/k}(N_{K'/k'}C_{K'}) = N_{k'/k}(C_{k'}(S')) \subset C_k(S).$$

Thus $C_k(S)$ contains $N_{K'/k}(C_{K'})$. Even if we do not even know whether K' is Galois over k, we can conclude using Lemma 15.8, (c).

(b) *The case of a global field k of characteristic p and a subgroup of C_k of index divisible by p.*

In this book we will not give a complete proof in this case (see also Remark 15.15 below) but in the next paragraph we will explain how this theorem can be deduced from the triviality of the universal norm group for such a field k (which is shown in [1], Sect. 6.5).   □

We will deduce from the existence theorem a description of the abelian Galois group of a number field.

Theorem 15.10 *Let k be a number field with Galois group G_k. Then we have an exact sequence of topological groups:*

$$0 \longrightarrow D_k \longrightarrow C_k \xrightarrow{\text{rec}} G_k^{\mathrm{ab}} \longrightarrow 0, \tag{15.1}$$

where D_k is the neutral connected component C_k. The group D_k is equal to the universal norm group

$$N_{G_k}C = \bigcap_K N_{K/k}C_K,$$

where K ranges through finite abelian extensions of k (or all finite extensions of k), it is also the kernel of the profinite completion morphism $C_k \to C_k^\wedge$, i.e., the intersection of open subgroups of finite index of C_k.

Proof As $\mathbf{R}_+^*$ does not have a non-trivial finite quotient, the image of $C_k = C_k^0 \times \mathbf{R}_+^*$ by the reciprocity map rec is the same as that of C_k^0. This image is thus dense and compact, i.e., it is the whole of G_k^{ab}. Besides we know that the kernel of rec is $N_{G_k}C$.

We define D_k as the neutral connected component of C_k. Observe that in the decomposition $C_k = C_k^0 \times \mathbf{R}_+^*$, we have $\mathbf{R}_+^* \subset D_k$ (because $\mathbf{R}_+^*$ is connected) hence we have a decomposition $D_k = D_k^0 \times \mathbf{R}_+^*$, where D_k^0 is the neutral connected component of D_k. The group C_k/D_k is compact since it is isomorphic to $\widetilde{C_k^0/D_k^0}$. It is profinite since it is completely disconnected. All its open subgroups $\widetilde{U}_i$ are thus of finite index and their intersection is trivial. Thus the intersection of their inverse images U_i in C_k is D_k and each open subgroup of C_k contains D_k (its intersection with the connected subgroup D_k is non-empty, open, and closed in D_k, hence this intersection is D_k). Finally D_k is the intersection of the open subgroups of finite index of C_k. Theorem 15.9 then gives that D_k is also the intersection of norm subgroups, i.e., $D_k = N_{G_k}C$ (that we can define by taking the intersection of $N_{K/k}C_K$ over all the finite extensions of K or over the finite abelian extensions, by Lemma 15.8c). Besides, the compactness of C_k/D_k implies that the topology on the profinite group G_k^{ab} corresponds to the quotient topology on C_k/D_k, which means that (15.1) is an exact sequence of topological groups. $\square$

For global duality theorems, we will need some supplementary properties of the neutral connected component D_k.

Theorem 15.11 *With the above notation:*

(a) *for any finite extension K of k, the norm $N_{K/k} : D_K^0 \to D_k^0$ is surjective, and the same holds for $N_{K/k} : D_K \to D_k$.*

(b) *the abelian group D_k is divisible and we also have*

$$D_k = \bigcap_{n>0} C_k^n,$$

where $C_k^n \subset C_k$ is the subgroup of nth powers in C_k.

(c) *the map rec induces an isomorphism between C_k modulo its maximal divisible subgroup[1] and G_k^{ab}.*

---

[1]Recall that an abelian group A always possesses one largest divisible subgroup D, which is a subgroup of $D' := \bigcap_{n>0} nA$. But it may happen that D is strictly contained in D'.

Proof (a) Theorem 15.10 and the decompositions

$$C_k = C_k^0 \times \mathbf{R}_+^* ; \quad D_k = D_k^0 \times \mathbf{R}_+^*$$

seen above, give that D_k^0 is the group of universal norms $N_{G_k} C^0 = \bigcap_K N_{K/k} C_K^0$ of C^0. Let us fix a finite extension K of k. Let $a \in D_k^0$. For any finite extension L of K, let $E(L)$ be the set of elements $b \in C_K^0$ such that $N_{K/k} b = a$ and such that furthermore there exists $c \in C_L^0$ such that $b = N_{L/K} c$. Then $E(L)$ is non-empty since as a is a universal norm of C^0, it can be written as $a = N_{L/k}(c)$ with $c \in C_L^0$ and the element $b = N_{L/K}(c)$ is then in $E(L)$ by transitivity of the norms. The $E(L)$ then form a filtered decreasing family of non-empty compacts (as C_K^0, C_L^0 and C_k^0 are compact and the norm is continuous). Their intersection is therefore non-empty (Lemma 4.9), and any element b in this intersection satisfies $N_{K/k} b = a$, as well as $b \in N_{G_K} C^0 = D_K^0$. Finally $N_{K/k} : D_K^0 \to D_k^0$ is surjective (for a similar statement in the case of class formations, see Proposition 16.33). The analogous assertion for $N_{K/k} : D_K \to D_k$ is deduced immediately, noting that the map induced by $N_{K/k}$ on $\mathbf{R}_+^*$ is $x \mapsto x^{[K:k]}$, which is also surjective.

(b) As $\mathbf{R}_+^*$ is divisible, it is enough to show that D_k^0 is divisible, or that for every prime number ℓ, the map $x \mapsto x^\ell$ is surjective from D_k^0 to itself. Let K be a finite extension of k containing all the ℓth roots of 1. For S a finite set of places of K (containing the archimedean places and places above ℓ), let us set $K' = K(\sqrt[\ell]{E_{K,S}})$, where $E_{K,S}$ is the group of S-units in K. We define, as in the proof of Theorem 15.9:

$$I_K(S) = \prod_{v \in S} K_v^{*\ell} \times \prod_{v \notin S} \mathcal{O}_v^*$$

and $C_K(S) = I_K(S) K^* / K^* \subset C_K$. For S large enough, we have $I_K = I_{K,S} K^*$ (Proposition 12.25), and then (Lemma 15.7) $C_K(S) = N_{K'/K}(C_{K'})$. As D_K is the universal norm group in C_K, we deduce that $D_K \subset C_K(S)$, hence $D_K \subset (C_K)^\ell U_S$, where U_S consists of idèle classes (u_v) with $u_v = 1$ if $v \in S$ and $u_v \in \mathcal{O}_v^*$ if $v \notin S$. As each element u of U_S satisfies $|u| = 1$, we thus obtain $D_K^0 \subset (C_K^0)^\ell U_S$. As U_S and $(C_K^0)^\ell$ are compact, Remark 4.11 implies that the intersection (over all the S large enough) of the $(C_K^0)^\ell U_S$ is $(C_K^0)^\ell \cdot (\bigcap_S U_S) = (C_K^0)^\ell$. Thus $D_K^0 \subset (C_K^0)^\ell$ and, by taking the norm, we find

$$D_k^0 \subset (N_{K/k} C_K^0)^\ell$$

by (a).

Let us fix $a \in D_k^0$. For any finite extension K of k containing the ℓth roots of 1, denote by X_K the set of the $x \in N_{K/k} C_K^0 \subset C_k^0$ such that $x^\ell = a$. We just saw that the X_K are non-empty and they are compact by compactness of C_k^0 and C_K^0 (and the continuity of the norm). Their intersection is then non-empty. But, if b is in this intersection, we have $b^\ell = a$ and b is also a universal norm, hence by

Theorem 15.10, we have $b \in C_k^0 \cap D_k = D_k^0$. Finally D_k^0 is ℓ-divisible.

Conversely, if $x \in \bigcap_{n>0} C_k^n$, then x is in the kernel of $C_k \to C_k^\wedge$ since it is in every subgroup of finite index in C_k, and hence $x \in D_k$ by Theorem 15.10.

(c) By (b), the group D_k is the maximal divisible subgroup of C_k. We conclude using Theorem 15.10. $\square$

Remark 15.12 For $k = \mathbf{Q}$, we easily see that $D_\mathbf{Q} \simeq \mathbf{R}_+^*$ and for k imaginary quadratic, we have $D_k \simeq \mathbf{C}^*$ (cf. Exercise 15.2), But in general the structure of D_k can be very complicated. For other properties of D_k, see [42], Chap. VIII, Sect. 2.

15.3   The Case of a Function Field

In this paragraph, assume that k is a global field of characteristic $p > 0$. To obtain the existence theorem, also for subgroups of C_k with index divisible par p, the key step is the following theorem, whose proof can be found in [1], Sect. 6.5:

Theorem 15.13 *For a global field k of characteristic $p > 0$, the reciprocity map* rec $: C_k \to G_k^{\mathrm{ab}}$ *is injective. In other words the group of universal norms $N_{G_k} C$ is trivial.*

Let us recall the following topological result ([22], Theorem 3.5 of Chap. I and Theorem 7.11 of Chap. II).

Proposition 15.14 *Let G be a locally compact topological group which is totally disconnected. Let H be a closed subgroup of G. Then the quotient G/H is also locally compact and totally disconnected.*

Remark 15.15 Proposition 15.14 implies that C_k is totally disconnected (as a quotient of a locally compact and totally disconnected group I_k by the closed subgroup k^*), hence C_k^0 is profinite. The quotient C_k/C_k^0 is isomorphic to $\mathbf{Z}$, which implies that $N_{G_k} C$ is contained in C_k^0, and therefore is profinite (it is closed in C_k^0). In order to prove Theorem 15.13, it is enough to know that $N_{G_k} C$ is ℓ-divisible for every prime number ℓ. The method of Theorem 15.11, (b) still works for $\ell \neq p$, but one needs specific arguments for the case $\ell = p$.

We would now like to deduce the existence theorem from Theorem 15.13, also for subgroups of index divisible by p. To do this, it is more convenient (unlike what we have done in the case of a number field) to first determine the structure of the abelianised Galois group of k. Consider k to be the field of functions of a curve (projective and smooth) C over a finite field $\kappa = \mathbf{F}_q$. We can assume κ to be algebraically closed in k (i.e., the curve C is geometrically integral over κ). For any place v of k, corresponding to a closed point of C, denote by $\kappa(v)$ the residue field of v; it is a finite extension of κ. We define the degree map

$$\deg : I_k \longrightarrow \mathbf{Z}, \quad \alpha = (\alpha_v)_{v \in \Omega_k} \longmapsto \sum_{v \in \Omega_k} v(\alpha_v) \cdot [\kappa(v) : \kappa].$$

Lemma 15.16 *(a) Let $\alpha \in I_k$. Let F be the canonical topological generator $x \mapsto x^q$ of $\mathrm{Gal}(\overline{\kappa}/\kappa)$. Then the image of $\mathrm{rec}(\alpha) \in G_k^{\mathrm{ab}}$ by the canonical surjection $\pi : G_k^{\mathrm{ab}} \to \mathrm{Gal}(\overline{\kappa}/\kappa) \simeq \widehat{\mathbf{Z}}$ is $F^{\deg \alpha}$. In other words, $\pi(\mathrm{rec}(\alpha))$ corresponds to the integer $(\deg \alpha) \in \mathbf{Z} \subset \widehat{\mathbf{Z}}$.*
(b) the map $\deg$ factors as a surjective map (still denoted by $\deg$) from C_k to $\mathbf{Z}$, with kernel C_k^0.

Proof (a) Let us set $\alpha = (\alpha_v)_{v \in \Omega_k}$. For any finite abelian extension K of k, let $\omega_{K/k} : I_k \to \mathrm{Gal}(K/k)$ be the norm residue symbol. We have, by formula (14.1):

$$\omega_{K/k}(\alpha) = \prod_{v \in \Omega_k} \omega_{K_v/k_v}(\alpha_v),$$

where $\omega_{K_v/k_v} : k_v^* \to \mathrm{Gal}(K_v/k_v) \subset \mathrm{Gal}(K/k)$ is the local reciprocity map associated to the extension K_v/k_v. Let us set $f_v = [\kappa(v) : \kappa]$. Let $F_{K/k}$ be the automorphism $x \mapsto x^q$ of $\mathrm{Gal}(K/k)$ and $\pi_{K/k} : \mathrm{Gal}(K/k) \to \mathrm{Gal}(K(v)/\kappa(v))$ the projection, where $K(v)$ is the residue field of K_v. By Lemma 9.8(a), we have in $\mathrm{Gal}(K/k)$:

$$\pi_{K/k}(\omega_{K_v/k_v}(\alpha_v)) = (F_{K/k}^{f_v})^{v(\alpha_v)}$$

as $F_{K/k}^{f_v}$ is the canonical generator of $\mathrm{Gal}(K(v)/\kappa(v))$. We deduce:

$$\pi_{K/k}(\omega_{K/k}(\alpha)) = \prod_{v \in \Omega_k} F_{K/k}^{f_v \cdot v(\alpha_v)}.$$

By passing to the limit over the K in the last equality, we obtain in $\mathrm{Gal}(\overline{\kappa}/\kappa)$:

$$\pi(\mathrm{rec}(\alpha)) = \prod_{v \in \Omega_k} F^{f_v v(\alpha_v)} = F^{\deg \alpha}.$$

(b) By point (a), the integer $\deg \alpha$ depends only on the image of α by rec. The global reciprocity law (Theorem 14.9) implies that $\deg$ factors as a map from C_k to $\mathbf{Z}$. Furthermore, as rec has dense image and π is surjective, the image of $\pi \circ \mathrm{rec}$ is dense in $\mathrm{Gal}(\overline{\kappa}/\kappa) \simeq \widehat{\mathbf{Z}}$ which, by formula from (a), shows that $\deg : I_k \to \mathbf{Z}$ is surjective. Lastly, the kernel of $\deg : I_k \to \mathbf{Z}$ is by definition I_k^0, which implies the last assertion since $C_k^0 = I_k^0/k^*$. $\qquad\square$

Remark 15.17 To prove (b), we can also use [21], Part. II, Corollary 6.10 and (for surjectivity of $\deg$) the fact that the curve C has points over $\mathbf{F}_{q^n}$ for n large enough (corollary of the Lang–Weil theorem [34]).

We deduce a description of the abelianised Galois group of k, which is in a certain sense simpler than the case of a number field since it is completely analogous to the local case.

Theorem 15.18 *Let k be a global field of characteristic $p > 0$. Then, we have an exact sequence*

$$0 \longrightarrow C_k \xrightarrow{\ \mathrm{rec}\ } G_k^{\mathrm{ab}} \longrightarrow \widehat{\mathbf{Z}}/\mathbf{Z} \longrightarrow 0.$$

Furthermore, the reciprocity map rec induces an isomorphism of profinite groups between C_k^0 and $G_k^0 := \ker[\pi : G_k^{\mathrm{ab}} \to \widehat{\mathbf{Z}}]$.

Proof By Lemma 15.16, we have a commutative diagram with exact rows:

$$
\begin{array}{ccccccccc}
0 & \longrightarrow & C_k^0 & \longrightarrow & C_k & \xrightarrow{\ \deg\ } & \mathbf{Z} & \longrightarrow & 0 \\
 & & \downarrow{\scriptstyle\mathrm{rec}} & & \downarrow{\scriptstyle\mathrm{rec}} & & \downarrow{\scriptstyle i} & & \\
0 & \longrightarrow & G_k^0 & \longrightarrow & G_k^{\mathrm{ab}} & \xrightarrow{\ \pi\ } & \widehat{\mathbf{Z}} & \longrightarrow & 0.
\end{array}
\qquad (15.2)
$$

Let us fix an element c of C_k with $\deg c = 1$. Theorem 15.13 implies that rec is injective. Its image is dense and C_k is the direct product of C_k^0 and the subgroup generated par c, which implies, using the diagram, that the image of C_k^0 by rec is dense in G_k^0. As C_k^0 is compact and G_k^0 is separated, we deduce that rec induces an isomorphism between C_k^0 and G_k^0. We immediately obtain the desired exact sequence. $\square$

Proof of Theorem 15.9 for a function field, from Theorem 15.18 We can now deal with the general case of the existence theorem for a global field of characteristic $p > 0$ (keeping in mind that we have not in this book given a proof of Theorem 15.13, which is the difficult point). As we have seen in Sect. 15.2, it remains to show that every finite index open subgroup of C_k is a norm subgroup (including the case where the index is divisible by the characteristic p of k).

Let U be an open finite index subgroup of C_k. The image of U by the map $\deg : C_k \to \mathbf{Z}$ is of the form $d\mathbf{Z}$ with $d > 0$. Let us set $U_0 = U \cap C_k^0$. Then U_0 is an open subgroup (hence compact) of C_k^0, and by Theorem 15.18, the map rec induces an isomorphism between U_0 and a compact subgroup $\widetilde{U}_0$ of G_k^0. Let $\widetilde{U}$ be the closure of $\mathrm{rec}(U)$ in G_k^{ab}. The diagram (15.2) implies that $\pi(\widetilde{U}) = d\widehat{\mathbf{Z}}$. The same argument as in the Proof of Theorem 15.18 (using the fact that U is the direct product of the compact subgroup U_0 and a subgroup isomorphic to $d\mathbf{Z}$) shows that $\mathrm{rec}(U_0)$ is dense in the kernel of $\pi : \widetilde{U} \to d\widehat{\mathbf{Z}}$, which means that this kernel is exactly $\widetilde{U}_0$. We thus have an exact sequence

$$0 \longrightarrow \widetilde{U}_0 \longrightarrow \widetilde{U} \xrightarrow{\ \pi\ } d\widehat{\mathbf{Z}} \longrightarrow 0.$$

This shows that $\widetilde{U}$ is a compact subgroup of finite index (hence also open) of G_k^{ab}. This implies that $\widetilde{U} = \mathrm{Gal}(k^{\mathrm{ab}}/K)$, where K is a finite abelian extension of k.

Let us show that $U = N_{K/k}C_k$. We have by construction $\mathrm{rec}(U) \subset \tilde{U}$, hence the image of U by the reciprocity map $\omega_{K/k} : C_k \to \mathrm{Gal}(K/k)$ is reduced to the neutral element, which shows that $U \subset N_{K/k}C_k$. On the other hand

$$[C_k : U] = [C_0 : U_0] \cdot d = [G_k^0 : \tilde{U}_0] \cdot d = [G_k^{\mathrm{ab}} : \tilde{U}] = [K : k] = [C_k : N_{K/k}C_K],$$

which allows to conclude that $U = N_{K/k}C_k$. $\qquad\qquad\square$

15.4 Ray Class Fields; Hilbert Class Field

In this paragraph, we will use the existence theorem to establish a relation between the abelian extensions of k and *ray class fields* which historically came about before the idelic formulation of theory. To simplify notation, we will assume that k is a number field. Similar results hold for a field of functions of a smooth projective curve C over $\mathbf{F}_q$ with a few alterations (in particular replacing cycles by cycles of degree 0, the ideal class group of $\mathcal{O}_k$ by $\mathrm{Pic}^0 C$, etc.).

Definition 15.19 Let k be a number field. A *cycle* $\mathcal{M}$ of k is a formal product $\mathcal{M} = \prod_{v \in \Omega_k} v^{n_v}$, where $n_v \in \mathbf{N}$, n_v is zero for almost all v, and $n_v \in \{0, 1\}$ if v is archimedean. We say that a place v_0 of k *divides* the cycle $\mathcal{M}$ if $n_{v_0} \geq 1$.

For a finite v and $n_v \geq 1$, we denote by $U_v^{n_v}$ the multiplicative group consisting of $x \in \mathcal{O}_v^*$ such that $v(x - 1) \geq n_v$, and we also set $U_v^0 = U_v = \mathcal{O}_v^*$. For a complex v, we set $U_v^{n_v} = k_v^* \simeq \mathbf{C}^*$. For a real v, we set $U_v^1 = \mathbf{R}_+^* \subset k_v^*$ and $U_v^0 = k_v^* \simeq \mathbf{R}^*$. The notation $\alpha_v \equiv 1 \bmod v^{n_v}$ will mean $v(\alpha_v - 1) \geq n_v$ if v is finite and $n_v \geq 1$ (by convention for $n_v = 0$ the condition is always satisfied). For a complex v or a real v with $n_v = 0$, this condition will by definition always be satisfied and for a real v with $n_v = 1$, will simply mean that $\alpha_v > 0$. Let $\mathcal{M} = \prod_v v^{n_v}$ be a cycle and let $\alpha \in I_k$. We will write $\alpha \equiv 1 \bmod \mathcal{M}$ to indicate: $\alpha_v \equiv 1 \bmod v^{n_v}$ for every place v of k.

Definition 15.20 Let $\mathcal{M}$ be a cycle of k. Let us set

$$I_k^{\mathcal{M}} = \{\alpha \in I_k, \alpha \equiv 1 \quad \bmod \mathcal{M}\} = \prod_{v \in \Omega_k} U_v^{n_v}.$$

The image $C_k^{\mathcal{M}} = I_k^{\mathcal{M}} \cdot k^*/k^*$ of the group $I_k^{\mathcal{M}}$ in C_k is called the *congruence subgroup modulo* $\mathcal{M}$ of C_k. The quotient $C_k/C_k^{\mathcal{M}}$ is the *ray class group* modulo $\mathcal{M}$.

The case where $\mathcal{M} = 1$ is the trivial cycle is particularly interesting: we then have $I_k^1 = \prod_{v \in \Omega_\infty} k_v^* \times \prod_{v \in \Omega_f} U_v$, hence we deduce that $C_k/C_k^1 \cong \mathcal{I}_k/\mathcal{P}_k$ (an isomorphism is obtained by sending the class of an idèle (x_v) to the class of the fractional ideal $\prod_{v \in \Omega_f} v^{\mathrm{val}(x_v)}$, cf. Remark 15.5) is the ideal class group of $\mathcal{O}_k$. Its cardinality is the class number $h(k)$.

Theorem 15.21 *A subgroup of C_k is a norm group if and only if it contains a congruence subgroup $C_k^{\mathcal{M}}$.*

Proof Let $\mathcal{M} = \prod_v v^{n_v}$ be a cycle. The group $C_k^{\mathcal{M}}$ is an open subgroup of C_k since $I_k^{\mathcal{M}} = \prod_{v \in \Omega_k} U_v^{n_v}$ is open in I_k. Besides, $C_k^{\mathcal{M}}$ is of finite index in C_k since $[C_k : C_k^1] = h(k)$ is finite and $[C_k^1 : C_k^{\mathcal{M}}]$ is bounded above by the order of $I_k^1 / I_k^{\mathcal{M}} = \prod_v U_v / U_v^{n_v}$ which is finite (recall that $n_v = 0$ for almost all places v and $U_v^0 := U_v$). Finally, $C_k^{\mathcal{M}}$ is a norm group by Theorem 15.9.

Conversely, let N be a norm group. Then its inverse image J in I_k is open, hence J contains a subset of the form $\prod_{v \in S} W_v \times \prod_{v \notin S} U_v$, where $S \supset \Omega_\infty$ is a finite set of places and W_v is an open neighbourhood of 1 in k_v^*. For $v \in S$ finite, we can assume that $W_v = U_v^{n_v}$ with $n_v \in \mathbf{N}$ since the U_v^m, $m \geq 0$ form a basis of neighbourhoods of 1. For v archimedean, the only subgroups of k_v^* generated by an open neighbourhood of 1 are k_v^* or $\mathbf{R}_+^*$ if v is real. We deduce that J contains a subgroup of the form $I_k^{\mathcal{M}}$, and hence that N contains a congruence subgroup $C_k^{\mathcal{M}}$.   □

Definition 15.22   The class field $k^{\mathcal{M}}$ associated to the congruence subgroup $C_k^{\mathcal{M}}$ is called the *ray class field* modulo $\mathcal{M}$. The field k^1 is called the *Hilbert class field* of k.

We thus have $\mathrm{Gal}(k^{\mathcal{M}}/k) \simeq C_k / C_k^{\mathcal{M}}$, and any finite abelian extension of k is contained in a ray class field. We also have

$$\mathcal{M} | \mathcal{M}' \implies k^{\mathcal{M}} \subset k^{\mathcal{M}'} \implies C_k^{\mathcal{M}} \supset C_k^{\mathcal{M}'}.$$

In particular, we have $\mathrm{Gal}(k^1/k) \simeq \mathcal{I}_k / \mathcal{P}_k$, which implies that the degree $[k^1 : k]$ is the class number $h(k)$ of k.

Definition 15.23   Let K be a local field and let L be a finite Galois extension of K with group G. The *conductor* $\mathcal{F}(L/K)$ of the extension L/K is the smallest integer $n \geq 0$ such that the reciprocity map $\omega_{L/K} : K^* \to G^{\mathrm{ab}}$ is trivial on U_K^n (with the usual convention that $U_K^0 = U_K$).

Note that as the kernel of $\omega_{L/K}$ is an open subgroup of K^*, this subgroup contains U_K^n for n large enough and the conductor is well defined. Besides, the computation of the reciprocity map via Lubin–Tate theory (Theorem 11.17) implies:

Proposition 15.24   *The conductor $\mathcal{F}(L/K)$ is trivial if and only if the extension L/K is unramified.*

We generalise the notion of conductor to $K = \mathbf{R}$ or $K = \mathbf{C}$, by taking $\mathcal{F}(L/K) = 0$ or $\mathcal{F}(L/K) = 1$ depending on whether the extension L is equal to K or not.

Definition 15.25   Let k be a number field. Let K be a finite abelian extension of k, associated to the norm group $N_K := N_{K/k} C_K$. The *conductor* $\mathcal{F}$ of the extension K/k (or of N_K) is the greatest common divisor of cycles $\mathcal{M}$ such that $K \subset k^{\mathcal{M}}$ (or that $C_k^{\mathcal{M}} \subset N_{K/k} C_K$).

In other words, $k^{\mathcal{F}}$ is the smallest ray class field containing K. Note that this does not imply that the conductor of $k^{\mathcal{M}}$ is $\mathcal{M}$ as $\mathcal{M} \mapsto C_K^{\mathcal{M}}$ is decreasing, but not in general injective (see Exercise 15.3).

Proposition 15.26 *Let K be a finite abelian extension of a number field k, with conductor $\mathcal{F}$. Then we have*

$$\mathcal{F} = \prod_{v \in \Omega_k} v^{\mathcal{F}(K_v/k_v)}.$$

Proof We start with a lemma:

Lemma 15.27 *For $x_v \in k_v^*$, denote by $[x_v]$ the idèle $(1, \ldots, 1, x_v, 1, \ldots,)$. Let K be a finite abelian extension of k. Then the class of $[x_v]$ in C_k is in $N_{K/k}C_K$ if and only if $x_v \in N_{K_v/k_v}K_v^*$. Furthermore, if α is an idèle such that the class of $[\alpha_v]$ in C_k is in $N_{K/k}C_K$ for every place v of k, then the class of α in C_k is also in $N_{K/k}C_K$.*

Proof of the lemma. Assume that $x_v \in N_{K_v/k_v}K_v^*$. Then we have by definition of the norm residue symbol:

$$([x_v], K/k) = (x_v, K_v/k_v) = 1$$

which shows that the class of $[x_v]$ belongs to $N_{K/k}C_K$ by Theorem 15.4. Conversely, assume that the class of $[x_v]$ is in $N_{K/k}C_K$. This means that there exists $\beta \in I_K$ and $a \in k^*$ such that $[x_v] \cdot a = N_{K/k}\beta$. This implies that a is a norm of K_u/k_u for any place u of k other than v. But then, we obtain that a is also a norm of K_v/k_v by reciprocity law (Theorem 14.9, b), which proves that $x_v \in N_{K_v/k_v}K_v^*$.

To prove the second assertion, we note that if $\alpha = (\alpha_v)$ is an idèle such that for every place v of k, we have $\alpha_v \in N_{K_v/k_v}K_v^*$, then $\alpha \in N_{K/k}I_K$ by Lemma 13.3, and hence a fortiori the class of α in C_k is in $N_{K/k}C_K$. $\qquad\square$

We can now prove Proposition 15.26. Let $N = N_{K/k}C_K$ and let $\mathcal{M} = \prod_v v^{n_v}$ be a cycle. For any idèle $\alpha \in I_k$, consider the following three conditions:

(i) $\alpha \equiv 1 \bmod \mathcal{M}$.

(ii) the class of $[\alpha_v]$ in C_k is in N for every place v of k.

(iii) the class of α in C_k is in N.

By second part of Lemma 15.27, (ii) implies (iii). Furthermore, if α satisfies (i), then $[\alpha_v]$ also satisfies (i) for any place v. Hence the two following conditions are equivalent:

(a) Each $\alpha \in I_k$ satisfying (i) also satisfies (ii).

(b) Each $\alpha \in I_k$ satisfying (i) also satisfies (iii).

Now, the condition $C_k^{\mathcal{M}} \subset N$ means precisely that (b) is satisfied. It is then equivalent to saying that (a) is satisfied, and with the first part of Lemma 15.27, it is equivalent to saying that:

$\alpha_v \equiv 1 \bmod v^{n_v}$ for every place v implies $\alpha_v \in N_{K_v/k_v}K_v^*$ for every place v,

or that:

for every place v, we have $U_v^{n_v} \subset N_{K_v/k_v}K_v^*$.

Finally, we have shown that $C_k^{\mathcal{M}} \subset N$ is equivalent to $\mathcal{F}(K_v/k_v) \leq n_v$ for every place v, which exactly means that $\mathcal{F} = \prod_v v^{\mathcal{F}(K_v/k_v)}$. $\qquad\square$

Remark 15.28 Note on the other hand that, as J. Riou pointed out, the above condition (iii) does not imply condition (ii). To see this, it is enough to take a principal idèle $\alpha \in k^*$ which is not a local norm at some place v. It is to overcome this difficulty that we have introduced idèles of the form $[\alpha_v]$ instead of working with the $\alpha_v \in k_v^*$.

Corollary 15.29 *Let K be a finite abelian extension of a number field k, with conductor $\mathcal{F}$. Then a place v is ramified in K/k if and only if v divides $\mathcal{F}$ (with the convention that for an archimedean place, unramified means totally split). The Hilbert class field is the maximal abelian extension of k which is unramified at all places of k (including archimedean places).*

Proof If v is a finite place, the previous proposition says that it divides $\mathcal{F}$ if and only if the local conductor $\mathcal{F}(K_v/k_v)$ is nonzero, in other words if and only if K_v/k_v is ramified (by Proposition 15.24). If v is complex, we automatically have $\mathcal{F}(K_v/k_v) = 0$, hence Proposition 15.26 says that v does not divide $\mathcal{F}$, while by definition K_v/k_v is a totally split extension. If v is real, we obtain that v divides $\mathcal{F}$ if and only if $\mathcal{F}(K_v/k_v) = 1$, i.e., if and only if K_v/k_v is a non-trivial extension, which is equivalent to v being ramified in K/k in the sense of the statement of this corollary.

Propositions 15.26 and 15.24 say that a finite abelian extension K of k is unramified at every place of k if and only if its conductor is 1, or in other words (by definition of the conductor) if and only if the Hilbert class field k^1 contains K. Thus k^1 is the maximal abelian extension of k unramified at every place of k. $\square$

We also have the following property of the Hilbert class field:

Proposition 15.30 *Let $\wp$ be a prime ideal of the ring of integers $\mathcal{O}_k$ of k. Then $\wp$ is a principal ideal of $\mathcal{O}_k$ if and only if it is totally split in the extension k^1/k.*

Proof We already know that $\wp$ is unramified in k^1/k. We also know that we have an isomorphism between $C_k/C_k^1 = C_k/N_{k^1/k}C_{k^1}$ and the ideal class group $\mathcal{I}_k/\mathcal{P}_k$, the isomorphism that sends the class of an idèle of the form $a := (1, 1, \ldots, \pi_v, 1, \ldots)$ (where π_v is a uniformiser of k_v at the place v that corresponds to $\wp$) to the class of $\wp$. The reciprocity map $\omega_{k^1/k} : C_k/C_k^1 \to \mathrm{Gal}(k^1/k)$ is an isomorphism, and we thus have that $\wp$ is principal if and only if $\omega_{k^1/k}(a) = 0$. As k^1/k is unramified at v, this means by the formula (14.1) and Lemma 9.8, (a) that the local extension k_v^1/k_v is trivial, which is that v is totally split in k^1/k. $\square$

The properties of the Hilbert class field have applications to questions about representation of an integer by a quadratic form: see the book by D. Cox [13] (Chap. 2). In Exercises 15.7 and 15.8 we give a foretaste of these applications.

The following statement is the famous *principal ideal theorem*. The reduction of this statement to a result from group theory that we have already encountered (Theorem 2.13) is due to Artin.

Theorem 15.31 *Let k be a number field. Let K_1 be its Hilbert class field. Then every ideal of $\mathcal{O}_k$ becomes principal in $\mathcal{O}_{K_1}$.*

Proof Let K_2 be the Hilbert class field of K_1, which is an abelian extension of K_1. The definition of the reciprocity map and the interpretation of transfer (resp. of the inclusion $i : C_k \subset C_{K_1}$) as the restriction

$$\widehat{H}^{-2}(\mathrm{Gal}(K_2/k), \mathbf{Z}) \longrightarrow \widehat{H}^{-2}(\mathrm{Gal}(K_2/K_1), \mathbf{Z})$$

$$H^0(\mathrm{Gal}(K_2/k), C_{K_2}) \longrightarrow H^0(\mathrm{Gal}(K_2/K_1), C_{K_2}) \qquad \text{resp.}$$

yields a commutative diagram[2]:

$$
\begin{array}{ccc}
C_k & \xrightarrow{\ \omega_{K_2/k}\ } & \mathrm{Gal}(K_2/k)^{\mathrm{ab}} \\
{\scriptstyle i}\downarrow & & \downarrow{\scriptstyle V} \\
C_{K_1} & \xrightarrow{\ \omega_{K_2/K_1}\ } & \mathrm{Gal}(K_2/K_1)^{\mathrm{ab}},
\end{array}
$$

where V is the transfer. We have seen that K_1 is the maximal abelian extension of k everywhere unramified and totally split at archimedean places. It is then the maximal abelian subextension of K_2/k. In other words, $\mathrm{Gal}(K_1/k)$ is the abelianisation of $\mathrm{Gal}(K_2/k)$ and $\mathrm{Gal}(K_2/K_1)$ is the derived subgroup of $\mathrm{Gal}(K_2/K)$. As K_2 is an abelian extension of K_1, we finally obtain a commutative diagram:

$$
\begin{array}{ccc}
\mathcal{I}_k/\mathcal{P}_k = C_k/N_{K_1/k}C_{K_1} & \xrightarrow{\ \omega_{K_1/k}\ } & \mathrm{Gal}(K_1/k) \\
{\scriptstyle i}\downarrow & & \downarrow{\scriptstyle V} \\
\mathcal{I}_{K_1}/\mathcal{P}_{K_1} = C_{K_1}/N_{K_2/K_1}C_{K_2} & \xrightarrow{\ \omega_{K_2/K_1}\ } & \mathrm{Gal}(K_2/K_1),
\end{array}
$$

where horizontal maps are isomorphisms. Theorem 2.13 then says that the map V of the diagram is zero, which yields the result. $\qquad\square$

Remark 15.32 When contemplating the last theorem, one can naturally wonder whether the tower of class fields $K_0 = k \subset K_1 \subset K_2 \subset \cdots$ (where, for $i \geq 0$, K_{i+1} is the Hilbert class field of K_i) is necessarily finite, which would imply that the last field in the tower has class number 1 (and not just that the ideals of $\mathcal{O}_k$ become principal there). This question, asked by Furtwängler, was finally solved in the negative by Golod and Shafarevich in 1964. See Exercises 15.5 and 15.6, which are taken from [49], Part. I, Sect. 4.4 and Ann. 3.

Remark 15.33 Let S be a non-empty set of places of k, containing the archimedean places. The same proof yields a result analogous to Theorem 15.31 (which corresponds to the case $S = \Omega_\infty$) for ideals of the ring $\mathcal{O}_{k,S}$ of S-integers: they all become principal in a finite extension of k, namely in the maximal abelian extension of k

---

[2] We have an analogous diagram for any class formation, see Proposition 16.30, and also Exercise 9.3 in the local case.

unramified outside S and totally split at places in S. This also holds for a function field (by assumption $S \neq \emptyset$, which ensures the finiteness of the class group of $\mathcal{O}_{k,S}$).

We conclude this paragraph with the Kronecker–Weber theorem (that can also be deduced from its version over $\mathbf{Q}_p$, cf. Exercise 12.6).

Theorem 15.34 *The maximal abelian extension of the field* $\mathbf{Q}$ *is the extension* $\mathbf{Q}(\mu_\infty)$ *generated by all the roots of unity.*

Let ζ_m be a primitive mth root of unity. Note that if a prime number p does not divide m, then the extension $\mathbf{Q}(\zeta_m)/\mathbf{Q}$ is unramified at p (indeed the reduction modulo p of the polynomial $X^m - 1$ is then a separable polynomial over $\mathbf{F}_p$). We already know that every cyclotomic extension $\mathbf{Q}(\zeta_m)$ is abelian. It is thus enough to show that every finite abelian extension of $\mathbf{Q}$ is contained in $\mathbf{Q}(\zeta_m)$ for a certain m. To do this, we will prove the following more precise form.

Theorem 15.35 *Let* $m = \prod p^{n_p} \in \mathbf{N}^*$. *Let* p_∞ *be the real place of* $\mathbf{Q}$. *Let* $\mathcal{M}$ *be the cycle* $m \cdot p_\infty$. *Then the ray class field modulo* $\mathcal{M}$ *of* $\mathbf{Q}$ *is* $\mathbf{Q}(\zeta_m)$.

Proof Let us set $m = m' \cdot p^{n_p}$ (for p fixed). Then $U_p^{n_p}$ is contained in the norm group of the unramified extension $\mathbf{Q}_p(\zeta_{m'})/\mathbf{Q}_p$, and also in that of $\mathbf{Q}_p(\zeta_{p^{n_p}})/\mathbf{Q}_p$ by Lubin–Tate (cf. Example 11.23). We deduce that $U_p^{n_p}$ is contained in the norm group of $\mathbf{Q}_p(\zeta_m)/\mathbf{Q}_p$, as by compatibility of the reciprocity maps, its image by $\omega_{\mathbf{Q}_p(\zeta_m)/\mathbf{Q}_p}$ is trivial. As $I_\mathbf{Q}^\mathcal{M} = \prod_{p \neq p_\infty} U_p^{n_p} \times \mathbf{R}_+^*$, we obtain $C_\mathbf{Q}^\mathcal{M} \subset N C_{\mathbf{Q}(\zeta_m)}$ by Lemma 13.3.
  Besides, we have

$$[C_\mathbf{Q} : C_\mathbf{Q}^\mathcal{M}] = [I_\mathbf{Q}^1 . \mathbf{Q}^* : I_\mathbf{Q}^\mathcal{M} . \mathbf{Q}^*] = [I_\mathbf{Q}^1 : I_\mathbf{Q}^\mathcal{M}] / [I_\mathbf{Q}^1 \cap \mathbf{Q}^* : I_\mathbf{Q}^\mathcal{M} \cap \mathbf{Q}^*]$$

since $[C_\mathbf{Q} : C_\mathbf{Q}^1] = 1$ as $\mathbf{Z} = \mathcal{O}_\mathbf{Q}$ is principal. But $I_\mathbf{Q}^\mathcal{M} \cap \mathbf{Q}^* = \{1\}$ and $I_\mathbf{Q}^1 \cap \mathbf{Q}^* = \{\pm 1\}$ hence

$$[C_\mathbf{Q} : C_\mathbf{Q}^\mathcal{M}] = 1/2 \cdot \prod_{p \neq p_\infty} [U_p : U_p^{n_p}] \cdot 2 = \prod_{p \neq p_\infty} \varphi(p^{n_p}) = \varphi(m).$$

Finally, we have $[C_\mathbf{Q} : C_\mathbf{Q}^\mathcal{M}] = [\mathbf{Q}(\zeta_m) : \mathbf{Q}] = [C_\mathbf{Q} : N C_{\mathbf{Q}(\zeta_m)}]$, which concludes. $\square$

15.5   Galois Groups with Restricted Ramification

It is often useful in applications to work not with the absolute Galois group G_k of a global field k, but with certain quotients G_S of G_k associated to non-empty subsets S of the set Ω_k of all places of k. This in particular will occur when we will be obtaining a relatively general version of the Poitou–Tate duality.
  In all that follows, we denote by S a *non-empty* subset of Ω_k, containing all the archimedean places if k is a number field. We attract the readers attention to the fact

that here (contrary to the conventions we have usually adopted before), the set S *is not assumed to be finite*. We denote by Ω_∞ the set of archimedean places of k and by $\Omega_{\mathbf{R}}$ the set of its real places. We fix an algebraic closure $\bar{k}$ of k and we denote by $\mathcal{O}_{k,S}$ the ring of S-integers of k. For example, if $S = \Omega_k$, we simply have $\mathcal{O}_{k,S} = k$; if k is a number field and $S = \Omega_\infty$, we have $\mathcal{O}_{k,S} = \mathcal{O}_k$. If v is a finite place of k, we, as usual, denote by $\mathcal{O}_v$ the ring of integers of k_v (and by convention we set $\mathcal{O}_v = k_v$ if v is archimedean).

Definition 15.36 We denote by k_S the maximal extension of k contained in $\bar{k}$ which is unramified outside S and we set $G_S = \mathrm{Gal}(k_S/k)$. We say that G_S is the *Galois group with ramification restricted to S* of k.

Example 15.37 If S is finite, the extension k_S is "mildly ramified". In the language of algebraic geometry, the group G_S is the étale fundamental group of the affine scheme $U = \mathrm{Spec}(\mathcal{O}_{k,S})$. Here U is a Zariski open set of $\mathrm{Spec}(\mathcal{O}_k)$ if k is a number field (resp. an affine open set of X if k is the function field of a smooth projective curve X over a finite field). If for example $S = \Omega_\infty$, then k_S is the maximal extension of k, unramified at every prime ideal of $\mathcal{O}_k$. If we furthermore assume that k does not have real places, the extension k_{Ω_∞} is the Hilbert class field of k (cf. Corollary 15.29) and G_{Ω_∞} is the ideal class group of $\mathcal{O}_k$.

If on the other hand, S contains almost all places of k (i.e., all places with the exception of a finite number of them), the extension k_S is "close" to $\bar{k}$. The case $S = \Omega_k$ corresponds to $k_S = \bar{k}$.

We would like to generalise some results of Sect. 15.2 to the restricted ramification group G_S. We will work with finite extensions F of k (that we will usually assume to be Galois). To simplify notations, we will often still denote by S the set of places of F dividing a place of k that belongs to S. It appears that we have an analogue of the idèle class group in this context, which is the following.

Definition 15.38 Let F be a finite extension of k, with idèle group I_F and idèle class group $C_F = I_F/F^*$. Let $U_{F,S} \simeq \prod_{w \notin S} \mathcal{O}_w^*$ be the subgroup of I_F consisting of families $(a_w)_{w \in \Omega_F}$ whose component at w is trivial (resp. invertible) if the place $w \in \Omega_F$ is in S (resp. is not in S). We set

$$C_S(F) := I_F/F^* U_{F,S}$$

(in other words $C_S(F)$ is the quotient of C_F by the image of $U_{F,S}$ modulo F^*), and we call C_S the inductive limit of the $C_S(F)$ when F runs through the finite Galois extensions of k contained in k_S.

Note that the assumption $S \neq \emptyset$ implies $F^* \cap U_{F,S} = \{1\}$, which allows us to view $U_{F,S}$ as a subgroup of C_F.

We begin with a lemma which compares $C_S(F)$ with a group $C_{F,S}$ which may seem more natural to define. The reason why $C_S(F)$ is more useful is that, as we will see (Proposition 15.40, c), it satisfies a good Galois descent property, which is not the case for $C_{F,S}$ in general.

Lemma 15.39 *We define $J_{F,S}$ as the subgroup of I_F consisting of families $(a_w)_{w \in \Omega_F}$ whose component at w is trivial for any place $w \in \Omega_F$ not in S. Let*

$$C_{F,S} := J_{F,S}/\mathcal{O}_{F,S}^*$$

be the quotient of $J_{F,S}$ by the group of units of $\mathcal{O}_{F,S}$. Then there exists an exact sequence

$$0 \longrightarrow C_{F,S} \longrightarrow C_S(F) \longrightarrow \mathrm{Cl}_{F,S} \longrightarrow 0,$$

where $\mathrm{Cl}_{F,S}$ is the ideal class group of $\mathcal{O}_{F,S}$. If S contains almost all places of k, then $C_{F,S} = C_S(F)$.

Note that $J_{F,S}$ is a kind of the "idèle group truncated at S": indeed it identifies with the restricted product of the F_w^* for w places of F above places of S. One should not confuse it with the group $I_{F,S}$ of S-idèles introduced in Proposition 12.25.

Proof The group $\mathcal{O}_{F,S}^*$ identifies with a subgroup of $J_{F,S}$ via the map $i : a \mapsto (a, a, \ldots, a, 1, 1, \ldots)$ (where the a appears at places of S and the 1 at places not in S). In I_F, we then have

$$J_{F,S} \cap (F^* \cdot U_{F,S}) = i(\mathcal{O}_{F,S}^*) \simeq \mathcal{O}_{F,S}^*.$$

We deduce an embedding

$$j : C_{F,S} \hookrightarrow C_F/U_{F,S} = C_S(F).$$

On the other hand the group $J_{F,S} \cdot U_{F,S}$ is the subgroup of I_F consisting of idèles whose component at each place $v \notin S$ is in $\mathcal{O}_v^*$. The cokernel of j identifies with $I_F/J_{F,S} \cdot U_{F,S} \cdot F^*$, i.e., to the quotient of $I_F/J_{F,S} \cdot U_{F,S} \simeq \bigoplus_{v \notin S} \mathbf{Z}$ by the image of F^* under the map

$$F^* \longrightarrow \bigoplus_{v \notin S} \mathbf{Z}, \quad a \longmapsto (w(a))_{w \notin S}.$$

Thus this cokernel is isomorphic to the ideal class group of $\mathcal{O}_{F,S}$ (note that the nonzero prime ideals of $\mathcal{O}_{F,S}$ correspond to places of F not in S; in geometric language $\mathrm{Cl}_{F,S}$ is the class group of divisors of $\mathrm{Spec}(\mathcal{O}_{F,S})$, cf. Theorem 12.24).

Finally if S contains almost all places of k, then $\mathcal{O}_{F,S}$ is a Dedekind ring with only finitely many prime ideals, it is thus a principal ideal domain (Proposition 12.5), which implies that its class group is trivial. $\qquad\square$

Next proposition will be useful later to define a P-class formation associated to G_S (Theorem 17.2).

Proposition 15.40 *(a) The group $C_S = \varinjlim_{F \subset k_s} C_S(F)$ is also the inductive limit of the $C_{F,S}$.*

(b) For any finite Galois extension F of k, the $\mathrm{Gal}(F/k)$-module $U_{F,S}$ is cohomologically trivial. The G_S-module $U_S := \varinjlim_{F \subset k_S} U_{F,S}$ is cohomologically trivial.

(c) We have $C_S^{G_S} = C_S(k)$.

Proof (a) By the last lemma, it is enough to check that $\varinjlim_{F \subset k_S} \mathrm{Cl}_{F,S} = 0$. This is an immediate consequence of the principal ideal theorem (Theorem 15.31 and Remark 15.33).

(b) The same argument as in Proposition 13.1 yields

$$\widehat{H}^i(\mathrm{Gal}(F/k), U_{F,S}) = \bigoplus_{v \notin S} \widehat{H}^i(\mathrm{Gal}(F_v/k_v), \mathcal{O}_v^*)$$

for any $i \in \mathbf{Z}$. With this, the result for $U_{F,S}$ follows from Proposition 8.3 since for $v \notin S$, the extension F_v/k_v is unramified. We deduce the result for U_S by passing to the limit, since $U_{F,S} = U_S^{\mathrm{Gal}(k_S/F)}$.

(c) Let $C(k_S)$ be the inductive limit of the C_F for F finite Galois extension of k contained in k_S. By passing to the limit in the definition of $C_S(F)$, we obtain an exact sequence

$$0 \longrightarrow U_S \longrightarrow C(k_S) \longrightarrow C_S \longrightarrow 0. \tag{15.3}$$

We then apply the cohomology exact sequence for the action of the group G_S (note that $U_S^{G_S} = U_{k,S}$ is obvious and we also have $C(k_S)^{G_S} = C_k$, cf. Proposition 13.4). As U_S is a cohomologically trivial G_S-module, we have $H^1(G_S, U_S) = 0$ and hence $C_S^{G_S} = C_k/U_{k,S} = C_S(k)$. $\qquad\qquad\qquad\qquad\qquad\qquad\qquad\qquad\qquad\quad\square$

Lemma 15.41 *Let A be a locally compact abelian group, with neutral connected component A^0. Let K be a compact subgroup of A. Then the image B of A^0 by the projection $\pi : A \to A/K$ is the neutral connected component of A/K (endowed with the quotient topology).*

Proof We already know that B is connected, since it is the image of the connected set A^0 by the continuous map π. It is then enough to check that the quotient $(A/K)/B = A/(K + A^0)$ is totally disconnected. But the topological group $A/(K + A^0)$ is the quotient of the locally compact and totally disconnected group A/A^0 by the compact (and hence closed) subgroup $(K + A^0)/A^0$, which allows us to conclude using Proposition 15.14. $\qquad\qquad\qquad\qquad\qquad\qquad\qquad\qquad\qquad\qquad\square$

Below are analogues of Theorems 15.11 and 15.18.

Proposition 15.42 *(a) Let k be a number field. Let $D_S(k)$ be the connected component of the identity in $C_S(k)$. Then we have $D_S(k) = D_k U_{k,S}/U_{k,S}$. The group $D_S(k)$ is divisible and we have an exact sequence*

$$0 \longrightarrow D_S(k) \longrightarrow C_S(k) \xrightarrow{\;\omega\;} G_S^{\mathrm{ab}} \longrightarrow 0,$$

where ω is the morphism induced by the reciprocity map $\mathrm{rec}_k : C_k \to G_k^{\mathrm{ab}}$ (cf. Theorem 15.4).

(b) Let k be a function field over $\mathbf{F}_q$. Then we have an exact sequence

$$0 \longrightarrow C_S(k) \xrightarrow{\;\omega\;} G_S^{\mathrm{ab}} \longrightarrow \widehat{\mathbf{Z}}/\mathbf{Z} \longrightarrow 0.$$

Proof (a) The group $C_S(k)$ is the quotient of C_k by the compact subgroup $U_{k,S}$. Lemma 15.41 then implies that the image of the neutral component D_k of C_k by p is the neutral component of $C_S(k)$, which proves the first assertion. Then $D_S(k)$ is divisible as it is a quotient of D_k, which is divisible by Theorem 15.11(b).

Now, we see that the image of $U_{k,S}$ by $\mathrm{rec}_k : C_k \to G_k^{\mathrm{ab}}$ is, by Lemma 9.8 and formula (14.1), the subgroup H of G_k^{ab} generated by the inertia subgroups I_v for $v \notin S$, which means that the fixed field of H is the maximal subextension of k^{ab} unramified outside of S, which is k_S^{ab} (indeed, a subgroup of G_k^{ab} contains I_v if and only if the corresponding extension of k is unramified at v). We thus have, by Theorem 15.10, a commutative diagram whose rows are exact and vertical arrows are injective:

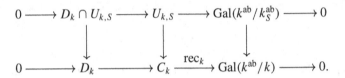

As $D_S(k)$, $C_S(k)$ and G_S^{ab} are the respective cokernels of the left, middle and right vertical arrows, the result follows from the snake lemma.

(b) The argument is exactly the same as in (a). Simply apply Theorem 15.18 instead of Theorem 15.10. □

15.6 Exercises

Exercise 15.1 Let k be a global field. Let F be a finite Galois non-trivial extension of k. Using Exercise 13.4, show that the group $k^*/N_{F/k}F^*$ is infinite (see Exercise 18.8 for a generalisation to the non-Galois case).

Exercise 15.2 Let $k = \mathbf{Q}(\sqrt{-d})$ (with d a strictly positive squarefree integer) a quadratic imaginary field. Let $C_k = I_k/k^*$ be the idèle class group of k and D_k be the neutral connected component of C_k.

(a) Show that k has one and only one archimedean place, which is complex.

(b) We denote by I_k^f the restricted product of the k_v^* for v a finite place of k (with respect to the $\mathcal{O}_v^*$). Show that there exists an injective group morphism $\mathbf{C}^* \to C_k$ such that the topological group quotient $C_k/\mathbf{C}^*$ is isomorphic to I_k^f/k^*.

(c) Deduce that D_k is isomorphic to $\mathbf{C}^*$.

Exercise 15.3 Let $k = \mathbf{Q}$. We denote by $\mathcal{M}$ the cycle p_∞, where p_∞ is the real place of $\mathbf{Q}$. Show that the conductor of $k^{\mathcal{M}}$ is 1, so it is not $\mathcal{M}$ (use Corollary 12.30).

Exercise 15.4 Let K be a cyclic extension of $\mathbf{Q}$ of degree ℓ^m with ℓ prime. Let $F \subset K$ be the intermediate extension such that $[F : \mathbf{Q}] = \ell$. Let $p \neq \ell$ be a prime number ramified in the extension $F/\mathbf{Q}$. We denote by K_p the completion of K at a place dividing p.

(a) Show that the extension $K_p/\mathbf{Q}_p$ is totally ramified.
(b) Show that every $x \in 1 + p\mathbf{Z}_p$ satisfies $x \in \mathbf{Q}_p^{*\ell^m}$.
(c) Deduce that the conductor of the extension $K_p/\mathbf{Q}_p$ is 1.
(d) Show that $p \equiv 1 \bmod \ell^m$ (you may first show that the reciprocity map induces a surjection from $\mathbf{Z}_p^*$ to $\mathrm{Gal}(K_p/\mathbf{Q}_p)$).
   (e) Let E be a quadratic extension of $\mathbf{Q}$, unramified and inert at 2. Show that E can be written as $E = \mathbf{Q}(\sqrt{m})$, where m is an integer with $m \equiv 5 \bmod 8$ (use the Example 7.11).
(f) Deduce from (d) and (e) that if K is a cyclic extension of $\mathbf{Q}$ of degree 8, then 2 cannot be unramified and inert in $K/\mathbf{Q}$ ("Wang's counter-example to Grunwald's theorem", 1948).

Exercise 15.5 Let k be a number field. In this exercise we will say that a finite Galois extension of k is *unramified* if it is unramified at every finite place of k, and totally split at every archimedean place of k. Fix a prime number p.

Let K be a finite Galois unramified extension of k, whose Galois group $G = \mathrm{Gal}(K/k)$ is a p-group. Assume that K has no extensions which are unramified and cyclic of order p at the same time. We denote by $\Omega_{K,\infty}$ (resp. $\Omega_{K,f}$) the set of archimedean (resp. finite) places of K.

(a) (i) Show that the cardinality $h(K)$ of the ideal class group $\mathrm{Cl}(K)$ of K is not divisible by p.
   (ii) Compute $\widehat{H}^q(G, \mathrm{Cl}(K))$ for any $q \in \mathbf{Z}$.
(b) Let I_K be the idèle group of K. We denote by I_K^1 the subgroup of I_K defined by

$$I_K^1 := \prod_{w \in \Omega_{K,\infty}} K_w^* \times \prod_{w \in \Omega_{K,f}} U_w,$$

where $U_w = \mathcal{O}_w^*$ is the group of units of the ring of integers of K_w^*.
   We denote by $C_K = I_K/K^*$ the idèle class group of K and $E_K = K^* \cap I_K^1$ the group of units of $\mathcal{O}_K$.
   (i) Show that $\widehat{H}^q(G, I_K^1) = 0$ for any $q \in \mathbf{Z}$ (you can draw inspiration from Sect. 13.1).
   (ii) Show that for any $q \in \mathbf{Z}$, we have an isomorphism $\widehat{H}^q(G, C_K) \simeq \widehat{H}^{q+1}(G, E_K)$.
(c) (i) Show that the groups $\widehat{H}^0(G, E_K)$ and $\widehat{H}^{-3}(G, \mathbf{Z})$ are isomorphic.
   (ii) Deduce that the rank of $\widehat{H}^{-3}(G, \mathbf{Z})$ (which is its minimal number of generators) is at most r, where r is the number of archimedean places of k. Show that the same is true of the group $H^3(G, \mathbf{Z})$ (use Exercise 2.1).
   (iii) With the notation of Exercise 1.10, show that

$$r(G) - n(G) \leq [k : \mathbf{Q}].$$

Exercise 15.6 (*sequel to the previous one*) Let p be a prime number. We admit the *Golod–Shafarevich inequality* ([49], Part. I, Ann. 3) which says that with the notations of Exercises 1.10 and 15.5, we have

$$r(G) > n(G)^2/4$$

for any finite p-group $G \neq \{1\}$.

(a) Let k be a number field, assumed to be totally imaginary if $p = 2$. Let $k(p)$ be the largest unramified Galois extension of k whose Galois group G is a pro-p-group. Show that there exists a constant $C(d)$ (depending only on the degree $d := [k : \mathbf{Q}]$) such that if the degree $[k(p) : k]$ is finite, we have

$$n(G) \leq C(d).$$

Deduce that the rank of the p-primary component of the ideal class group $\mathrm{Cl}(k)$ is at most $C(d)$.

(b) We take $p = 2$.

We consider the imaginary quadratic fields $k = \mathbf{Q}(\sqrt{-p_1 \cdots p_N})$, where the p_i are pairwise distinct prime numbers congruent to 1 modulo 4, and their extensions $k_i = k(\sqrt{p_i})$. Show that the extension k_i/k is unramified at every place of k (you can use Exercise 7.2). Show that there exists such a field k such that the degree $[k(2) : k]$ is infinite (analogous examples exist for any prime number p).

(c) Deduce that for such a field k, every finite extension of k has class number divisible by 2. In particular, the "tower of class fields", obtained by $k_0 = k$ and $k_{i+1} :=$ the Hilbert class field of k_i for $i \geq 0$, is infinite. It is the famous negative answer to the class field tower problem posed by Furtwängler, which remained open for about forty years until it was solved in 1964 by Golod and Shafarevich, see also article IX in [9].

Exercise 15.7 (*After D. Cox*, [13], *Theorem 5.1 et 5.26*) Let n be a squarefree natural number, not congruent to 3 modulo 4. Set $K = \mathbf{Q}(\sqrt{-n})$ and denote by h the class number of K. Recall that the ring of integers of K is then $\mathcal{O}_K = \mathbf{Z}[\sqrt{-n}]$. We denote by L the Hilbert class field of K. For any odd prime number p not dividing n, we say that p *verifies property (E)* if there exists $x, y \in \mathbf{Z}$ such that $p = x^2 + ny^2$.

(a) Show that $\mathcal{O}_K^* = \{\pm 1\}$.
(b) For any prime ideal $\wp$ of $\mathcal{O}_K$, we denote by $\overline{\wp}$ the prime ideal consisting of $\overline{x}$ with $x \in \wp$, where $\overline{x}$ is the conjugate of the complex number x. Show that p verifies (E) if and only if there exists a principal prime ideal $\wp$ of $\mathcal{O}_K$ with $p\mathcal{O}_K = \wp\overline{\wp}$ and $\wp \neq \overline{\wp}$.
(c) Let τ be the complex conjugation. Show that $\tau(L) = L$ (observe that L is an unramified extension of $\tau(K) = K$), then show that L is a finite Galois extension of $\mathbf{Q}$.
(d) Deduce that p verifies (E) if and only if it is totally split in the extension $L/\mathbf{Q}$.

(e) We view L as a subfield of $\mathbf{C}$ and set $L_1 = L \cap \mathbf{R}$. Show that L_1 is an extension of $\mathbf{Q}$ of degree $h = [L : K]$, then show that there exists a real algebraic integer α satisfying: $L_1 = \mathbf{Q}(\alpha)$ and $L = K(\alpha)$.

(f) We denote by f the minimal polynomial of α over $\mathbf{Q}$ (by e), it is a monic polynomial in $\mathbf{Z}[X]$ of degree h). Let $\mathrm{disc}(f) \in \mathbf{Z}$ be the discriminant of f (which is the resultant of polynomials f and f'). Let p be an odd prime number not dividing $n.\mathrm{disc}(f)$.

Prove the equivalence of the two assertions: (i) p verifies (E); (b) the Legendre symbol $(\frac{-n}{p})$ is 1 and the equation $f(x) = 0$ mod p has a solution in $\mathbf{Z}/p\mathbf{Z}$. You can use Example 12.15.

Exercise 15.8 (*sequel to the previous exercise*) Let $K = \mathbf{Q}(\sqrt{-14})$. We admit (cf. For example [13], Theorem 5.30, where the link is established between the class group of a quadratic imaginary field and the theory of integral binary quadratic forms due to Gauss) that the class number of K is 4. We set $\beta = 2\sqrt{2} - 1$ and $\beta' = -2\sqrt{2} - 1$, then $\alpha = \sqrt{\beta}$. Consider the fields $K_1 = K(\sqrt{2})$ and $L = K(\alpha) = K(\sqrt{2\sqrt{2} - 1})$.

(a) Show that $K_1 = K(\sqrt{-7})$ and $L = K_1(\alpha) = K_1(\sqrt{\beta'})$.
(b) Deduce that L/K is a finite Galois extension with group $\mathbf{Z}/4\mathbf{Z}$.
(c) Using Exercise 12.7, show that the extension L/K is unramified at every prime ideal of $\mathcal{O}_K$, and deduce that L is the Hilbert class field of K.
(d) Let p a prime number different from 2 and 7. Using Exercise 15.7, show that p can be written as $p = x^2 + 14y^2$ (with $x, y \in \mathbf{Z}$) if and only if: $(\frac{-14}{p}) = 1$ and the equation $(x^2 + 1)^2 = 8$ has a solution in $\mathbf{Z}/p\mathbf{Z}$.

Part IV
Duality Theorems

In this last part, we are interested in the arithmetic duality theorems. Our aim is to prove the Poitou–Tate theorems, which are essential tools in many modern number theoretic questions. We attain this objective in Chap. 17, after a chapter devoted to class formations in which we present an abstract duality theorem which will be a crucial step in the proof of the Poitou–Tate theorems and also allow us to recover and extend the Tate local duality theorem of Chap. 10. We end this book with a chapter devoted to a few classical applications, such as Grunwald–Wang theorem and a calculation of the strict cohomological dimension of a number field.

There is a number of generalisations of duality theorems covered in this book. One finds a very complete exposition, within particular many extensions to étale and flat cohomology in [39]. Let us mention that the étale cohomology is a powerful tool which provides for example a proof of the Poitou–Tate theorems which is in a certain way more natural, via what we call the *Artin-Verdier duality theorem* (loc. cit., paragraph II.3). A complete proof of Theorem 17.13 using this approach is covered in [25]. Duality theorems and their applications are still objects of active research; see for example papers [10, 17, 19] and [20].

Chapter 16
Class Formations

In this chapter, we develop a kind of "abstract" class field theory, by introducing the general formalism of class formation. The main goal is to prove the general duality Theorem 16.21 (due to Tate), which allows to recover and extend the local duality Theorem 10.9, and will be an important tool in the proof of Poitou–Tate duality theorems for global fields. The main examples of class formations are associated to the absolute Galois group of a local or global field, using properties proved in parts II and III (cf. Example 16.5).

16.1 Notion of Class Formation

Let G be a profinite group. Let C be a discrete G-module. We consider a family[1] of isomorphisms $\mathrm{inv}_U : H^2(U, C) \simeq \mathbf{Q}/\mathbf{Z}$ indexed by the open subgroups U of G.

Definition 16.1 We say that the system $(C, \{\mathrm{inv}_U\})$ (that we will also denote by (C, G)) as above is a *class formation* (associated to G) if the two following properties are satisfied:

(a) for every open subgroup U of G, we have $H^1(U, C) = 0$.
(b) for all open subgroups U, V of G with $V \subset U$, the following diagram is commutative:

$$
\begin{array}{ccc}
H^2(U, C) & \xrightarrow{\ \mathrm{Res}\ } & H^2(V, C) \\
{\scriptstyle \mathrm{inv}_U} \downarrow & & \downarrow {\scriptstyle \mathrm{inv}_V} \\
\mathbf{Q}/\mathbf{Z} & \xrightarrow{\ \cdot n\ } & \mathbf{Q}/\mathbf{Z}
\end{array}
\qquad (16.1)
$$

where $n := [U : V]$.

---

[1] The definition that we adopt here is that of [39], §I.1. It is slightly less general than that of Artin and Tate ([1], Chap. 14; [47], Chap. XI), but it will be sufficient for applications.

© Springer Nature Switzerland AG 2020
D. Harari, *Galois Cohomology and Class Field Theory*, Universitext,
https://doi.org/10.1007/978-3-030-43901-9_16

Let $(C, \{\mathrm{inv}_U\})$ be a class formation. Definition and Corollary 1.45 yield, for any pair of open subgroups U and V of G with V normal of index n in U, a commutative diagram with exact rows:

$$\begin{array}{ccccccccc}
0 & \longrightarrow & H^2(U/V, C^V) & \longrightarrow & H^2(U, C) & \xrightarrow{\ \mathrm{Res}\ } & H^2(V, C) & \longrightarrow & 0 \\
& & \downarrow{\scriptstyle \mathrm{inv}_{U/V}} & & \downarrow{\scriptstyle \mathrm{inv}_U} & & \downarrow{\scriptstyle \mathrm{inv}_V} & & \\
0 & \longrightarrow & \frac{1}{n}\mathbf{Z}/\mathbf{Z} & \longrightarrow & \mathbf{Q}/\mathbf{Z} & \xrightarrow{\ \cdot n\ } & \mathbf{Q}/\mathbf{Z} & \longrightarrow & 0.
\end{array}$$

In particular, if U is an open normal subgroup of G of index n, we have an isomorphism $\mathrm{inv}_{G/U} : H^2(G/U, C^U) \simeq \frac{1}{n}\mathbf{Z}/\mathbf{Z}$.

Definition 16.2 We denote by $u_{G/U}$ the unique element of $H^2(G/U, C^U)$ which is sent to $1/n$ by $\mathrm{inv}_{G/U}$; We say that $u_{G/U}$ is the *fundamental class* of G/U.

Remark 16.3 We note that with our definition, the existence of a class formation associated to G implies that the order of G (as a supernatural number) is divisible by each element of $\mathbf{N}^*$. Assume indeed there was a prime number p such that this order is not divisible by p^{r+1} with $r \in \mathbf{N}$. Then for U open and normal in G, $H^2(G/U, C^U)\{p\}$ would be killed by p^r (since each element of $H^2(G/U, C^U)\{p\}$ is killed by the order of G/U and by a power of p, hence by their gcd). Thus $H^2(G, C)\{p\}$ would be killed by p^r and hence would not be isomorphic to $(\mathbf{Q}/\mathbf{Z})\{p\}$.

Proposition 16.4 *Let $(C, \{\mathrm{inv}_U\})$ be a class formation. Let U, V be open subgroups of G with $V \subset U$. Then:*

(a) We have a commutative diagram:

$$\begin{array}{ccc}
H^2(V, C) & \xrightarrow{\ \mathrm{Cores}\ } & H^2(U, C) \\
{\scriptstyle \mathrm{inv}_V}\downarrow & & \downarrow{\scriptstyle \mathrm{inv}_U} \\
\mathbf{Q}/\mathbf{Z} & \xrightarrow{\ \mathrm{Id}\ } & \mathbf{Q}/\mathbf{Z}.
\end{array}$$

(b) Assume furthermore U and V to be normal in G. Then the image of the fundamental class $u_{G/U} \in H^2(G/U, C^U)$ by the inflation map $H^2(G/U, C^U) \to H^2(G/V, C^V)$ is $[U : V] \cdot u_{G/V}$.

Proof (a) follows immediately from the axioms of class formation and the formula $\mathrm{Cores} \circ \mathrm{Res} = \cdot [U : V]$. For (b), we write a commutative diagram:

$$\begin{array}{ccccc}
H^2(G/U, C^U) & \xrightarrow{\ \mathrm{Inf}\ } & H^2(G, C) & \xrightarrow{\ \mathrm{inv}_G\ } & \mathbf{Q}/\mathbf{Z} \\
{\scriptstyle \mathrm{Inf}}\downarrow & & \| & & \| \\
H^2(G/V, C^V) & \xrightarrow{\ \mathrm{Inf}\ } & H^2(G, C) & \xrightarrow{\ \mathrm{inv}_G\ } & \mathbf{Q}/\mathbf{Z}
\end{array}$$

and the result then follows from the definitions of $u_{G/U}$ and $u_{G/V}$, that are sent to $1/[G : U]$ and $1/[G : V]$ (respectively) in $\mathbf{Q}/\mathbf{Z}$.   □

Example 16.5 (a) Let $G = \text{Gal}(\overline{K}/K)$ be the absolute Galois group of a p-adic field K. Let us set $C = \overline{K}^*$. Then by Hilbert 90 and Proposition 8.4, the isomorphisms $\text{inv}_L : \text{Br } L \to \mathbf{Q}/\mathbf{Z}$ (defined for any open subgroup $U = \text{Gal}(\overline{K}/L)$ of G, i.e., for any finite extension L of K) define a class formation.

(b) Let $G = \widehat{\mathbf{Z}}$ (for example G may be the Galois group of a finite field). Take $C = \mathbf{Z}$, endowed with the trivial action of G. Let σ a topological generator of G. For $m > 0$, the unique subgroup U of index m of G is topologically generated by σ^m. The exact sequence

$$0 \longrightarrow \mathbf{Z} \longrightarrow \mathbf{Q} \longrightarrow \mathbf{Q}/\mathbf{Z} \longrightarrow 0$$

induces an isomorphism $\delta : H^1(U, \mathbf{Q}/\mathbf{Z}) \to H^2(U, \mathbf{Z})$. We obtain a class formation by taking $\text{inv}_U : H^2(U, \mathbf{Z}) \to \mathbf{Q}/\mathbf{Z}$ as the composite of $\delta^{-1} :$ $H^2(U, \mathbf{Z}) \to H^1(U, \mathbf{Q}/\mathbf{Z})$ with the evaluation $\chi \mapsto \chi(\sigma^m)$ of $H^1(U, \mathbf{Q}/\mathbf{Z})$ in $\mathbf{Q}/\mathbf{Z}$.

(c) Let k be a global field with absolute Galois group G_k. Let C be the inductive limit (for K/k finite separable extension) of the idèle class groups $C_K = I_K/K^*$. The properties we saw in Sect. 15.1 tell us that the isomorphisms

$$\text{inv}_K : H^2(G_K, C) \longrightarrow \mathbf{Q}/\mathbf{Z}$$

constitute a class formation on the G_k-module C.

Proposition 16.6 *Let $(C, \{\text{inv}_U\})$ be a class formation. Let U be an open normal subgroup of G. Then for any torsion-free G/U-module M, the cup-product with $u_{G/U}$:*

$$\widehat{H}^r(G/U, M) \longrightarrow \widehat{H}^{r+2}(G/U, M \otimes C^U)$$

is an isomorphism for any $r \in \mathbf{Z}$.

Proof It is the Tate-Nakayama theorem (Theorem 3.14).   □

Proposition 16.7 *Let $(C, \{\text{inv}_U\})$ be a class formation. Then we have a canonical homomorphism*

$$\omega_G : C^G \longrightarrow G^{\text{ab}}$$

with dense image, and such that the kernel is the group of universal norms $\bigcap_U N_{G/U} C^U$ (the intersection is taken over open normal subgroups U of G). We say that ω_G is the reciprocity map associated to the class formation $(C, \{\text{inv}_U\})$.

Proof It is exactly the same argument as the one we have used in Theorem 9.2 and Definition 9.5. We take $r = -2$ and $M = \mathbf{Z}$ in the previous proposition. We obtain isomorphisms $(G/U)^{\text{ab}} \to C^G/N_{G/U} C^U$, and converse isomorphisms

$$\omega_{G/U} : C^G/N_{G/U}C^U \longrightarrow (G/U)^{\mathrm{ab}}$$

which are compatible with each other in an obvious way (cf. Corollary 9.4, a), by the same calculation as in Proposition 9.3. This allows us to pass to the limit to obtain an isomorphism

$$\varprojlim_U C^G/N_{G/U}C^U \longrightarrow G^{\mathrm{ab}}.$$

We then obtain ω_G by composing the canonical homomorphism

$$C^G \longrightarrow \varprojlim_U C^G/N_{G/U}C^U$$

with this isomorphism. The assertion about the kernel of ω_G is then evident and the density of $\mathrm{Im}\,\omega_G$ follows from the fact that its composite with the projection $G^{\mathrm{ab}} \to (G/U)^{\mathrm{ab}}$ is surjective for any U.   □

Remark 16.8 The same argument as in Corollary 9.6 gives, for any c in C^G and for any character χ of G^{ab}, the formula

$$\mathrm{inv}_G(c \cup \chi) = \chi(\omega_G(c)),$$

where, to form the cup-product, c is viewed as an element of $H^0(G, C)$ and χ of $H^2(G, \mathbf{Z})$.

Remark 16.9 To define the universal norm group, we can limit ourselves to forming the intersection of the $N_{G/U}C^U$ for the U which satisfy the additional condition that the G/U are abelian. This will follow from Proposition 16.29 (b) below (also see Exercise 9.1). In the case of class formation associated to a local field (Example 16.5, a), the universal norm group is trivial as a corollary of the existence theorem (cf. Exercise 11.2 or Corollary 9.15 for the p-adic case).

16.2   The Spectral Sequence of the Ext

In this paragraph, we will recall some standard results on a spectral sequence which often occurs in group cohomology. We always denote by G a profinite group and by C_G the category of discrete G-modules. The category of abelian groups will be denoted $\mathcal{A}b$. Consider objects M and N of C_G. A difficulty is that if M is not assumed to be of finite type the G-module $\mathrm{Hom}(M, N) = \mathrm{Hom}_{\mathbf{Z}}(M, N)$ (the action of G is defined by the usual formula, cf. Example 1.3, d) is in general not discrete (in other words, the stabiliser of an element is not always open), cf. Exercise 16.1.

Definition 16.10 Let M and N be discrete G-modules. We define the discrete G-module $\mathcal{H}om(M, N)$ by the formula

$$\mathcal{H}om(M, N) = \bigcup_U \mathrm{Hom}(M, N)^U,$$

where U runs through the set of open subgroups of G (or that of normal open subgroups, which is the same as these form a basis of neighbourhoods of 1). For any normal closed subgroup H of G, we similarly denote

$$\mathcal{H}om_H(M, N) = \mathcal{H}om(M, N)^H = \bigcup_{U \supset H} \mathrm{Hom}(M, N)^U,$$

where the union is taken over the set of open subgroups of G containing H. It is a discrete G/H-module.

Definition 16.11 Let H be a closed subgroup of a profinite group G. We assume H to be normal in G. Let M be a discrete G-module. For any $r \geqslant 0$, we define $\mathcal{E}xt_H^r(M, N)$ to be the rth right derived functor of the functor

$$N \longmapsto \mathcal{H}om_H(M, N) : C_G \longrightarrow C_{G/H}.$$

Thus $\mathcal{E}xt_H^r(M, N)$ is a discrete G/H-module.

Remark 16.12 The G/H-module $\mathcal{H}om_H(M, N)$ corresponds to the *internal $\mathcal{H}om$* in the category $C_{G/H}$, for the G/H-modules M^H and N^H, and the $\mathcal{E}xt_H(M, N)$ to the corresponding *internal $\mathcal{E}xt$*. When G is the Galois group of a field k, we can identify M and N to *étale sheaves* on Spec k ([36], Part. II) and we recover the distinction between Hom (resp. Ext) and $\mathcal{H}om$ (resp. $\mathcal{E}xt$) for sheaves.

Remark 16.13 Recall that we also have the $\mathrm{Ext}_H^r(M, .)$, defined as derived functors of $\mathrm{Hom}_H(M, .)$ (from the category C_H to the category $\mathcal{A}b$ of abelian groups). By Proposition 4.25, we see[2] that if we restrict these functors $\mathrm{Ext}_H^r(M, .)$ to the category C_G (viewing the discrete G-modules as discrete H-modules), we also obtain derived functors of $\mathrm{Hom}_H(M, .)$ from the category C_G to $\mathcal{A}b$ (or to $C_{G/H}$). If furthermore M is of finite type, we have $\mathrm{Hom}_H(M, N) = \mathcal{H}om_H(M, N)$ for any $N \in C_G$, and there is thus no need to distinguish between $\mathrm{Ext}_H^{r(M,N)}$ and $\mathcal{E}xt_H^r(M, N)$. Recall also that Remark 1.37 still applies to a profinite G, and so does Proposition 1.39 if the subgroup H is assumed to be open. Lastly, we prove in exactly the same way as Shapiro's lemma (cf. also Exercise 1.5) that we have, for any discrete G-module A and any discrete H-module B, canonical isomorphisms

$$\mathrm{Ext}_G^i(A, I_G^H(B)) \simeq \mathrm{Ext}_H^i(A, B).$$

When $H = \{1\}$, we will write $\mathcal{E}xt^r(M, N)$ instead of $\mathcal{E}xt_{\{1\}}^r(M, N)$. We also denote by $\mathrm{Ext}^{r(M,N)}$ or $\mathrm{Ext}_{\mathbf{Z}}^r(M, N)$ the Ext in the category of abelian groups (corresponding to $\mathrm{Ext}_H^{r(M,N)}$ for $H = \{1\}$).

---

[2]The importance to know here Proposition 4.25 does not seem to have been observed in the earlier texts.

Theorem 16.14 *Let H be a closed normal subgroup of a profinite group G. Let $N, P \in C_G$ and let M be a discrete G/H-module without torsion. Then we have a spectral sequence*

$$E_2^{r,s} = \mathrm{Ext}_{G/H}^r(M, \mathcal{E}xt_H^s(N, P)) \implies \mathrm{Ext}_G^{r+s}(M \otimes N, P).$$

As usual $\mathrm{Tor}_{\mathbf{Z}}(M, N) = 0$ (instead of M torsion-free) would be enough.

Proof Let M be any discrete G/H-module (for now we do not assume it to be torsion-free). We can consider M as a discrete G-module with a trivial action of H. We then have

$$\begin{aligned}\mathrm{Hom}_{G/H}(M, \mathcal{H}om_H(N, P)) &= \mathrm{Hom}_G(M, \mathcal{H}om_H(N, P))\\ &= \mathrm{Hom}_G(M \otimes N, P)\end{aligned} \qquad (16.2)$$

since

$$\mathrm{Hom}_G(M, \mathcal{H}om_H(N, P)) = \mathrm{Hom}_G(M, \mathcal{H}om(N, P))$$

(the action of H on M is trivial) and

$$\mathrm{Hom}_G(M, \mathcal{H}om(N, P)) = \mathrm{Hom}_G(M, \mathrm{Hom}(N, P))$$

(since M is a discrete G-module, union of the M^U for U open subgroup of G). We deduce that for any injective I of C_G and any $N \in C_G$ torsion-free, the G/H-module $Q := \mathcal{H}om_H(N, I)$ is injective in $C_{G/H}$ since $\mathrm{Hom}_{G/H}(., Q)$ is the composite of three exact functors: $M \mapsto M$ from $C_{G/H}$ to C_G, $. \otimes N$ and $\mathrm{Hom}_G(., I)$.

Assume now that M is torsion-free and N is any module in C_G. To apply the Grothendieck spectral sequence of composed functors (Appendix, Theorem A.67), it is enough to check that for every injective I of C_G, the G/H-module $\mathcal{H}om_H(N, I)$ is acyclic in $C_{G/H}$ for the functor $\mathrm{Hom}_{G/H}(M, .)$. To do this, we write a resolution of N by discrete torsion-free G-modules N_1 and N_0

$$0 \longrightarrow N_1 \longrightarrow N_0 \longrightarrow N \longrightarrow 0.$$

This is possible since every x of N generates a G-module which is a quotient of $\mathbf{Z}[G/U]$ for a certain open subgroup U of G, which implies that N is a quotient of a direct sum of such modules (which are torsion-free). Then I remains injective as a U-module for any open subgroup U of G (it is a special case of Proposition 4.25, which by the way can be dealt with directly as the special case where G is finite, cf. Lemma 1.38). For such a U, we thus obtain an exact sequence

$$0 \longrightarrow \mathrm{Hom}_U(N, I) \longrightarrow \mathrm{Hom}_U(N_0, I) \longrightarrow \mathrm{Hom}_U(N_1, I) \longrightarrow 0.$$

By passing to the inductive limit over the U which contain H, we obtain an exact sequence

$$0 \longrightarrow \mathcal{H}om_H(N, I) \longrightarrow \mathcal{H}om_H(N_0, I) \longrightarrow \mathcal{H}om_H(N_1, I) \longrightarrow 0$$

which is then (by what has been seen above regarding Q) an injective resolution of the G/H-module $\mathcal{H}om_H(N, I)$. This already implies that if $r \geqslant 2$, we have $\mathrm{Ext}^r_{G/H}(M, \mathcal{H}om_H(N, I)) = 0$. For $r = 1$, we just need to check that the map

$$\mathrm{Hom}_{G/H}(M, \mathcal{H}om_H(N_0, I)) \longrightarrow \mathrm{Hom}_{G/H}(M, \mathcal{H}om_H(N_1, I))$$

remains surjective. By (16.2), this map identifies with to the natural map

$$\mathrm{Hom}_G(M \otimes N_0, I) \longrightarrow \mathrm{Hom}_G(M \otimes N_1, I)$$

which is surjective since I is injective and the map $M \otimes N_1 \to M \otimes N_0$ remains injective since M is torsion-free (the kernel is $\mathrm{Tor}_{\mathbf{Z}}(M, N)$). Finally we have

$$\mathrm{Ext}^r_{G/H}(M, \mathcal{H}om_H(N, I)) = 0$$

for any $r \geqslant 1$. In other words $\mathcal{H}om_H(N, I)$ is acyclic for the functor $\mathrm{Hom}_{G/H}(M, .)$.   □

Example 16.15 (a) For $M = N = \mathbf{Z}$, we recover the Hochschild-Serre sequence since $\mathcal{E}xt^r_H(\mathbf{Z}, .) = \mathrm{Ext}^r_H(\mathbf{Z}, .) = H^r(H, .)$ for any $r \geqslant 0$.
(b) Let $M = \mathbf{Z}$ and $H = \{1\}$. We obtain a spectral sequence

$$H^r(G, \mathcal{E}xt^s(N, P)) \implies \mathrm{Ext}^{r+s}_G(N, P).$$

Proposition 16.16 *Let N and P be discrete G-modules. Assume N to be of finite type.*

(a) We have an exact sequence

$$0 \longrightarrow H^1(G, \mathrm{Hom}(N, P)) \longrightarrow \mathrm{Ext}^1_G(N, P) \longrightarrow H^0(G, \mathrm{Ext}^1_{\mathbf{Z}}(N, P))$$
$$\longrightarrow H^2(G, \mathrm{Hom}(N, P)) \longrightarrow \cdots$$
$$(16.3)$$

Furthermore, the $\mathrm{Ext}^r_G(N, P)$ are torsion for any $r \geqslant 1$.
(b) If furthermore N is torsion-free, we have

$$H^r(G, \mathrm{Hom}(N, P)) = \mathrm{Ext}^r_G(N, P)$$

for any $r \geqslant 0$.
(c) Let (G_i) be a projective system of profinite groups. Let (P_i) be an inductive system of discrete G_i-modules, the transition maps being compatible with those of the (G_i). Let $G = \varprojlim G_i$ and $P = \varinjlim P_i$. Then we have isomorphisms

$$\varinjlim_i \operatorname{Ext}^r_{G_i}(N, P_i) \simeq \operatorname{Ext}^r_G(N, P).$$

Proof (a) As N is of finite type, the spectral sequence of the Example 16.15 (b) becomes

$$H^r(G, \operatorname{Ext}^s(N, P)) \implies \operatorname{Ext}^{r+s}_G(N, P).$$

Using the fact that $\operatorname{Ext}^s_{\mathbf{Z}}(N, P)$ is zero for $s \geqslant 2$ (Appendix, Proposition A.54), we obtain the desired exact sequence by Proposition A.65 (c) of the appendix.

As the abelian group N is the direct sum of copies of $\mathbf{Z}/n$ and of $\mathbf{Z}$, and $\operatorname{Ext}^1_{\mathbf{Z}}(\mathbf{Z}, P) = 0$, this implies that there exists an integer $m > 0$ such that $\operatorname{Ext}^1_{\mathbf{Z}}(N, P)$ is m-torsion; using the exact sequence, all $\operatorname{Ext}^r_G(N, P)$ are also m-torsion for $r \geqslant 1$.

(b) Note that $\operatorname{Ext}^1_{\mathbf{Z}}(N, P) = 0$ is zero if N is furthermore torsion-free, since as an abelian group N is then isomorphic to the direct sum of a finite number of copies of $\mathbf{Z}$. Then (a) yields

$$H^r(G, \operatorname{Hom}(N, P)) = \operatorname{Ext}^r_G(N, P)$$

for any $r \geqslant 0$.

(c) We write the exact sequence (16.3) for G_i and P_i, then we pass to the inductive limit. As N is of finite type, the functors $\operatorname{Hom}(N, .)$ and $\operatorname{Ext}^1_{\mathbf{Z}}(N, .)$ commute with inductive limits. Proposition 4.18 then yields for any $r \geqslant 0$ isomorphisms

$$\varinjlim_i H^r(G_i, \operatorname{Hom}(N, P_i)) \simeq H^r(G, \operatorname{Hom}(N, P)),$$

$$\varinjlim_i H^r(G_i, \operatorname{Ext}^1_{\mathbf{Z}}(N, P_i)) \simeq H^r(G, \operatorname{Ext}^1_{\mathbf{Z}}(N, P)).$$

We conclude with the five lemma.   □

Let now M, N and P be discrete G-modules. We have a pairing (cf. appendix, Definition A.57, see also [39], p. 4)

$$\operatorname{Ext}^r_G(M, N) \times \operatorname{Ext}^s_G(P, M) \longrightarrow \operatorname{Ext}^{r+s}_G(P, N).$$

As $H^s(G, .) = \operatorname{Ext}^s_G(\mathbf{Z}, .)$ for any integer s, we deduce (taking $P = \mathbf{Z}$) a pairing

$$\operatorname{Ext}_{M,N} : \operatorname{Ext}^r_G(M, N) \times H^s(G, M) \longrightarrow H^{r+s}(G, N) \qquad (16.4)$$

which, for $r = 0$, is the map $(f, a) \mapsto f_*(a)$ of $\operatorname{Hom}_G(M, N) \times H^s(G, M)$ in $H^s(G, N)$. This pairing is functorial in M and N. It is also compatible (in an obvious way) with the long exact sequences associated to the exact sequences

$$0 \longrightarrow M \longrightarrow M' \longrightarrow M'' \longrightarrow 0$$

("left" compatibility) and to the exact sequences

$$0 \longrightarrow N \longrightarrow N' \longrightarrow N'' \longrightarrow 0$$

("right" compatibility).

This pairing also admits the following compatibility with the cup-product:

Proposition 16.17 *Let M, N, P be discrete G-modules. Let*

$$\varphi : M \times N \longrightarrow P$$

be a pairing of G-modules (i.e., bilinear map compatible with the action of G). Denote by u : M → Hom(N, P) the corresponding homomorphism of G-modules. Let

$$v : H^r(G, M) \longrightarrow \mathrm{Ext}^r_G(N, P)$$

be the composite of $u_* : H^r(G, M) \to H^r(G, \mathcal{H}om(N, P))$ with the map

$$H^r(G, \mathcal{H}om(N, P)) \longrightarrow \mathrm{Ext}^r_G(N, P)$$

coming from the spectral sequence of Theorem 16.14. Then the following diagram is commutative:

$$
\begin{array}{ccc}
H^r(G, M) \times H^s(G, N) & \xrightarrow{\ \cup\ } & H^{r+s}(G, P) \\
{\scriptstyle (v,\, \mathrm{Id})} \downarrow & & \downarrow {\scriptstyle \mathrm{Id}} \\
\mathrm{Ext}^r_G(N, P) \times H^s(G, N) & \xrightarrow{\ \mathrm{Ext}_{N,P}\ } & H^{r+s}(G, P).
\end{array}
$$

For a proof[3] (in a more general case), see [36], Prop. V.1.20.

We will also need the following compatibility between the above pairings, Shapiro isomorphisms, and those of the Proposition 1.39 (in their version where G is profinite, cf. Remark 16.13):

Proposition 16.18 *Let G be a profinite group. Let U be an open subgroup of G. Let M and N be discrete G-modules. Let $M_* = I^U_G(M)$. Then we have a commutative diagram:*

---

[3]This type of verification is much simpler when one knows the formalism of derived categories, see Remark A.58 of the appendix.

$$\begin{array}{ccc}
\mathrm{Ext}^p_U(M, N) \times H^q(U, M) & \longrightarrow & H^{p+q}(U, N) \\
\wr \downarrow i & \wr \uparrow \mathrm{Sh} & \downarrow \mathrm{Cores} \\
\mathrm{Ext}^p_G(M_*, N) \times H^q(G, M_*) & \longrightarrow & H^{p+q}(G, N),
\end{array}$$

where the left vertical map is the isomorphism given by the profinite version of the Proposition 1.39, the middle one is the isomorphism of Shapiro's lemma and the right one is the corestriction.

Proof We start with the case $q = 0$. We need to show that the diagram

$$\begin{array}{ccc}
\mathrm{Ext}^p_U(M, N) \times M^U & \longrightarrow & H^p(U, N) \\
\wr \downarrow i & \| & \downarrow \mathrm{Cores} \\
\mathrm{Ext}^p_G(M_*, N) \times (M_*)^G & \longrightarrow & H^p(G, N)
\end{array}$$

commutes. For $p = 0$, this follows from the explicit definition of the left isomorphism, given by Proposition 1.15 (which extends immediately to the case of an open subgroup of a profinite group), and from the definition of the corestriction in degree zero: indeed we find that for all $\varphi \in \mathrm{Hom}_U(M, N)$ and any $a \in M^U = (M_*)^G$, the pairing at the top $\varphi \cdot a$ is $\varphi(a) \in N^U$; while the image of φ in $\mathrm{Hom}_G(M_*, N)$ is $f \mapsto \sum_{g \in G/U} g \cdot \varphi(f(g^{-1}))$ which, paired at the bottom with the constant function $f = a$, gives $\sum_{g \in G/U} g \cdot \varphi(a) \in N^G$. But this last term is $\mathrm{Cores}(\varphi(a))$. To deal with the case of an arbitrary p (still in the case $q = 0$), we proceed by shifting, writing an exact sequence

$$0 \longrightarrow N \longrightarrow I \longrightarrow N_1 \longrightarrow 0$$

with I injective in C_G. Assume the result to be true for p, and let us prove it for $p + 1$. The induction assumption and the compatibility of the pairing (16.4) with the "right" coboundaries yield a commutative diagram:

$$\begin{array}{ccc}
\mathrm{Ext}^{p+1}_G(M_*, N) \times (M_*)^G & \longrightarrow & H^{p+1}(G, N) \\
\partial \uparrow & \| & \uparrow \partial \\
\mathrm{Ext}^p_G(M_*, N_1) \times (M_*)^G & \longrightarrow & H^p(G, N_1) \\
i \uparrow \wr & \| & \uparrow \mathrm{Cores} \\
\mathrm{Ext}^p_U(M, N_1) \times M^U & \longrightarrow & H^p(U, N_1) \\
\partial \downarrow & \| & \downarrow \partial \\
\mathrm{Ext}^{p+1}_U(M, N) \times M^U & \longrightarrow & H^{p+1}(U, N).
\end{array} \qquad (16.5)$$

Besides, the compatibility with the coboundaries of the corestriction and of isomorphisms i (Proposition 1.39, in profinite version) yield commutative diagrams:

$$
\begin{array}{ccc}
H^p(G, N_1) \xrightarrow{\ \partial\ } H^{p+1}(G, N) & \quad & \mathrm{Ext}_U^p(M, N_1) \xrightarrow{\ \partial\ } \mathrm{Ext}_U^{p+1}(M, N) \\
\Big\uparrow \text{Cores} \qquad\qquad \Big\uparrow \text{Cores} & \quad & \Big\downarrow i \qquad\qquad\qquad \Big\downarrow i \\
H^p(U, N_1) \xrightarrow{\ \partial\ } H^{p+1}(U, N) & \quad & \mathrm{Ext}_G^p(M_*, N_1) \xrightarrow{\ \partial\ } \mathrm{Ext}_G^{p+1}(M_*, N).
\end{array}
$$

In addition, the coboundary $\partial : \mathrm{Ext}_U^p(M, N_1) \to \mathrm{Ext}_U^{p+1}(M, N)$ is surjective (and even bijective if $p > 1$) as $\mathrm{Ext}_U^{p+1}(M, I) = 0$ since I is injective in C_G (hence also in C_U, Proposition 4.25 in the special case of an open subgroup). We obtain the desired commutative diagram (which corresponds to the rectangle relating the first row of the diagram (16.5) to the last) for $q = 0$, by induction on p.

The case where q is arbitrary is dealt with using exactly the same shifting method (i.e., by induction on q) writing an exact sequence

$$ 0 \longrightarrow M \longrightarrow I \longrightarrow M_1 \longrightarrow 0 $$

with I injective of C_G. The sequence

$$ 0 \longrightarrow M_* \longrightarrow I_* \longrightarrow (M_1)_* \longrightarrow 0 $$

then remains exact. Consider the diagram

$$
\begin{array}{ccc}
\mathrm{Ext}_U^p(M, N) \times H^{q+1}(U, M) & \longrightarrow & H^{p+q+1}(U, N) \\
\partial \big\downarrow\wr \qquad \wr\big\uparrow \partial & & \big\| \\
\mathrm{Ext}_U^{p+1}(M_1, N) \times H^q(U, M_1) & \longrightarrow & H^{p+q+1}(U, N) \\
i \big\downarrow\wr \qquad \wr\big\uparrow \mathrm{Sh} & & \big\downarrow \text{Cores} \\
\mathrm{Ext}_G^{p+1}((M_1)_*, N) \times H^q(G, (M_1)_*) & \longrightarrow & H^{p+q+1}(G, N) \\
\partial \big\uparrow \qquad\qquad \big\downarrow \partial & & \big\| \\
\mathrm{Ext}_G^p(M_*, N) \times H^{q+1}(G, M_*) & \longrightarrow & H^{p+q+1}(G, N).
\end{array}
$$

It is commutative by compatibility of the "left" pairing (16.4) with the coboundaries and the induction assumption. The coboundary $H^q(U, M_1) \to H^{q+1}(U, M)$ is surjective (since I is an injective of C_G, hence of C_U), and we also have compatibility of Shapiro isomorphisms with the coboundaries of the long exact cohomology sequences and the compatibility of isomorphisms i with the boundaries of long exact homology sequences (cf. the last assertion of the Proposition 1.39). We deduce the desired commutative diagram (with $q + 1$), which again corresponds to the rectangle relating the first row to the last. $\qquad\square$

16.3   The Duality Theorem for a Class Formation

In this paragraph, G is a profinite group and $(C, \{\text{inv}_U\})$ is a class formation associated to G. The expression "G-module" as usual refers to a discrete G-module. To simplify the notation, we set $H^i(G, \ldots) = 0$ if i is a strictly negative integer.

For any G-module M, we have (for $r \in \mathbf{N}$) a pairing

$$\text{Ext}^r_G(M, C) \times H^{2-r}(G, M) \longrightarrow \mathbf{Q}/\mathbf{Z}$$

obtained by composing $\text{Ext}_{M,C} : \text{Ext}^r_G(M, C) \times H^{2-r}(G, M) \to H^2(G, C)$ with $\text{inv}_G : H^2(G, C) \to \mathbf{Q}/\mathbf{Z}$. We deduce, for $r \in \mathbf{N}$, a homomorphisms of abelian groups (functorial and compatible with coboundaries associated with short exact sequences):

$$\alpha^r(G, M) : \text{Ext}^r_G(M, C) \longrightarrow H^{2-r}(G, M)^*.$$

For $r = 0$ and $M = \mathbf{Z}$, the homomorphism $\alpha^0(G, \mathbf{Z})$ goes from C^G to $G^{\text{ab}} = H^2(G, \mathbf{Z})^*$. We also have $\text{Ext}^1_G(\mathbf{Z}/m, C) = C^G/m$. This follows from the exact sequence

$$0 \longrightarrow \mathbf{Z} \xrightarrow{\cdot m} \mathbf{Z} \longrightarrow \mathbf{Z}/m \longrightarrow 0 \tag{16.6}$$

and the axiom $H^1(G, C) = 0$. Similarly, $\text{Ext}^2_G(\mathbf{Z}/m, C) = H^2(G, C)[m]$.

We begin by writing explicitly the $\alpha^r(G, M)$ for $M = \mathbf{Z}$ and $M = \mathbf{Z}/m$.

Lemma 16.19 (a) *The map $\alpha^0(G, \mathbf{Z})$ is the reciprocity map $\omega_G : C^G \to G^{\text{ab}}$. The source and the target of the homomorphism $\alpha^1(G, \mathbf{Z})$ are zero and $\alpha^2(G, \mathbf{Z}) : H^2(G, C) \to \mathbf{Q}/\mathbf{Z}$ is equal to inv_G.*

(b) *The composite of the map $\alpha^0(G, \mathbf{Z}/m) : (C^G)[m] \to H^2(G, \mathbf{Z}/m)^*$ with the natural map $H^2(G, \mathbf{Z}/m)^* \to H^2(G, \mathbf{Z})^*[m] = G^{\text{ab}}[m]$ is the restriction $C^G[m] \to G^{\text{ab}}[m]$ of the map ω_G. The map $\alpha^1(G, \mathbf{Z}/m) : C^G/m \to H^1(G, \mathbf{Z}/m)^* = G^{\text{ab}}/m$ is induced by ω_G and the map*

$$\alpha^2(G, \mathbf{Z}/m) : H^2(G, C)[m] \longrightarrow \tfrac{1}{m}\mathbf{Z}/\mathbf{Z}$$

is induced by inv_G.

Proof The statements about the maps $\alpha^1(G, \mathbf{Z})$ and $\alpha^2(G, \mathbf{Z})$ are straightforward. Proposition 16.17 along with the fact that the homomorphism $\omega_G : H^0(G, C) \to H^2(G, \mathbf{Z})^*$ is induced by the cup-product (Remark 16.8) imply the assertion about $\alpha^0(G, \mathbf{Z})$. The assertions about the $\alpha^r(G, \mathbf{Z}/m)$ then follow immediately from those on the $\alpha^r(G, \mathbf{Z})$ from identifications $\text{Ext}^1_G(\mathbf{Z}/m, C) = C^G/m$ and $\text{Ext}^2_G(\mathbf{Z}/m, C) = H^2(G, C)[m]$, which result from long exact sequences associated to the exact sequence (16.6). $\qquad\square$

Lemma 16.20 *Let M be a finite type G-module. Then $\text{Ext}^r_G(M, C) = 0$ if $r \geqslant 4$. If we assume furthermore M to be torsion-free, then $\text{Ext}^3_G(M, C) = 0$.*

Proof We can find an exact sequence of G-modules

$$0 \longrightarrow M_1 \longrightarrow M_0 \longrightarrow M \longrightarrow 0$$

with M_0 of finite type and torsion-free, and hence M_1 is also of finite type (as it is a submodule of a finite type module over $\mathbf{Z}$) and torsion-free. Using the long exact sequence, we are thus reduced to showing that if $r \geqslant 3$, we have $\mathrm{Ext}_G^r(M, C) = 0$ for M of finite type and torsion-free. Let then N be the G-module $N = \mathrm{Hom}(M, \mathbf{Z})$. The G-modules $N \otimes C$ and $\mathrm{Hom}(M, C)$ are isomorphic, via the map

$$\varphi \otimes c \longmapsto (m \longmapsto \varphi(m)c), \quad \varphi \in \mathrm{Hom}(M, \mathbf{Z}), c \in C, m \in M,$$

hence we get an isomorphism

$$\mathrm{Ext}_G^r(M, C) \simeq H^r(G, N \otimes C)$$

by Proposition 16.16 (b). As N is of finite type, we have $N = N^U$ (and hence the U-module N is isomorphic to $\mathbf{Z}^m$ with $m \in \mathbf{N}$) for U an open normal small enough subgroup of G, which means that we have

$$H^r(G, N \otimes C) = \varinjlim_U H^r(G/U, N \otimes C^U),$$

where the limit is taken over open normal subgroups U of G such that $N = N^U$. Proposition 16.6 (immediate consequence of the Tate-Nakayama theorem) implies then that for any $r \geqslant 3$ (this assumption is necessary to guarantee that the Tate modified groups coincide with the usual cohomology groups), the cup-product with the fundamental class $u_{G/U}$ provides an isomorphism

$$H^{r-2}(G/U, N) \simeq H^r(G/U, N \otimes C^U).$$

We then have, for V open and normal contained in U, a commutative diagram, whose horizontal arrows are isomorphisms:

$$
\begin{array}{ccc}
H^{r-2}(G/U, N) & \xrightarrow{\ \cup\, u_{G/U}\ } & H^r(G/U, N \otimes C^U) \\
{\scriptstyle [U\,:\,V]\,\cdot\,\mathrm{Inf}} \downarrow & & \downarrow {\scriptstyle \mathrm{Inf}} \\
H^{r-2}(G/V, N) & \xrightarrow{\ \cup\, u_{G/V}\ } & H^r(G/V, N \otimes C^V).
\end{array}
$$

The fact that the diagram is commutative follows from the formulas

$$\mathrm{Inf}(u_{G/U}) = [U : V] \cdot u_{G/V}$$

(Proposition 16.4) and

$$\mathrm{Inf}(a \cup b) = \mathrm{Inf}(a) \cup \mathrm{Inf}(b)$$

(Proposition 2.27, d). On the other hand for $r \geqslant 3$, the group $H^{r-2}(G/U, N)$ is torsion and the order of U (as a supernatural number) is divisible by any integer n (Remark 16.3). We deduce that the inductive limit of the $H^{r-2}(G/U, N)$ (for transition maps $[U : V] \cdot \mathrm{Inf}$) is zero: indeed, if α is an m-torsion element (with $m > 0$) in an $H^{r-2}(G/U, N)$, we can find an open normal subgroup V of G contained in U such that the index $[U : V]$ is divisible by m, and the image of α in $H^{r-2}(G/V, N)$ is then zero. Finally $H^r(G, N \otimes C)$ is zero, as desired. $\square$

We are now reaching the main theorem of this chapter, which is the general duality theorem for a class formation.

Theorem 16.21 (Tate) *Let M be a finite type G-module.*

(a) The homomorphism

$$\alpha^r(G, M) : \mathrm{Ext}^r_G(M, C) \longrightarrow H^{2-r}(G, M)^*$$

is bijective for any $r \geqslant 2$ and if M is torsion-free, $\alpha^1(G, M)$ is also bijective. In particular, $\mathrm{Ext}^r_G(M, C) = 0$ for any $r \geqslant 3$.

(b) Assume $\alpha^1(U, \mathbf{Z}/m)$ to be bijective for every $m > 0$ and every open (normal) subgroup U of G. Then $\alpha^1(G, M)$ is bijective.

(c) Keep the assumptions of (b) and assume furthermore M to be finite. We also make the assumption that $\alpha^0(U, \mathbf{Z}/m)$ is surjective (resp. bijective) for every $m > 0$ and every open (normal) subgroup U of G. Then $\alpha^0(G, M)$ is surjective (resp. bijective).

Proof By Lemma 16.20, we have the result if $r \geqslant 4$. Assume then $r \leqslant 3$. Let us first prove the theorem when G acts trivially on M. In this case, M is a direct sum of G-modules isomorphic to either $\mathbf{Z}$ or to $\mathbf{Z}/n$ and it is enough to treat these two cases. For $M = \mathbf{Z}$ (where we limit ourselves to $1 \leqslant r \leqslant 3$ as $\mathbf{Z}$ is not finite), this is Lemma 16.19 (a) and Lemma 16.20. For $M = \mathbf{Z}/n$, this is Lemma 16.19 (b) if $r = 2$ and assumption (b) (resp. c) of the theorem if $r = 1$ (resp. $r = 0$). Lastly, for $r = 3$, we deduce $\mathrm{Ext}^3_G(\mathbf{Z}/m, C) = 0$ from $\mathrm{Ext}^3_G(\mathbf{Z}, C) = 0$ (Lemma 16.20) using the exact sequence

$$0 \longrightarrow \mathrm{Ext}^2_G(\mathbf{Z}, C)/m \longrightarrow \mathrm{Ext}^3_G(\mathbf{Z}/m, C) \longrightarrow \mathrm{Ext}^3_G(\mathbf{Z}, C)$$

(which follows from the exact sequence (16.6)) and the fact that $\mathrm{Ext}^2_G(\mathbf{Z}, C) \simeq \mathbf{Q}/\mathbf{Z}$ is divisible. We obtain that the theorem is valid when the action of G on M is trivial.

Let us now treat the general case. As M is of finite type, we can find an open normal subgroup U with G acting trivially on M. Write the exact sequence

$$0 \longrightarrow M \longrightarrow M_* \longrightarrow M_1 \longrightarrow 0$$

with $M_* = I_G^U(M)$. Shapiro's lemma implies $H^r(G, M_*) \simeq H^r(U, M)$. On the other hand we have $\mathrm{Ext}_U^r(M, C) \simeq \mathrm{Ext}_G^r(M_*, C)$ (profinite version of Proposition 1.39). We obtain a commutative diagram (the non-obvious commutations[4] come from Proposition 16.18 and the fact that the corestriction $H^2(U, C) \to H^2(G, C)$ induces via inv_U and inv_G the identity on $\mathbf{Q}/\mathbf{Z}$ by Proposition 16.4 (a)) with exact rows:

$$\cdots \to \mathrm{Ext}_G^r(M_1, C) \longrightarrow \mathrm{Ext}_U^r(M, C) \longrightarrow \mathrm{Ext}_G^r(M, C) \to \mathrm{Ext}_G^{r+1}(M_1, C) \to \cdots$$

$$\alpha^r(G, M_1) \Big\downarrow \qquad \alpha^r(U, M) \Big\downarrow \qquad \alpha^r(G, M) \Big\downarrow \qquad \alpha^{r+1}(G, M_1) \Big\downarrow$$

$$\cdots \to H^{2-r}(G, M_1)^* \to H^{2-r}(U, M)^* \to H^{2-r}(G, M)^* \to H^{1-r}(G, M_1)^* \to \cdots$$

For $r = 3$, we already know, by Lemma 16.20 and the case where the action of G is trivial, that $\mathrm{Ext}_U^3(M, C)$ and $\mathrm{Ext}_G^4(M_1, C)$ are zero. The same then holds true for $\mathrm{Ext}_G^3(M, C)$. As this is true for any G-module of finite type M, it also applies to M_1 hence $\mathrm{Ext}_G^3(M_1, C) = 0$. Finally all the terms in the diagram are zero for $r = 3$. For $r = 2$, we now know that $\alpha^3(G, M_1)$ has zero as both source and target. On the other hand $\alpha^2(U, M)$ is an isomorphism (the action of U on M is trivial). Thus $\alpha^2(G, M)$ is surjective (for any M), hence also $\alpha^2(G, M_1)$. The five lemma then implies that $\alpha^2(G, M)$ is an isomorphism. If furthermore M is torsion-free, it is also the case for M_* and M_1 and we show in the same way that $\alpha^1(G, M)$ is an isomorphism. Lastly the proof of assertions (b) and (c) follows exactly the same scheme. $\qquad\square$

Remark 16.22 We easily verify that in the assertions (b) and (c), it is enough to assume that for a fixed m, the assumptions on $\alpha^1(U, \mathbf{Z}/m)$ and $\alpha^0(U, \mathbf{Z}/m)$ are valid *for U small enough*. Indeed, in the proof of the general case, we can always replace the subgroup U by any open subgroup contained in U.

Example 16.23 (a) Let G be a group isomorphic to $\widehat{\mathbf{Z}}$ (for example the absolute Galois group of a finite field) and let C be the class formation associated to a topological generator σ of G (cf. Example 16.5, b). Here the reciprocity map is the inclusion $n \mapsto \sigma^n$ of $\mathbf{Z}$ into G. By Lemma 16.19, the source and the target of the map $\alpha^0(U, \mathbf{Z}/m)$ are zero for any integer m and any open subgroup U of G (indeed, $\mathrm{cd}(G) = 1$, hence also $\mathrm{cd}(U) \leqslant 1$, which implies that $H^2(U, \mathbf{Z}/m) = 0$). The map $\alpha^1(U, \mathbf{Z}/m)$ is also an isomorphism since it is the natural map $\mathbf{Z}/m \to \widehat{\mathbf{Z}}/m$. The theorem then implies that if $r \geqslant 1$, then $\alpha^r(G, M)$ is an isomorphism for any G-module of finite type M, and the same property is valid when $r = 0$ if M is finite.

(b) Let K be a field with the absolute Galois group $G = \mathrm{Gal}(\overline{K}/K)$ and assume we have a class formation (G, C) associated to G. Let M be a finite type torsion-free G-module, and let N be the G-module $N = \mathrm{Hom}(M, \mathbf{Z})$ (M can be seen as the Galois module of *characters* of an *algebraic K-torus* T and N as its module of *co-characters*). By Theorem 16.21, we have isomorphisms

$$\mathrm{Ext}_G^r(M, C) \longrightarrow H^{2-r}(G, M)^*$$

---

[4]Thanks to Clément Gomez for pointing out this difficulty.

for any $r \geqslant 1$. But on the other hand $\mathrm{Ext}_G^r(M, C) = H^r(G, \mathrm{Hom}_{\mathbf{Z}}(M, C))$ (Proposition 16.16). As $\mathrm{Hom}_{\mathbf{Z}}(M, C) = N \otimes C$, Proposition 16.17 yields that the cup-product

$$H^r(G, N \otimes C) \times H^{2-r}(G, M) \longrightarrow H^2(G, C) \simeq \mathbf{Q}/\mathbf{Z}$$

induces an isomorphism $H^r(G, N \otimes C) \to H^{2-r}(G, M)^*$ for $r \geqslant 1$. For $r = 1$, we even obtain a perfect duality of finite groups (use the fact that $H^1(U, \mathbf{Z}) = 0$, and hence by restriction-inflation sequence we have $H^1(G, M) = H^1(G/U, M)$, where U is a open normal subgroup G such that $M = M^U$. Then apply Corollary 1.50). For $r \in \{0, 2\}$, one has to be slightly careful since if for example we take for K a p-adic field (with the usual class formation associated to $C = \overline{K}^*$) and $M = \mathbf{Z}$, we obtain $N = \mathbf{Z}$ hence

$$H^0(G, N \otimes C) = K^*; \quad H^2(G, M) = (G^{\mathrm{ab}})^*$$

but the dual of the discrete group $(G^{\mathrm{ab}})^*$ is the profinite group $G^{\mathrm{ab}} = \widehat{K}^*$ (Corollary 9.15), profinite completion of K^*. Similarly $H^2(G, N \otimes C) = \mathrm{Br}\, K \simeq \mathbf{Q}/\mathbf{Z}$, whose dual is $\widehat{\mathbf{Z}}$, the completion of $H^0(G, M) = \mathbf{Z}$. The fact that we have to "complete the H^0" to obtain good duality theorems involving these groups is a frequent problem when we no longer work with finite modules.

A bit later we will see how Theorem 16.21 applies to the Galois group of a local field to recover (and extend) the Tate duality theorem (Tate's initial proof used the cohomology of tori, mentioned in Example 16.23 b). At the end of this book we will apply Theorem 16.21 to prove the Poitou–Tate duality over global fields.

16.4 *P*-Class Formation

It is sometimes useful to consider a slightly more general notion than that of class formation as we have defined it. For a profinite group G and a set P of prime numbers, we define a *P-class formation* $(C, \{\mathrm{inv}_U\})$ by the same axioms as a class formation except that instead of asking for the inv_U to be isomorphisms, we are only asking for them to be injections satisfying the following two axioms:

(a) for open subgroups U and V of G with V normal in U, the map

$$\mathrm{inv}_{U/V} : H^2(U/V, C^V) \longrightarrow [U : V]^{-1}\mathbf{Z}/\mathbf{Z}$$

   is an isomorphism.
(b) for every open subgroup U of G and every $\ell \in P$, the restriction

$$\mathrm{inv}_U : H^2(U, C)\{\ell\} \longrightarrow \mathbf{Q}_\ell/\mathbf{Z}_\ell$$

   is an isomorphism.

Remark 16.24 (a) If P is the set of all primes, a P-class formation is then a class formation as we have defined it previously, and if $P = \varnothing$ it is a class formation as defined by Artin and Tate (cf. [47], Chap. XI).

(b) If all the other axioms are satisfied, the axiom (b) is equivalent to the requirement that the order of G is divisible by ℓ^∞ for every $\ell \in P$. Note also that if (G, C) is a class formation and H is a closed normal subgroup of G, then $(G/H, C^H)$ is a P-class formation, where P is the set of prime numbers ℓ such that ℓ^∞ divides $[G : H]$.

Theorem 16.21 extends immediately to the P-class formations provided one restricts everywhere to ℓ-primary components of groups under consideration with $\ell \in P$ (recall that, for $r \geqslant 1$, the groups $\mathrm{Ext}_G^r(M, N)$ are always torsion when M is a G-module of finite type by Proposition 16.16. For $r = 0$, we limit ourselves to M finite). This will be useful for the Poitou–Tate duality, when we will have defined the P-class formation associated to a restricted ramification group (Theorem 17.2).

16.5 From the Existence Theorem to the Duality Theorem for a p-adic Field

In this paragraph, we follow a somehow reverse procedure to that of Chap. 10: assuming known the existence theorem, we will reprove (and even slightly generalise) Theorem 10.9 (Tate local duality) from the general duality Theorem 16.21 for class formations.

In all that follows, we denote by K a p-adic field and we set $\Gamma = \mathrm{Gal}(\overline{K}/K)$. We have defined in Sect. 16.3 the maps α^r, for which we have the duality Theorem 16.21, that we will now apply.

Proposition 16.25 *Let M be a discrete Γ-module of finite type. Then the maps*

$$\alpha^r(\Gamma, M) : \mathrm{Ext}_\Gamma^r(M, \overline{K}^*) \longrightarrow H^{2-r}(\Gamma, M)^*$$

are isomorphisms for every $r \geqslant 1$. If furthermore M is finite, then α^0 is also an isomorphism (of finite groups).

Proof We apply the duality Theorem 16.21 to the class formation $(\Gamma, \overline{K}^*)$. Let $U = \mathrm{Gal}(\overline{K}/L)$ be an open subgroup of Γ. Then $\alpha^1(U, \mathbf{Z}/n)$ is the map

$$L^*/L^{*n} \longrightarrow U^{\mathrm{ab}}/n$$

induced by the reciprocity map ω_L (Lemma 16.19, b). The existence theorem (Theorems 9.13 and 11.20) identifies the profinite completion $(L^*)^\wedge = \varprojlim_n L^*/L^{*n}$ with the projective limit of the $L^*/N_{K/L}F^*$, the limit being taken over the finite abelian extensions F of L. It follows that the reciprocity map ω_L induces an isomorphism $\hat{\omega}_L$ between $(L^*)^\wedge$ and U^{ab}, and hence (viewing this isomorphism modulo n) that

$\alpha^1(U, \mathbf{Z}/n)$ is an isomorphism. Theorem 16.21 then says that $\alpha^r(\Gamma, M)$ is an isomorphism for $r \geqslant 1$.

If now M is finite of order s it is enough to check that $\alpha^0(U, \mathbf{Z}/n)$ is bijective for any n dividing s and any U sufficiently small (see Remark 16.22). We can thus assume that L contains the nth roots of unity. Then we have maps

$$\mu_n(L) \longrightarrow H^2(L, \mathbf{Z}/n)^* \longrightarrow (U^{\mathrm{ab}})[n]$$

and we know that the composed map (induced by the reciprocity map via Lemma 16.19, b) is an isomorphism, since $\hat{\omega}_L$ is an isomorphism, while L^* and its profinite completion have the same n-torsion subgroup.

We deduce that the first map (which is $\alpha^0(U, \mathbf{Z}/n)$) is injective, but as $H^2(L, \mathbf{Z}/n) = H^2(L, \mu_n) = (\mathrm{Br}\, L)[n] \simeq \mathbf{Z}/n$ is of cardinality n, we deduce that $\alpha^0(U, \mathbf{Z}/n)$ is also surjective. $\qquad\square$

We obtain a new proof (and an extension to modules of finite type) of the Tate local duality theorem:

Theorem 16.26 (Tate) *Let K be a p-adic field. Let M be a finite Γ-module with dual $M' = \mathrm{Hom}(M, \overline{K}^*)$. Then for any $r \geqslant 0$ the cup-product*

$$H^r(K, M) \times H^{2-r}(K, M') \longrightarrow \mathrm{Br}\, K = \mathbf{Q}/\mathbf{Z}$$

is a perfect pairing of finite groups. For $r = 1$, the same statement is true if we only assume M to be of finite type.

Proof Assume first M to be finite. We already know that all the intervening groups are finite (Corollary 8.15). Besides, we have the compatibility of pairings defined by the Ext and the cup-products (Proposition 16.17). It is then enough to apply the preceding result noticing that

$$\mathrm{Ext}^r_\Gamma(M, \overline{K}^*) = H^r(K, M')$$

for any $r \geqslant 0$ by Proposition 16.16, since $\mathrm{Ext}^i_{\mathbf{Z}}(M, \overline{K}^*) = 0$ for $i > 0$ by divisibility of $\overline{K}^*$ (this point is essential and has no analogues in the case of a totally imaginary number field, cf. [49], Part. I, Ann. 1).

The proof for M of finite type when $r = 1$ is exactly the same, provided we check that $H^1(K, M)$ remains finite (cf. Exercise 8.3). To do this, we first reduce to the case where M is free by noticing that M_{tors} is finite and M/M_{tors} is free of finite type. If L is a finite Galois extension of K such that the action of $\mathrm{Gal}(\overline{K}/L)$ on M is trivial, we obtain $H^1(K, M) = H^1(\mathrm{Gal}(L/K), M)$ by restriction-inflation, and we conclude with Corollary 1.50. $\qquad\square$

Remark 16.27 For M of finite type (but not finite), we need, for $r = 0$ or $r = 2$, to replace the H^0 by their profinite completions to find a good duality theorem, see Exercise 16.3 or [49], Part. II, Sect. 5.8.

16.6 Complements

In this paragraph, we provide some complements on class formations, which generalise the properties already seen in the special case of a class formation associated to a local field or to a global field.

In what follows, we will denote by (C, G) a class formation (one can check that all results in this paragraph remain valid for a P-class formation, where P is any set of prime numbers, for example $P = \varnothing$ which is the minimal assumption. The point being, Proposition 16.6 remains valid as stated).

For any pair (U, V) of open subgroups of G with $V \subset U$, we have a *norm* homomorphism

$$N_{U/V} : C^V \longrightarrow C^U \quad x \longmapsto \sum_{s \in U/V} s \cdot x.$$

We also have the transitivity of the norms: if $W \subset V \subset U$, then $N_{U/W}$ is the composite of $N_{V/W}$ and $N_{U/V}$.

Definition 16.28 A *norm subgroup* of C^U is a subgroup of the form $N_{U/V}(C^V)$ for a certain open subgroup V of U.

The next proposition generalises Proposition 9.7 (in the global case, we have also seen it in the proof of the existence theorem). Its proof is very similar.

Proposition 16.29 *Let* (C, G) *be a class formation. Let* (U, V) *be a pair of open subgroups of* G *with* $V \subset U$.

(a) *Assume that* V *is normal in* U. *Then we have a (reciprocity) isomorphism*

$$C^U / N_{U/V} C^V \longrightarrow (U/V)^{\mathrm{ab}}.$$

In particular, if U/V *is abelian (i.e., if* V *contains the profinite derived subgroup* $D(U)$ *of* U*), we obtain an isomorphism between* $C^U / N_{U/V} C^V$ *and* U/V.

(b) *In the general case, let* W *be the smallest closed subgroup of* U *containing* V, *normal in* U, *and such that* U/W *is abelian (*W *is the subgroup generated by* V *and* $D(U)$*). Then we have*

$$N_{U/V}(C^V) = N_{U/W}(C^W).$$

(c) *The group of norms* $N_{U/V}(C^V)$ *is of finite index in* C^U. *This index divides* $[U : V]$, *and is equal to it only if* V *is normal in* U *with* U/V *abelian.*

Note that if U is the absolute Galois group of a field K and V corresponds to an extension L of K, then W corresponds to the largest abelian extension of K contained in L. When V is a normal subgroup of U, we denote by $(x, U/V)$ the image of an element x of C^U by the reciprocity map $C^U \to (U/V)^{\mathrm{ab}}$ of (a).

Proof (a) This is Proposition 16.6 (which is a corollary of Tate-Nakayama) applied to the class formation (U, C) and to the subgroup V of U, with $r = -2$ and $M = \mathbf{Z}$.

(b) The inclusion $N_{U/V}(C^V) \subset N_{U/W}(C^W)$ follows from the transitivity of the norms. Let $a \in N_{U/W}(C^W)$, let us show that $a \in N_{U/V}(C^V)$. Let Z be an open subgroup of U contained in V and normal in U. Let us set $J = U/Z$ and $H = V/Z$. Then $W/Z = J'.H$, where J' is the derived subgroup of J. The inclusion $H \to J$ induces a canonical homomorphism $i : H^{ab} = H/H' \to J^{ab} = J/J'$. The image $(a, U/W)$ of a in $U/W = J/J'H$ is trivial, which means that $(a, U/Z) \in (J/J')$ is coming from H/H'. We have a commutative diagram, which follows from the definition of the reciprocity map as the inverse of the map "cup-product with the fundamental class", and from the compatibility of the cup-product with the corestrictions (Proposition 2.27, e):

$$
\begin{array}{ccc}
C^V & \xrightarrow{\;N_{U/V}\;} & C^U \\
{\scriptstyle(.,\,V/Z)}\Big\downarrow & & \Big\downarrow{\scriptstyle(.,\,U/Z)} \\
H/H' & \xrightarrow{\;i\;} & J/J'.
\end{array}
\tag{16.7}
$$

As C^V is surjective onto H/H' by the reciprocity map, the diagram then provides the existence of an a' in C^V such that

$$(a, U/Z) = (N_{U/V}a', U/Z).$$

Thus $N_{U/V}a' - a$ can be written as $N_{U/Z}(a'')$ with $a'' \in C^Z$, hence

$$a = N_{U/V}(a' - N_{Z/V}(a'')).$$

(c) By (b), we have $C^U/N_{U/V}(C^V) = C^U/N_{U/W}(C^W)$, which is finite with cardinality $[U : W]$ by (a). The result follows. $\qquad\qquad\square$

We also have the following compatibility formula (that we will compare with that of diagram (16.7)), which in the global case appears in the Proof of Theorem 15.31.

Proposition 16.30 *Let U and V be open subgroups of G with $V \subset U$ and V normal in G. Then we have a commutative diagram:*

$$
\begin{array}{ccc}
C^G & \xrightarrow{\;i\;} & C^U \\
{\scriptstyle\omega_{G/V}}\Big\downarrow & & \Big\downarrow{\scriptstyle\omega_{U/V}} \\
(G/V)^{ab} & \xrightarrow{\;v\;} & (U/V)^{ab}
\end{array}
$$

where i is the inclusion, $\omega_{G/V}$ and $\omega_{U/V}$ are the reciprocity maps and $v : (G/V)^{ab} \to (U/V)^{ab}$ is the transfer (cf. Definition 2.11) associated to the inclusion $U/V \to G/V$.

Proof This follows immediately from the definition of the reciprocity map and from the fact that the transfer (resp. the inclusion i) corresponds to the restriction $\widehat{H}^{-2}(G/V, \mathbf{Z}) \to \widehat{H}^{-2}(U/V, \mathbf{Z})$ (resp. to the restriction $\widehat{H}^{0}(G/V, C^V) \to \widehat{H}^{0}(U/V, C^V)$). $\qquad\square$

We also have a generalisation of the Lemma 9.9, (a) and (b) (and in the global case these arguments have also been used in the proof of the existence theorem):

Proposition 16.31 *Let U be an open subgroup of the group G. We denote by Ab_U the set of open subgroups V of U such that V is normal in U and U/V is abelian (i.e., such that V contains the derived subgroup $D(U)$ of U). For V in Ab_U, set $I_V = N_{U/V}C^V$. Then the map $V \mapsto I_V$ is an increasing bijection from Ab_U to the set of norm subgroups of C^U. We furthermore have, for all V, W in Ab_U:*

$$I_{V \cap W} = I_V \cap I_W$$

and any subgroup of C^U which contains a norm group is a norm group.

Proof Note that if V and W are in Ab_U, then $V \cap W$ is too. The fact that $V \mapsto I_V$ is increasing follows from the transitivity of the norms, which in particular gives the inclusion $I_{V \cap W} \subset I_V \cap I_W$. Conversely, if $a \in I_V \cap I_W$, then the element $(a, U/(V \cap W))$ of $U/(V \cap W)$ has a trivial image in U/V and U/W, hence is trivial, which implies that $a \in I_{V \cap W}$.

If now $I_V \supset I_W$, then $I_{V \cap W} = I_W$ hence $V \cap W$ and W have the same index in U, which gives $V \supset W$. Using Proposition 16.29 (b), we finally obtain that the correspondence is bijective.

Let I be a subgroup of C^U which contains a subgroup of norms $N_{U/V}C^V$. We can assume (always by Proposition 16.29, b) that U/V is an abelian group. Then the image of I by the reciprocity map $C^U \to U/V$ is a subgroup W/V, where W is an open subgroup of U which contains V. We deduce that I is the kernel of the reciprocity map $C^U \to U/W$, hence $I = N_{U/W}C^W$ is a norm group. $\qquad\square$

Assume furthermore that we have a structure of a Hausdorff topological group on each C^U, such that for $V \subset U$ the topology on C^U is that induced by $C^V \supset C^U$, and for every s of G the map $C^U \to C^{sUs^{-1}}$, $a \mapsto s \cdot a$ is continuous (which implies that $N_{U/V}: C^V \to C^U$ is continuous). A typical example is where $G = \mathrm{Gal}(\overline{K}/K)$ is the Galois group of local field K and $C = \overline{K}^*$. Then the open subgroups U correspond to finite extensions L of K and we endow $C^U = L^*$ with the usual topology associated to its valuation. The map $N_{U/V}$ is the norm (in the usual sense) $N_{L/K}: L^* \to K^*$. Another example is that of the class formation (C, G_k) associated to a global field k (Example 16.5, c), by endowing each $C_K = I_K/K^*$ with the quotient topology.

Let us make one further assumption, which is satisfied in the two previous examples (cf. Proof of the Lemma 9.9, c) and the Proof of Theorem 15.11).

Hypothesis (H) *For every pair $V \subset U$ of open subgroups of G, the norm $N_{U/V}:$ $C^V \to C^U$ has a closed image and a compact kernel (if we work with locally compact countable at infinity groups, this means that $N_{U/V}$ is a proper map).*

This assumption implies that $N_{U/V}C^V$ is an open subgroup of C^U: indeed, it is of finite index by Proposition 16.29, (c).

Definition 16.32 Let U be an open subgroup of G. We denote by D_U the intersection of all the subgroups of norms of U. This is also the kernel of the reciprocity map $C^U \to U^{Ab}$.

For $U = G$, we have already encountered D_G, the group of universal norms of C^G, in Sect. 16.1.

Proposition 16.33 *Assume (H) to be satisfied. Then for any pair $V \subset U$ of open subgroups of G, we have $N_{U/V}D_V = D_U$.*

Proof The inclusion $N_{U/V}D_V \subset D_U$ follows from the transitivity of the norms. Let conversely $a \in D_U$. For any open subgroup W of V, denote by $K(W)$ the set of b of C^V of norm a (in C^U), and which are the norm of some element of W. The assumption (H) implies that $K(W)$ is compact. It is non-empty since $a \in D_U$, hence a is the norm of an element c of C^W and we can take $b = N_{V/W}(c)$ via the transitivity of the norms. As the family of the $K(W)$ is filtering and decreasing, its intersection over all of the W is non-empty. But an element b of this intersection clearly satisfies to $b \in D_V$ and $N_{U/V}(b) = a$. $\qquad\square$

To obtain the general existence theorem for a class formation , it is necessary to make a few further assumptions. This is the object of [47], Chap. XI, Sect. 5, that we have taken up in the Exercises 16.5 and 16.6.

16.7 Exercises

Exercise 16.1 Give an example of a profinite group G and of discrete G-modules M, N, such that the G-module $\mathrm{Hom}_{\mathbf{Z}}(M, N)$ is not discrete (we can take for G the absolute Galois group of a field and for M an infinite direct sum of copies of $\mathbf{Z}$).

Exercise 16.2 (a) Let A be a finite abelian group. Show that $\mathrm{Ext}^r_{\mathbf{Z}}(A, \mathbf{Z}) = 0$ if $r \neq 1$ and $\mathrm{Ext}^1_{\mathbf{Z}}(A, \mathbf{Z}) = A^*$, where $A^* = \mathrm{Hom}(A, \mathbf{Q}/\mathbf{Z})$ is the dual of A.
(b) Let $G = \widehat{\mathbf{Z}}$, endowed with its usual class formation. Show that for $r = 0, 1$, the cup-product

$$H^r(G, M) \times H^{1-r}(G, M^*) \longrightarrow H^1(G, \mathbf{Q}/\mathbf{Z}) = \mathbf{Q}/\mathbf{Z}$$

is non degenerate if M is a finite G-module.

Exercise 16.3 (after [39], pages 26–28) Let $G = \mathrm{Gal}(\overline{K}/K)$ be the Galois group of a p-adic field K. Let M be a discrete G-module of finite type. We denote by $M' = \mathrm{Hom}(M, \overline{K}^*)$ the Cartier dual of M. You can use the fact that if

$$0 \longrightarrow A \longrightarrow B \longrightarrow C \longrightarrow 0$$

is an exact sequence of abelian topological groups with B locally compact, totally disconnected, generated by a compact subset, such that $A \to B$ is a strict morphism (this notion is recalled in the notations and conventions at the beginning of the book) with closed image and the map $B \to C$ is open, then the sequence of profinite completions

$$0 \longrightarrow A^\wedge \longrightarrow B^\wedge \longrightarrow C^\wedge \longrightarrow 0$$

remains exact (cf. [20], Appendix).

(a) With the notation of the Proposition 16.25, show that $\alpha^0(G, M)$ induces an isomorphism between the profinite completion $\mathrm{Hom}_G(M, \overline{K}^*)^\wedge$ and $H^2(G, M)^*$ (you may first deal with the case where G acts trivially on M).

(b) Show that the cup-product induces a Pontryagin duality between the profinite group $H^0(K, M)^\wedge$ and the discrete group $H^2(K, M')$ and also between the profinite group $H^0(K, M')^\wedge$ and the discrete group $H^2(K, M)$ (you can use Proposition 16.16 and (a) of this exercise).

(c) Suppose furthermore that M is unramified and of prime to p torsion. Let N be the submodule $\mathrm{Hom}(M, \mathcal{O}_K^{\mathrm{nr}*})$ of M'. We denote by $I \subset G$ the inertia group of K. Show that in the duality

$$H^1(G, M) \times H^1(G, M') \longrightarrow \mathbf{Q}/\mathbf{Z},$$

the groups $H^1(G/I, M)$ and $H^1(G/I, N)$ are orthogonal to each other (you may reduce to the finite case by showing first that if M is torsion-free, then N is cohomologically trivial and $H^1(G/I, M) = H^1(G, M)$).

Exercise 16.4 Let (C, G) be a class formation. In this exercise, we have adopted the notation of Sect. 16.6. Let U and V be open subgroups of G with V a normal subgroup of U. Let $s \in G$. Show that we have a commutative diagram:

$$
\begin{array}{ccc}
C^U & \xrightarrow{\ s_1\ } & C^{sUs^{-1}} \\
{\scriptstyle \omega_{U/V}} \downarrow & & \downarrow {\scriptstyle \omega_{sUs^{-1}/sVs^{-1}}} \\
(U/V)^{\mathrm{ab}} & \xrightarrow{\ s_2\ } & (sUs^{-1}/sVs^{-1})^{\mathrm{ab}},
\end{array}
$$

where s_1 sends every $c \in C^U$ to $s \cdot c$ and s_2 sends $g \in (U/V)^{\mathrm{ab}}$ to sgs^{-1}.

Exercise 16.5 Consider a class formation (C, G). Assume that the C^U are endowed with a structure of topological group as at the end of Sect. 16.6 and that the assumption (H) is satisfied. Furthermore assume that for every prime number p and every open subgroup U of G, the kernel of the multiplication by p in C^U is compact. Lastly, assume that G contains an open subgroup U_p such that for every open subgroup V

of U_p, the image of the multiplication by p in C^V contains D_V (intersection of all the subgroups of norms of V).

(a) Show that if p is a prime number, then $D_U = pD_U$ (let $a \in D_U$. You may consider, for any open subgroup V of $U \cap U_p$, the set $L(V)$ of $b \in C^U$ such that $pb = a$ and $b \in N_{U/V}C^V$).

(b) Show that $D_U = \bigcap_{n>0} nC^U$.

We now additionally assume that for every open subgroup U of G, there exists a compact subgroup H of C^U such that every closed subgroup of finite index of C^U which contains H is a norm group. Let I be a closed subgroup of finite index n of C^U.

(c) Show that $D_U \subset I$.

(d) Let I' be the complement of I in C^U. Show that there exists a subgroup of norms N of C^U such that $I' \cap N \cap H = \varnothing$, i.e., such that $I \supset N \cap H$.

(e) Show that $N \cap (H + (N \cap I)) \subset I$.

(f) Show that if A is a Hausdorff topological abelian group, F is a closed subgroup of A, and G is a compact subgroup of A, then $F + G$ is a closed subgroup of A.

(g) Deduce that $H + (N \cap I)$ is a closed subgroup of finite index in C^U, then that I is a norm group.

Exercise 16.6 *(sequel to Exercise 16.5)* Let K be a p-adic field with absolute Galois group Γ. Let $(\overline{K}^*, \Gamma)$ be the associated class formation. The aim of this exercise is to recover the existence theorem for K directly from the general existence theorem of Exercise 16.5. You may not assume known Theorem 9.13 or its corollaries.

(a) Let ℓ be a prime number (distinct or not from p). Let K_ℓ be the field obtained by adjoining to K all the ℓth roots of unity, we set $U_\ell = \mathrm{Gal}(\overline{K}/K_\ell)$. Let $V = \mathrm{Gal}(\overline{K}/L)$ be an open subgroup of U_ℓ. Show that if $x \in L^*$ is a universal norm, then $x \in L^{*^\ell}$ (you may use Proposition 9.12).

(b) Let $H = U_K$ be the group of units of $\mathcal{O}_K$. Let I be a subgroup of finite index of K^* that contains H. Show that there exists an integer n such that I is the preimage of $n\mathbf{Z}$ by the valuation. Then show that if K' denotes the unramified extension of K of degree n, the group I is the kernel of the reciprocity map $\omega_{K'/K} : K^* \to \mathrm{Gal}(K'/K)$.

(c) Apply the result of the Exercise 16.5 (g) to recover the existence theorem for K.

Chapter 17
Poitou–Tate Duality

This chapter is devoted to the statement and to the proof of the difficult Poitou–Tate theorems, which are the duality theorems for the Galois cohomology of global fields. Even though totally imaginary number fields and function fields are of cohomological dimension 2 (cf. Exercise 14.1, or Corollary 17.14 and Remark 17.15), the method using the dualising module (that we have used in Chap. 10 for the local fields) does not work very well here. The problem is that the dualising module is not divisible. The same obstacle makes it difficult to obtain the Poitou–Tate duality as a direct application of the duality theorem for class formations, unlike in the local case (Theorem 16.26). Instead, we will therefore follow the method of [39], Chap. I.4, which uses in an essential way the duality Theorem 16.21 for a (P)-class formation. For another method (relying on more elaborate notions, namely étale cohomology, but more readily generalisable), one can consult [25].

17.1 The P-Class Formation Associated to a Galois Group with Restricted Ramification

We use notations and conventions of Sect. 15.5, which will be used throughout this chapter. In particular, k is a global field and S is a non-empty set of places of k, containing all the archimedean places if k is a number field. The readers wishing to familiarise themselves with the Poitou–Tate duality are invited to skip this paragraph in the first reading and to assume for the rest of this chapter that S is the set Ω_k of all the places of k. This avoids some complications due to restricted ramification.

We have the maximal extension k_S of k unramified outside S and the Galois group $G_S = \mathrm{Gal}(k_S/k)$. We denote by P the set of prime numbers ℓ such that ℓ^∞ divides the order of G_S (as a supernatural number).

© Springer Nature Switzerland AG 2020
D. Harari, *Galois Cohomology and Class Field Theory*, Universitext,
https://doi.org/10.1007/978-3-030-43901-9_17

Remark 17.1 If k is a function field over $\mathbf{F}_q$, the set P contains all the prime numbers since $\mathrm{Gal}(\overline{\mathbf{F}_q}/\mathbf{F}_q) = \widehat{\mathbf{Z}}$ is a quotient of G_S. If k is a number field, we have at least that P contains all the ℓ invertible in $\mathcal{O}_{k,S}$ (i.e., such that S contains all the places dividing ℓ) since for such an ℓ, the field k_S contains all the roots of unity of order a power of ℓ and we know that if ζ_{ℓ^m} is a primitive ℓ^mth root of unity, then $\mathbf{Q}(\zeta_{\ell^m})$ is of degree $\ell^{m-1}(\ell - 1)$ over $\mathbf{Q}$. Besides, if S contains almost all places of k, then (Exercise 17.1) P contains all the prime numbers.[1]

For any finite extension F of k, we have (Definition 15.38) the group $C_S(F)$, quotient of the idèle class group C_F by $U_{F,S} \simeq \prod_{w \notin S} \mathcal{O}_w^*$. We have seen (Proposition 15.40) the properties of the limit C_S of the $C_S(F)$ and of the limit U_S of the $U_{F,S}$, when F ranges through all finite Galois extensions of k contained in k_S. The following theorem says that we can define a P-class formation (G_S, C_S), as defined in Sect. 16.4:

Theorem 17.2 *The class formation (G_k, C) from Example 16.5, (c) induces a P-class formation (G_S, C_S).*

Proof Let $H_S := \mathrm{Gal}(\overline{k}/k_S)$. We know (given the definition of P and Remark 16.24, b) that the class formation (G_k, C) induces a P-class formation (G_S, C^{H_S}). Here C^{H_S} is the inductive limit $C(k_S)$ of the C_F (for F a finite Galois extension of k contained in k_S). As (by Proposition 15.40) the G_S-module U_S is cohomologically trivial, the exact sequence (15.3) shows that the cohomology groups $H^i(G_S, C^{H_S})$ identify with the $H^i(G_S, C_S)$ for $i \geqslant 1$, the result follows. $\qquad\square$

Let us finish this paragraph with a consequence of Remark 17.1, which will be useful in the proof of the Poitou–Tate duality (specifically in Proposition 17.25):

Lemma 17.3 *Let ℓ be a prime number, invertible in $\mathcal{O}_{k,S}$. Let v be a place of k (which may or may not be in S). For any finite Galois extension F of k contained in k_S, let F_v be the completion of F at a place above v. Let $r \geqslant 1$ be an integer. Then*

$$\varinjlim_{F \subset k_S} H^r(\mathrm{Gal}(\overline{k}_v/F_v), \overline{k}_v^*)\{\ell\} = 0.$$

Proof As by assumption $\ell \in \mathcal{O}_{k,S}^*$, we know (Remark 17.1) that ℓ^∞ then divides the order of G_S. We deduce, by Proposition 8.4, that:

$$\varinjlim_{F \subset k_S} (\mathrm{Br}\, F_v)\{\ell\} = 0,$$

that is,

$$\varinjlim_{F \subset k_S} H^2(\mathrm{Gal}(\overline{k}_v/F_v), \overline{k}_v^*)\{\ell\} = 0.$$

---

[1] We actually have much better: G. Chenevier and L. Clozel have in [11] proved that if there exist at least two primes p, q such that S contains all the places above p and q, then the set P contains all the prime numbers.

We also have on the other hand

$$\varinjlim_{F \subset k_S} H^r(\mathrm{Gal}(\overline{k}_v/F_v), \overline{k}_v^*)\{\ell\} = 0$$

for $r = 1$ (Hilbert 90), and also for $r \geqslant 3$: indeed, if v is finite it follows from the fact that the absolute Galois group of F_v is of strict cohomological dimension at most 2 (Theorem 10.6, Remark 10.7 and Proposition 6.15). If v is real, this follows from Theorem 2.16 along with the cases $r = 1$ and $r = 2$.   □

17.2 The Groups $\mathbf{P}^i_S(M)$

In this section, we consider discrete G_S-modules (as usual we will simply say "G_S-module" instead of "discrete G_S-module") of finite type M, *whose torsion subgroup* M_{tors} *is of cardinality invertible in* $\mathcal{O}_{k,S}$. This condition is always satisfied in the case of a number field if $S = \Omega_k$, and is equivalent to the fact that the cardinality of M_{tors} is prime to the characteristic of k in the case of a function field. In this paragraph we define groups obtained from "local" Galois cohomology groups $H^i(k_v, M)$.

If v is a place of k, we denote by $G_v \subset G_k$ the decomposition subgroup at v, which identifies with the absolute Galois group of the completion k_v (cf. Sect. 13.1). If v is finite, we denote by $k(v)$ the residue field at v, and by $G(v)$ the absolute Galois group of $k(v)$. We have restriction maps $H^i(G_S, M) \to H^i(G_v, M)$ (defined for any place v of k; we will mainly use them for $v \in S$).

Definition 17.4 Caution. In this chapter we make the convention that $H^i(k_v, M)$ denotes $H^i(G_v, M)$ *except for* $i = 0$ *and* v *archimedean*. In this case, by convention it will be the *modified* Tate group $\widehat{H}^0(G_v, M)$, which is $\{0\}$ if v is complex and the quotient of $M^{G_{\mathbf{R}}}$ by the group of norms $N_{G_{\mathbf{R}}} M$ if v is real, where $G_{\mathbf{R}} := \mathrm{Gal}(\mathbf{C}/\mathbf{R})$. The reason for this convention is that it is adapted to duality theorems, see Lemma 17.9 below.

Let now v be a finite place of a global field k and assume furthermore that M is *non ramified at* v (i.e., unramified for the induced action of G_v: cf. Definition 10.15 and Remark 10.18). For $i \geqslant 0$, we have the groups of unramified cohomology $H^i_{\mathrm{nr}}(k_v, M)$. For simplicity we also write $H^i_{\mathrm{nr}}(k_v, M) = H^i(k_v, M)$ if v is an archimedean place. Note that M is unramified outside a finite number of places (as it is of finite type, and any element of M is fixed by an open finite index subgroup of G_S. But a finite extension of k is ramified only at finitely many places, see background material of Sect. 12.1). This justifies the following definition.

Definition 17.5 Let M be a G_S-module of finite type such that $\#M_{\mathrm{tors}}$ is invertible in $\mathcal{O}_{k,S}$. We define $\mathbf{P}^i_S(k, M)$ (or $\mathbf{P}^i_S(M)$ if it is clear what k is) as the *restricted product* for v in S of the $H^i(k_v, M)$ with respect to the $H^i_{\mathrm{nr}}(k_v, M)$.

Thus $\mathbf{P}_S^i(k, M)$ consists of families $(x_v)_{v \in S}$ with $x_v \in H^i(k_v, M)$ for every place $v \in S$, and $x_v \in H_{nr}^i(k_v, M)$ for almost every v. When $S = \Omega_k$, we will shorten $\mathbf{P}_{\Omega_k}^i(k, M)$ to $\mathbf{P}^i(k, M)$. The definitions and the fact that, for v finite, the absolute Galois group of the field k_v is of strict cohomological dimension 2 (Theorem 10.6, Remark 10.7 and Proposition 6.15) immediately imply:

Proposition 17.6 We have $\mathbf{P}_S^0(k, M) = \prod_{v \in S} H^0(k_v, M)$. If furthermore M is finite, this group is compact and we also have $\mathbf{P}_S^2(k, M) = \bigoplus_{v \in S} H^2(k_v, M)$, which is then a discrete group. For $i \geqslant 3$, we have $\mathbf{P}_S^i(k, M) = \bigoplus_{v \in \Omega_{\mathbf{R}}} H^i(k_v, M)$, where $\Omega_{\mathbf{R}}$ is the set of real places of k.

Remark 17.7 (a) If we do not assume that M is of finite type, the definition of the $\mathbf{P}^i$ is more complicated, and is better interpreted in the language of the étale cohomology. We therefore will not be using this notion in this book (nevertheless, see Exercise 17.8).

(b) If M is a G_S-module of finite type (with $\#M_{\mathrm{tors}}$ invertible in $\mathcal{O}_{k,S}$) and $v \notin S$, then M is unramified at v as k_S is unramified outside S.

(c) Assume M to be finite (and always of order invertible in $\mathcal{O}_{k,S}$). The group $\mathbf{P}_S^1(k, M)$, endowed with its restricted product topology (each $H^1(k_v, M)$ is considered as discrete), is then locally compact. Let indeed T be a part of S with $S - T$ finite and M non ramified at places of T. Let us set $P_T = \prod_{v \in S-T} H^1(k_v, M) \times \prod_{v \in T} H_{nr}^1(k_v, M)$. Then all the $H^1(k_v, M)$ (and hence also the $H_{nr}^1(k_v, M)$) are finite by Corollary 8.15 and by the Remark 8.14, which shows that P_T is compact (product of compacts). This argument extends easily to the case where M is of finite type (cf. Exercise 8.3).

Lemma 17.8 Let M be a G_S-module of finite type such that $\#M_{\mathrm{tors}}$ is invertible in $\mathcal{O}_{k,S}$. Let $i \geqslant 0$. The image of the diagonal map (induced by the restrictions)

$$\beta^i : H^i(G_S, M) \longrightarrow \prod_{v \in S} H^i(k_v, M)$$

is contained in $\mathbf{P}_S^i(k, M)$.

Proof Let $x \in H^i(G_S, M)$. There exists a finite Galois extension $F \subset k_S$ of k such that the action of $\mathrm{Gal}(k_S/F)$ on M is trivial and x is in the image of $H^i(\mathrm{Gal}(F/k), M)$. The restriction of x to $H^i(k_v, M)$ is then in $H_{nr}^i(k_v, M)$ as soon as v is unramified in the extension F/k, hence for almost all places v of k. $\qquad\square$

The following lemma is the analogue of the local duality Theorem 10.9 for the field of real numbers (for the field of complex numbers, the assertion is trivial, because of our convention on $H^0(k_v, M)$).

Lemma 17.9 Let M be a finite $G_{\mathbf{R}}$-module with Cartier dual M'. Then the cup-product

$$\widehat{H}^i(G_{\mathbf{R}}, M) \times \widehat{H}^{2-i}(G_{\mathbf{R}}, M') \longrightarrow \mathrm{Br}\,\mathbf{R} \simeq \mathbf{Z}/2\mathbf{Z}$$

is a perfect duality of 2-torsion finite groups for all $i \in \mathbf{Z}$.

Note that by Theorem 2.16 (and Remark 2.29, which describes explicitly the periodicity isomorphism as a cup-product), the modified cohomology of the cyclic group $G_\mathbf{R}$ is 2-periodic. It is then enough to prove the result for $i \in \{0, 1, 2\}$ to deduce it for any $i \in \mathbf{Z}$.

Proof As the Galois group $G_\mathbf{R}$ of $\mathbf{R}$ is of order 2, we can assume (by Corollary 1.49) that M is 2-primary torsion. We can also assume M to be simple using induction on the order of M and the five lemma (along with the compatibility of the cup-product with exact sequences, Proposition 2.28) to show that the cup-product induces an isomorphism between $\widehat{H}^i(G_\mathbf{R}, M)$ and the dual of $\widehat{H}^{2-i}(G_\mathbf{R}, M')$. Now, the only simple $G_\mathbf{R}$-module is $\mathbf{Z}/2\mathbf{Z}$ with the trivial action (use Lemma 3.1 to first see that the action is trivial) as $G_\mathbf{R}$ is a 2-group. Finally for $M = \mathbf{Z}/2\mathbf{Z}$ with trivial action, all the groups under consideration are $\mathbf{Z}/2\mathbf{Z}$ and it is enough to show that the pairing is not zero. This is straightforward for $i = 0$ and $i = 2$ by Remark 2.22. For $i = 1$, the cup-product of the classes of $a, b \in \mathbf{R}^*$ in $H^1(G_\mathbf{R}, \mathbf{Z}/2\mathbf{Z}) \simeq \mathbf{R}^*/\mathbf{R}^{*^2}$ is the Hilbert symbol (a, b), which is nonzero if one chooses a and b both negative (Example 9.11, see also Exercise 9.4). $\qquad\square$

Proposition 17.10 *Let M be a finite G_S-module whose cardinality n is invertible in $\mathcal{O}_{k,S}$. Then the map*

$$\beta^1 : H^1(G_S, M) \longrightarrow \mathbf{P}^1_S(k, M)$$

is proper (i.e., the preimage of a compact subset of $\mathbf{P}^1_S(k, M)$ is finite).

Proof Let T be a part of S with $S - T$ finite and M unramified at the places of T. We set $P_T = \prod_{v \in S-T} H^1(k_v, M) \times \prod_{v \in T} H^1_{\mathrm{nr}}(k_v, M)$. It is compact (Remark 17.7 c) and any compact subset P of $\mathbf{P}^1_S(k, M)$ is contained in a P_T (cover P by the open sets $P \cap P_T$, and extract a finite covering noticing that $P_T \subset P_{T'}$ if $T' \subset T$). It is then enough to show that the preimage X_T of P_T by β^1 is finite. There exists a finite Galois extension $F \subset k_S$ of k such that $\mathrm{Gal}(k_S/F)$ acts trivially on M and F contains the nth roots of unity. For $v \in T$, the image of X_T in $H^1(F, M)$ is unramified at every place of F above v. But $H^1(G_S, M)$ is a subgroup of $H^1(k, M)$ and the kernel of $H^1(k, M) \to H^1(F, M)$ is finite (Theorem 1.42). It is therefore enough to show that the image of X_T in $H^1(F, M)$ is finite. We are thus reduced to the case where the action of G_S on M is trivial and where k contains the nth roots of unity. One can restrict to the case where $M = \mu_n$. An element of $H^1(G_S, \mu_n) \subset H^1(k, \mu_n)$ whose image in $H^1(k_v, \mu_n)$ is unramified for every place $v \in T$ also has an unramified image for every place $v \notin S$ (by definition of G_S), hence for every $v \notin S - T$. As $S - T$ is finite, the result follows from Proposition 12.28 (one can also directly use the unramified extension Theorem, cf. [49], Part. II, Sect. 6, Lemma 6). $\qquad\square$

For a finite G_S-module M whose order n is invertible in $\mathcal{O}_{k,S}$, the nth roots of unity of $\overline{k}$ are in k_S. Thus the G_S-module $M' = \mathrm{Hom}(M, \overline{k}^*)$ (the Cartier dual of M) is also $\mathrm{Hom}(M, k_S^*)$. Recall that the action of G_S on M is defined by the formula:

$$(\gamma \cdot \varphi)(x) := \gamma(\varphi(\gamma^{-1} \cdot x))$$

for $\gamma \in G_S$, $\varphi \in M'$, $x \in M$ (Example 4.16, d). We will also use in what follows the notion of dual $A^* = \mathrm{Hom}_c(A, \mathbf{Q}/\mathbf{Z})$ of an abelian topological group A (cf. Remark 10.8).

Proposition 17.11 *Let M be a finite G_S-module with $\#M$ invertible in $\mathcal{O}_{k,S}$. Then Tate local duality induces isomorphisms between:*

(a) *the dual $\mathbf{P}^0_S(k, M)^*$ of the compact group $\mathbf{P}^0_S(k, M)$ and the discrete group $\mathbf{P}^2_S(k, M')$;*

(b) *the dual $\mathbf{P}^1_S(k, M)^*$ of the locally compact group $\mathbf{P}^1_S(k, M)$ and the locally compact group $\mathbf{P}^1_S(k, M')$.*

Proof We apply the classical formula which describes the dual of a restricted product (cf. [42], Proposition 1.1.13). The proposition then follows from Theorems 10.9 and 10.17 (note that for almost all places v of k, the cardinality of M is prime to the residual characteristic of $\mathcal{O}_v$), combined with Lemma 17.9, Remarks 10.10 and 10.18.   □

17.3 Statement of the Poitou–Tate Theorems

This paragraph is devoted to the statement of the main duality theorem for the Galois cohomology of a global field. More precisely, fix a non-empty set S of places of k (containing the archimedean places if k is a number field) and in what follows consider a G_S-module M which is *finite and whose order is invertible in* $\mathcal{O}_{k,S}$. In particular, all primes ℓ dividing the order of M belong to the set P associated to S as in Sect. 17.1 (Remark 17.1). One may in the first reading only consider the case $S = \Omega_k$. In this case there is no restriction on the order of M and P is the set of all prime numbers.

Using Proposition 17.11, we have for $i = 0, 1, 2$ continuous maps

$$\gamma^i : \mathbf{P}^i_S(k, M) \longrightarrow H^{2-i}(G_S, M')^*$$

obtained by dualising the maps β^{2-i} (associated to M') of the Lemma 17.8. For $i = 1, 2$, we define

$$\mathrm{III}^i_S(k, M) = \mathrm{Ker}[\beta^i : H^i(G_S, M) \longrightarrow \mathbf{P}^i_S(k, M)]$$

(we sometimes write $\mathrm{III}^i_S(M)$ for $\mathrm{III}^i_S(k, M)$ if k is understood).

Remark 17.12 For $S = \Omega_\infty$, the group $\mathrm{III}^i_S(M)$ is the subgroup of $H^i(k, M)$ consisting of elements which are "everywhere locally " zero. The Cyrillic letter III (pronounced "Sha") refers to the Russian mathematician I.R. Shafarevich. The group $\mathrm{III}^1(k, M)$ is analogous to the classical Tate–Shafarevich group of an abelian variety (cf. for Example [39], Chap. I.6).

We can now state the main result of this chapter (and possibly of this whole book!).

Theorem 17.13 (Poitou–Tate) *Let M be a finite G_S-module whose cardinality is invertible[2] in $\mathcal{O}_{k,S}$. Let $M' = \mathrm{Hom}(M, \bar{k}^*)$ be the Cartier dual of M. Then:*

(a) For $r \geqslant 3$, we have $H^r(G_S, M) \simeq \bigoplus_{v \in \Omega_\mathbf{R}} H^r(k_v, M)$.

(b) The groups $\mathrm{III}^1_S(k, M')$ and $\mathrm{III}^2_S(k, M)$ are finite and dual to each other.

(c) we have a 9-term exact sequence of topological groups (called the Poitou–Tate *exact sequence):*

$$0 \longrightarrow H^0(G_S, M) \xrightarrow{\beta^0} \prod_{v \in S} H^0(k_v, M) \xrightarrow{\gamma^0} H^2(G_S, M')^*$$

$$\hookrightarrow H^1(G_S, M) \xrightarrow{\beta^1} \mathbf{P}^1_S(k, M) \xrightarrow{\gamma^1} H^1(G_S, M')^*$$

$$\hookrightarrow H^2(G_S, M) \xrightarrow{\beta^2} \bigoplus_{v \in S} H^2(k_v, M) \xrightarrow{\gamma^2} H^0(G_S, M')^* \longrightarrow 0,$$

where the "unnamed" maps come from the assertion (b).

Note that the first term is finite, the next two terms are compact, and the fourth term is discrete. Symmetrically, the last term is finite, the two preceding terms are discrete, and the sixth term is compact. The "middle term" $\mathbf{P}^1_S(k, M)$ is only locally compact. If we dualise the sequence, we obtain the same sequence with M and M' swapped (by Proposition 17.11).

Corollary 17.14 *Let p be a prime number invertible in $\mathcal{O}_{k,S}$, that we assume different from 2 if k has real places. Then G_S is of p-cohomological dimension $\leqslant 2$.*

Proof This follows from assertion (a) of Theorem 17.13. □

Remark 17.15 Let p be a prime number, assumed to be odd if K is a number field which is not totally imaginary. For $S = \Omega_k$, we have more precisely $\mathrm{cd}_p(G_k) = 2$ if k is a number field since $H^2(G_k, \mu_p) = (\mathrm{Br}\, k)[p]$ is nonzero (Brauer–Hasse–Noether theorem), and similarly for a function field of characteristic $\neq p$. We thus recover the results of Exercise 14.1. On the other hand, the computation of $\mathrm{scd}_p(G_S)$ (which is 2 or 3 by Corollary 17.14 and Proposition 5.8) in the case of a number field is a very difficult problem, which is still open in general and is related to the *Leopoldt conjecture* ([42], Chap. X, Sect. 3). In the next chapter we will see (Theorem 18.15) that if S contains almost all places of k, then $\mathrm{scd}_p(G_S) = 2$ (always under the assumptions of the previous corollary). For a function field, we have $\mathrm{scd}_\ell(G_S) = 2$ for any prime number ℓ as soon as S is non-empty, but this result does not seem to (for an arbitrary S) be easily deduced from the class field theory, even if one knows the theorems of Poitou–Tate; see [42], Theorem 8.3.17 or [39], Remark I.1.12.

---

[2]K. Česnavičius [10] has obtained a version of the Poitou–Tate exact sequence without this assumption; see also González-Avilés [17] for the case of a function field.

Example 17.16 If k is a number field and p is not invertible in $\mathcal{O}_{k,S}$, the question of deciding whether $\mathrm{cd}_p(G_S)$ is finite is also difficult, see [42], Chap. X. If $S = \Omega_\infty$, it can happen that the group G_S is finite and non-trivial even for a quadratic imaginary field, see Yamamura [56]. For example, if $k = \mathbf{Q}(\sqrt{-771})$, we obtain $G_S = \mathcal{S}_4 \times \mathbf{Z}/3\mathbf{Z}$. When G_S is finite, Exercise 5.1 shows that $\mathrm{cd}_p(G_S)$ is infinite for any prime number p dividing the cardinality of G_S.

Corollary 17.17 *Assume S to be finite. Let M be a finite G_S-module of order invertible in $\mathcal{O}_{k,S}$. Then the groups $H^r(G_S, M)$ are finite for any $r \geqslant 0$.*

Proof Here the groups $\mathbf{P}_S^r(k, M)$ are finite by Corollary 8.15 and Remark 8.14. The result follows from the finiteness of $\mathrm{III}_S^r(k, M)$ for $r = 1, 2$, and assertion (a) of Theorem 17.13 for $r \geqslant 3$. $\qquad\qquad\qquad\qquad\qquad\qquad\qquad\qquad\qquad\qquad\qquad\square$

17.4 Proof of Poitou–Tate Theorems (I): Computation of the Ext Groups

In this paragraph we start proving Theorem 17.13. We will work with finite Galois extensions F of k contained in k_S. In Lemma 15.39 we have defined the group $J_{F,S}$ (the idèle group truncated at S) as the subgroup of the idèle group I_F consisting of families $(a_w)_{w \in \Omega_F}$ whose component at w is trivial for every place $w \in \Omega_F$ not in S. Let $E_{F,S} := \mathcal{O}_{F,S}^*$ be the group of invertible elements of $\mathcal{O}_{F,S}$. Set $C_{F,S} = J_{F,S}/E_{F,S}$. We denote by I_S the inductive limit (over the $F \subset k_S$) of the $J_{F,S}$ and E_S that of the $E_{F,S}$. The G_S-module C_S of the P-class formation of Theorem 17.2 satisfies (Proposition 15.40, a):

$$C_S = \varinjlim_F C_{F,S}.$$

The exact sequence

$$0 \longrightarrow E_{F,S} \longrightarrow J_{F,S} \longrightarrow C_{F,S} \longrightarrow 0$$

then gives, by passing to the limit, the exact sequence

$$0 \longrightarrow E_S \longrightarrow I_S \longrightarrow C_S \longrightarrow 0. \tag{17.1}$$

Theorem 17.13 will be proved by writing the long exact sequence obtained by applying the functor $\mathrm{Hom}_{G_S}(M', .)$ to this exact sequence, and by calculating the Ext that occur. In this paragraph, we deal with the Ext with values in C_S and E_S. Those with values in I_S (which are more complicated) will be dealt with in the next paragraph.

Theorem 17.18 *Let M be a G_S-module of finite type. Let $\ell \in P$. Then the homomorphism*

$$\alpha^r(G_S, M)\{\ell\} : \operatorname{Ext}^r_{G_S}(M, C_S)\{\ell\} \longrightarrow H^{2-r}(G_S, M)^*\{\ell\}$$

(cf. Theorem 16.21) is an isomorphism for any $r \geqslant 1$. In particular, the groups $\operatorname{Ext}^r_{G_S}(M, C_S)\{\ell\}$ are zero for $r \geqslant 3$.

Note that the assumption on M in this statement is very slightly more general than that of the Poitou–Tate theorems (we allow M to be of finite type, possibly infinite). For a finite M whose cardinality is invertible in $\mathcal{O}_{k,S}$, the set P contains all the prime numbers dividing $\#M$ by Remark 17.1, and Theorem 17.18 applies to an arbitrary ℓ.

Proof We apply the version "P-class formation" of Theorem 16.21. After if necessary passing to a finite extension of k contained in k_S (corresponding to an open subgroup U of G_S), it is enough to check that for every $m > 0$, the map $\alpha^1(G_S, \mathbf{Z}/\ell^m)$ is bijective. Note that $H^0(G_S, C_S) = C_S(k)$ (Proposition 15.40, c). Then, by Lemma 16.19 (b), the map $\alpha^1(G_S, \mathbf{Z}/\ell^m) : C_S(k)/\ell^m \to G_S^{ab}/\ell^m$ is induced by the reciprocity map rec_k and Proposition 15.42 shows that it is an isomorphism (note the importance of the fact, which was proved in Proposition 15.42, that in the case of a number field, the kernel $D_S(k)$ of the map $\omega : C_S(k) \to G_S^{ab}$ induced by rec_k is divisible. In the case of a function field, we just use the fact that $\widehat{\mathbf{Z}}/\mathbf{Z}$ is uniquely divisible). $\qquad\square$

Remark 17.19 For $v \in S$ finite, we can also consider M as a G_v-module. The isomorphisms $\alpha^r(G_S, M)$ of Theorem 17.18 are compatible with the isomorphisms $\alpha^r(G_v, M)$ of Proposition 16.25, which give the local Tate duality. Indeed the $\alpha^r(G_v, M)$ also come from the duality Theorem 16.21 (applied to the class formation associated to G_v). We have an analogous compatibility for the archimedean places provided we consider the modified groups.

Remark 17.20 It looks likely that $\alpha^0(G_S, \mathbf{Z}/\ell^m)$ is surjective for a totally imaginary number field (or if $\ell \neq 2$), but this seems conditional on the assumption that $H^2(G_S, \mathbf{Q}_\ell/\mathbf{Z}_\ell) = 0$ (cf. Exercise 17.2). Unfortunately this result (which is a version of the Leopoldt conjecture) is not known in general. It is true over a function field, but does not seem to be easily deduced from class field theory. The cardinality argument used in the local case (Proposition 16.25) does not work here. See also Remark 17.15.

Let us look at the Ext with values in E_S. Recall that if v is a place of k, we denote by $G_v \simeq \operatorname{Gal}(\overline{k}_v/k_v)$ the decomposition group at v. If v is finite, we denote by $k(v)$ the residue field at v and $G(v)$ the absolute Galois group of $k(v)$.

Lemma 17.21 *Let M be a finite G_S-module, whose cardinality is invertible in $\mathcal{O}_{k,S}$. Let $r \geqslant 0$ be an integer. Then:*

(a) We have $\operatorname{Ext}^r_{G_S}(M, E_S) \simeq H^r(G_S, M')$.

(b) For any place v not in S, we have $\operatorname{Ext}^r_{G(v)}(M, \mathcal{O}_v^{nr^*}) \simeq H^r(k(v), M')$, and both groups are zero for $r \geqslant 2$.

(c) Let v be any place of k. Then

$$\mathrm{Ext}^r_{G_v}(M, \overline{k}^*_v) \simeq H^r(G_v, M')$$

(recall that this last group is also denoted by $H^r(k_v, M')$ if $r \geq 1$ or if v is non-archimedean).

Note that in assertion (a), M' denotes the G_S-module $\mathrm{Hom}(M, k^*_S)$, which is also $\mathrm{Hom}(M, \overline{k}^*)$ or $\mathrm{Hom}(M, E_S)$ since $\#M$ is invertible in $\mathcal{O}_{k,S}$. In assertion (b), we view M as a $G(v)$-module (since $v \notin S$, M is unramified at v) and M' designates its dual $\mathrm{Hom}(M, \overline{k(v)}^*)$. The notation $\mathcal{O}^{\mathrm{nr}}_v$ means that we consider the maximal unramified extension of $\mathcal{O}_v$, i.e., the ring of integers of k^{nr}_v. Lastly, in assertion (c), we view M as a G_v-module, the dual of $M' = \mathrm{Hom}(M, \overline{k}^*) = \mathrm{Hom}(M, \overline{k}^*_v)$.

Proof (a) Let us first show that E_S is ℓ-divisible for any prime number ℓ invertible in $\mathcal{O}_{k,S}$. The assumption implies that $k(\mu_\ell)/k$ is unramified outside S. If $a \in E_S$, let $K = k(\mu_\ell, a)$. Then K/k is unramified outside S, and $K(\sqrt[\ell]{a})$ is a Kummer extension of K, unramified at all places which are not above those in S, by Lemma 13.17 and the assumption that $a \in E_S \cap K = \mathcal{O}^*_{K,S}$. Thus $\sqrt[\ell]{a} \in k_S$, hence we conclude by noticing that $\sqrt[\ell]{a}$ remains of valuation zero at each place $v \notin S$.

Now, the multiplication by such an ℓ is surjective in $\mathrm{Ext}^1_{\mathbb{Z}}(M, E_S)$ via the exact sequence

$$0 \longrightarrow E_S[\ell] \longrightarrow E_S \xrightarrow{\cdot \ell} E_S \longrightarrow 0$$

and the fact that $\mathrm{Ext}^2_{\mathbb{Z}}$ are always zero (Appendix, Proposition A.54). We deduce that the multiplication by $\#M$ is surjective and zero in $\mathrm{Ext}^1_{\mathbb{Z}}(M, E_S)$, hence finally $\mathrm{Ext}^1_{\mathbb{Z}}(M, E_S) = 0$. The exact sequence (16.3) of the Proposition 16.16 then yields the result.

(b) Let ℓ be a prime number dividing $\#M$. As $v \notin S$, v does not divide ℓ because of the assumption that $\#M$ is invertible in $\mathcal{O}_{k,S}$. We deduce that $\mathcal{O}^{\mathrm{nr}*}_v$ is ℓ-divisible by applying the Hensel lemma to finite unramified extensions of k_v (cf. Example 10.19). The argument is then similar to (a). On the other hand $H^r(k(v), M') = 0$ if $r \geq 2$ because the cohomological dimension of the finite field $k(v)$ is 1.

(c) The proof is completely analogous to that of (a) since $\overline{k}^*_v$ is a divisible group.

□

17.5 Proof of the Poitou–Tate Theorems (II): Computation of the Ext with Values in I_S and End of the Proof

We complete the Proof of Theorem 17.13. We always denote by M a finite G_S-module whose cardinality is invertible in $\mathcal{O}_{k,S}$. The most difficult step is to compute $\mathrm{Ext}^r_{G_S}(M, I_S)$. We can then use the exact sequence (17.1).

Definition 17.22 We denote by $\mathcal{E}$ the set of pairs (F, T), where:

(a) T is a finite subset of S, containing all the archimedean places, the places where M is ramified, and the places above the prime numbers dividing $\#M$ (in particular M is a G_T-module whose order is invertible in $\mathcal{O}_{k,T}$ and we also have $k_T \subset k_S$),
(b) F is a finite Galois extension of k with $F \subset k_T$, such that the action of $\mathrm{Gal}(k_S/F)$ on M is trivial.

For $(F, T) \in \mathcal{E}$, we set

$$J^T_F = \prod_{w \in T_F} F^*_w \times \prod_{w \in S_F - T_F} \mathcal{O}^*_w,$$

where T_F (resp. S_F) is the set of places of F above T (resp. above S), F_w is the completion of F at w and $\mathcal{O}_w$ its ring of integers. We can view J^T_F (which is a subgroup of the idèle group I_F of F) as a $\mathrm{Gal}(F/k)$-module.

Note that the pairs (F, T) in $\mathcal{E}$ form an ordered filtered set for the relation: $(F, T) \leqslant (F', T')$ if $F \subset F'$ and $T \subset T'$.

The next lemma allows us to identify $\mathrm{Ext}^r_{G_S}(M, I_S)$ as an inductive limit over the $\mathcal{E}$ of the Ext "defined at finite level". For simplicity, as usual, we denote by F_v the completion of F at a place above v.

Lemma 17.23 *With the above notations, we have for any integer $r \geqslant 0$:*

$$\mathrm{Ext}^r_{G_S}(M, I_S)$$
$$\simeq \varinjlim_{(F,T) \in \mathcal{E}} \left[\prod_{v \in T} \mathrm{Ext}^r_{\mathrm{Gal}(F_v/k_v)}(M, F^*_v) \times \prod_{v \in S-T} \mathrm{Ext}^r_{\mathrm{Gal}(F_v/k_v)}(M, \mathcal{O}^*_{F_v}) \right]. \quad (17.2)$$

These groups also identify with the inductive limit (taken over the pairs $(F, T) \in \mathcal{E}$) of the $\mathrm{Ext}^r_{\mathrm{Gal}(F/k)}(M, J^T_F)$.

Proof Recall that for any fixed finite extension $F \subset k_S$ of k, the group $J_{F,S}$ (defined in Lemma 15.39) identifies with the restricted product of the F^*_w for w a place of S_F, or with the inductive limit (over the T satisfying the condition (a) of the Definition 17.22) of the J^T_F. Such an extension F of k is unramified outside a finite number of places, it is then contained in k_T for T finite large enough. As I_S is by definition the inductive limit of the $J_{F,S}$ for F as above, we obtain:

$$I_S = \varinjlim_{(F,T)\in\mathcal{E}} J_F^T.$$

We also have $G_S = \varprojlim_{(F,T)\in\mathcal{E}} \mathrm{Gal}(F/k)$. Proposition 16.16 (c) then implies

$$\mathrm{Ext}^r_{G_S}(M, I_S) \simeq \varinjlim_{(F,T)\in\mathcal{E}} \mathrm{Ext}^r_{\mathrm{Gal}(F/k)}(M, J_F^T).$$

Now, we use the fact that the $\mathrm{Ext}^r_{\mathrm{Gal}(F/k)}(M, .)$ commute with products (Appendix, Proposition A.53). On the other hand if w_1 is a place of F above $v \in S$, the group $\mathrm{Gal}(F_{w_1}/k_v)$ identifies with a decomposition subgroup of $\mathrm{Gal}(F/k)$, and the $\mathrm{Gal}(F/k)$-module $\prod_{w|v} F_w^*$ (resp. $\prod_{w|v} \mathcal{O}_w^*$) with the induced module of $F_{w_1}^*$ (resp. $\mathcal{O}_{w_1}^*$) relatively to this subgroup (cf. Propositions 12.14 and 13.1). The result follows by "Shapiro's lemma for the Ext" (see the end of Remark 16.13). $\qquad\Box$

We now need to compute local terms in the formula (17.2). The second one is easier:

Lemma 17.24 *Let $(F, T) \in \mathcal{E}$. Let $v \in S - T$. Then*

$$\mathrm{Ext}^r_{\mathrm{Gal}(F_v/k_v)}(M, \mathcal{O}_{F_v}^*) \simeq H^r(G(v), M')$$

for any integer $r \geqslant 0$, and this group is zero if $r \geqslant 2$.

Recall that here M' denotes the dual $\mathrm{Hom}(M, \overline{k(v)}^*)$ of the $G(v)$-module M, which is meaningful as M is unramified outside T.

Proof By definition of $\mathcal{E}$, the extension F/k is unramified outside T hence $F_v \subset k_v^{\mathrm{nr}}$ and $\mathcal{O}_{F_v}^* \subset \mathcal{O}_v^{\mathrm{nr}*}$. Write the spectral sequence of the Ext (Theorem 16.14) with $N = \mathbf{Z}$, $G = G(v) = \mathrm{Gal}(k_v^{\mathrm{nr}}/k_v)$, $H = \mathrm{Gal}(k_v^{\mathrm{nr}}/F_v)$ and $P = \mathcal{O}_v^{\mathrm{nr}*}$, we obtain a spectral sequence

$$\mathrm{Ext}^r_{\mathrm{Gal}(F_v/k_v)}(M, H^s(H, \mathcal{O}_v^{\mathrm{nr}*})) \implies \mathrm{Ext}^{r+s}_{G(v)}(M, \mathcal{O}_v^{\mathrm{nr}*}).$$

The G-module $\mathcal{O}_v^{\mathrm{nr}*}$ is cohomologically trivial by Proposition 8.3; besides $\mathbf{Z}$ is of finite type, hence all the terms

$$\mathcal{E}xt_H^s(\mathbf{Z}, \mathcal{O}_v^{\mathrm{nr}*}) = \mathrm{Ext}_H^s(\mathbf{Z}, \mathcal{O}_v^{\mathrm{nr}*}) = H^s(H, \mathcal{O}_v^{\mathrm{nr}*})$$

are zero for $s > 0$. For $s = 0$, this term is equal to $H^0(H, \mathcal{O}_v^{\mathrm{nr}*}) = \mathcal{O}_{F_v}^*$. We thus obtain

$$\mathrm{Ext}^r_{\mathrm{Gal}(F_v/k_v)}(M, \mathcal{O}_{F_v}^*) \simeq \mathrm{Ext}^r_{G(v)}(M, \mathcal{O}_v^{\mathrm{nr}*})$$

which equals $H^r(G(v), M')$ by Lemma 17.21 (b), which is applicable since $v \notin T$ and M is a G_T-module of order invertible in $\mathcal{O}_{k,T}$. If furthermore $r \geqslant 2$, then $H^r(G(v), M')$ is zero since $G(v)$ (absolute Galois group of a finite field) is of cohomological dimension 1. $\qquad\Box$

To compute $\mathrm{Ext}^r_{G_S}(M, I_S)$, we consider the cases $r \geqslant 2$ and $r = 1$ separately:

Proposition 17.25 *Let $r \geqslant 2$. Then*

$$\mathrm{Ext}^r_{G_S}(M, I_S) \simeq \bigoplus_{v \in S} H^r(k_v, M').$$

Observe that $\bigoplus_{v \in S} H^r(k_v, M')$ is here $\mathbf{P}^r_S(k, M')$ as $r \geqslant 2$. In this statement, we again denote by M' the Cartier dual $\mathrm{Hom}(M, \bar{k}^*)$ of the G_S-module M.

Proof By Lemmas 17.23 and 17.24, we have

$$\mathrm{Ext}^r_{G_S}(M, I_S) \simeq \varinjlim_{(F,T) \in \mathcal{E}} \prod_{v \in T} \mathrm{Ext}^r_{\mathrm{Gal}(F_v/k_v)}(M, F^*_v).$$

We observe (in particular using that inductive limits commute with direct sums) that this inductive limit is also

$$\varinjlim_{F \subset k_S} \bigoplus_{v \in S} \mathrm{Ext}^r_{\mathrm{Gal}(F_v/k_v)}(M, F^*_v) \simeq \bigoplus_{v \in S} \varinjlim_{F \subset k_S} \mathrm{Ext}^r_{\mathrm{Gal}(F_v/k_v)}(M, F^*_v),$$

the limit taken over the finite Galois extensions $F \subset k_S$ of k. Let ℓ be a prime divisor of the cardinality of M and let v be a place of S. As by assumption $\ell \in \mathcal{O}^*_{k,S}$, Lemma 17.3 gives

$$\varinjlim_{F \subset k_S} H^s(\mathrm{Gal}(\bar{k}_v/F_v), \bar{k}^*_v)\{\ell\} = 0$$

for any $s \geqslant 1$. As this is valid for any prime number ℓ dividing the order of M, the spectral sequence of the Ext (Theorem 16.14, again applied with $N = \mathbf{Z}$) then gives, by passing to the limit over the finite extensions $F \subset k_S$:

$$\varinjlim_{F \subset k_S} \mathrm{Ext}^r_{\mathrm{Gal}(F_v/k_v)}(M, F^*_v) \simeq \mathrm{Ext}^r_{G_v}(M, \bar{k}^*_v)$$

and this last term is $H^r(G_v, M') = H^r(k_v, M')$, by Lemma 17.21 c). The result follows. □

Proposition 17.26 *We have $\mathrm{Ext}^1_{G_S}(M, I_S) \simeq \mathbf{P}^1_S(k, M')$.*

Proof By Lemmas 17.23 and 17.24, we know that the group $\mathrm{Ext}^1_{G_S}(M, I_S)$ is isomorphic to

$$\varinjlim_{(F,T) \in \mathcal{E}} \left[\prod_{v \in T} \mathrm{Ext}^1_{\mathrm{Gal}(F_v/k_v)}(M, F^*_v) \times \prod_{v \in S-T} H^1(G(v), M') \right].$$

The sequence of terms of low degree of the spectral sequence of the Ext and Hilbert 90 give, for $v \in T$:

$$\mathrm{Ext}^1_{\mathrm{Gal}(F_v/k_v)}(M, F_v^*) \simeq \mathrm{Ext}^1_{G_v}(M, \bar{k}_v^*)$$

which is again $H^1(G_v, M')$ by Lemma 17.21 (c). This term is independent of F. On the other hand for $v \in S - T$, the group $H^1(G(v), M') = H^1_{\mathrm{nr}}(k_v, M')$ is again independent of F. This implies that the term to be computed is

$$\varinjlim_{T} \left[\prod_{v \in T} H^1(k_v, M') \times \prod_{v \in S-T} H^1_{\mathrm{nr}}(k_v, M') \right],$$

the limit taken over the finite $T \subset S$. By definition of the restricted product, this limit is precisely $\mathbf{P}^1_S(k, M')$. $\square$

Proof of Theorem 17.3 We use the short exact sequence

$$0 \longrightarrow E_S \longrightarrow I_S \longrightarrow C_S \longrightarrow 0 \tag{17.3}$$

and previous results to calculate the terms of the long exact sequence $\mathrm{Ext}^r_{G_S}(M', -)$ associated to (17.3).

Let us prove (a). Assume first that $r \geq 4$. Theorem 17.18 says that $\mathrm{Ext}^{r-1}_{G_S}(M', C_S)$ and $\mathrm{Ext}^r_{G_S}(M', C_S)$ are zero, hence we have an isomorphism between $\mathrm{Ext}^r_{G_S}(M', E_S)$ and $\mathrm{Ext}^r_{G_S}(M', I_S)$, which implies the result using Proposition 17.25 and Lemma 17.21.

Applying Proposition 17.25 with $r = 2, 3$, Lemma 17.21 with $r = 3$, and Theorem 17.18 with $r = 2, 3$, we obtain an exact sequence

$$\mathbf{P}^2_S(k, M) \xrightarrow{\gamma^2} H^0(G_S, M')^* \longrightarrow H^3(G_S, M) \xrightarrow{\beta^3} \mathbf{P}^3_S(k, M) \longrightarrow 0.$$

The map $\beta^0_{M'} : H^0(G_S, M') \to \prod_{v \in S} H^0(k_v, M')$ is injective as for M nonzero, S contains at last one finite place (Recall that the cardinality of M is invertible in $\mathcal{O}_{k,S}$ by assumption). As it is a map between profinite groups (the first one is actually finite), its dual map is surjective. By Proposition 17.11, this dual map is precisely the map $\gamma^2 : \mathbf{P}^2_S(k, M) \to H^0(G_S, M')^*$, which is thus surjective. Finally the map $\beta^3 : H^3(G_S, M) \to \mathbf{P}^3_S(k, M)$ is an isomorphism, which finishes the proof of (a).

Let us show (b). We apply Proposition 17.26, Theorem 17.18 for $r = 1$, and Lemma 17.21 for $r = 1, 2$. Using what we have already seen, this gives the last six terms of the Poitou–Tate exact sequence. We have, using Proposition 17.11, an exact sequence

$$\mathbf{P}^1_S(k, M')^* \longrightarrow H^1(G_S, M')^* \longrightarrow \text{Ш}^2_S(k, M) \longrightarrow 0,$$

which shows that $\text{Ш}^2_S(k, M)$ is the dual of $\text{Ш}^1_S(k, M')$, which is finite by Proposition 17.10. Whence (b).

Let us show (c). In view of the above, it remains to prove the exactness of

$$0 \to H^0(G_S, M) \longrightarrow \prod_{v \in S} H^0(k_v, M) \longrightarrow H^2(G_S, M')^* \longrightarrow \text{Ш}^1_S(k, M) \to 0.$$

But this is obtained (with b) and Proposition 17.11) by dualising the end of the following sequence applied to M':

$$0 \to \text{III}^2_S(k, M') \longrightarrow H^2(G_S, M') \longrightarrow \bigoplus_{v \in S} H^2(k_v, M') \longrightarrow H^0(G_S, M)^* \to 0,$$

which is not problematic since we are dualising a sequence of discrete torsion groups.   □

Remark 17.27 (a) The identification of the maps β^i in the Poitou–Tate theorems is immediate (they are induced by the inclusions $F^* \to F_v^*$, just like the map $E_S \to I_S$).

The fact that it is the maps γ^i that appear in the sequence is the consequence of their definitions, of the compatibility between the pairings given by the cup-products and those given by the Ext (Proposition 16.17), and from Remark 17.19.

The "unnamed" maps are more difficult to write down. We can make them precise using an explicit description of the pairing between $\text{III}^1_S(M)$ and $\text{III}^2_S(M')$, cf. [39], p. 65.

(b) The finiteness of $\text{III}^2_S(M)$ allows to immediately deduce the analogue of the Proposition 17.10 in degree 2 as in this case $\mathbf{P}^2_S(M)$ is discrete. It seems difficult (impossible?) to prove this finiteness statement without using the Poitou–Tate duality!

(c) The Poitou–Tate theorems have many generalisations to more general modules M than the finite ones: modules of finite type and tori (Exercise 17.8 and [42], Chap. VIII, Sect. 6), abelian varieties ([39], Part. I, Chap. 6), or even "1-motives" ([17, 20]).

d) Instead of obtaining the beginning of the Poitou–Tate sequence by duality from the last six terms, it is also possible (As in [39], pages 61–62) to compute them directly by applying $\text{Hom}_{G_S}(M', .)$ to the exact sequence (17.3). This approach has the advantage to give the whole of the Poitou–Tate sequence from the $\text{Ext}^*_{G_S}(M', .)$ applied to (17.3), but the computation of the third term is significantly more difficult (cf. Remark 17.20).

17.6 Exercises

Exercise 17.1 Let k be a number field. Let S be a subset of Ω_k, containing Ω_∞. Assume that S contains almost all places of k. Recall the weak version of the Dirichlet arithmetic progression theorem: for any integer $b \geqslant 1$, there are infinitely many prime numbers congruent to 1 modulo b.

(a) Assume that $k = \mathbf{Q}$. Show that the set P of prime numbers associated to S (cf. Remark 17.1) contains all the primes (consider cyclotomic extensions $\mathbf{Q}(\zeta_n)$ with well chosen n).

(b) Using (a), generalise to an arbitrary number field.

Exercise 17.2 Let k be a number field. Let S be a subset of Ω_k containing the archimedean places. Let ℓ be a prime number invertible in $\mathcal{O}_{k,S}$. We suppose that $H^2(G_S, \mathbf{Q}_\ell/\mathbf{Z}_\ell) = 0$ (with the notation of the text). Show that the homomorphism α^0 of Theorem 17.18 is surjective (it is an open question as to whether $H^2(G_S, \mathbf{Q}_\ell/\mathbf{Z}_\ell) = 0$ for an arbitrary S, under the usual assumption that $\ell \neq 2$ if k has real places).

Exercise 17.3 Extend Proposition 17.10 to the case where M is only assumed to be of finite type. Same question with Lemma 17.9.

Exercise 17.4 Using the notations of this book, let k be a global field and M be a finite G_S-module whose cardinality is invertible in $\mathcal{O}_{k,S}$. Let T be a finite subset of S such that there is at least one finite place of S which is not in T. Fix such a place w.

(a) Let $\chi : H^0(k, M') \to \mathbf{Q}/\mathbf{Z}$ be a character of $H^0(k, M')$. Show that there exists $a_w \in H^2(k_w, M)$ such that for any b of $H^0(k, M')$, we have

$$-\mathrm{inv}_w(a_w \cup b_w) = \chi(b),$$

where b_w is the image of b in $H^0(k_w, M')$.

(b) Deduce that the diagonal map

$$H^2(G_S, M) \longrightarrow \bigoplus_{v \in T} H^2(k_v, M)$$

is surjective.

Exercise 17.5 Let k be a global field with absolute Galois group G_k. Let M be a finite G_k-module, whose cardinality is prime to the characteristic of k if k is a function field. We denote by M' the Cartier dual of M. Let T be a finite set of places of k. We set

$$\mathrm{III}^1(T, M) := \mathrm{Ker}[H^1(k, M) \longrightarrow \bigoplus_{v \notin T} H^1(k_v, M)].$$

(a) Using the Poitou–Tate exact sequence, show that there is an exact sequence

$$\mathrm{III}^1(T, M') \longrightarrow \bigoplus_{v \in T} H^1(k_v, M') \xrightarrow{\ \theta\ } H^1(k, M)^*.$$

Give a description of the map θ.

(b) Deduce from it an exact sequence

$$H^1(k, M) \longrightarrow \bigoplus_{v \in T} H^1(k_v, M) \longrightarrow \mathrm{III}^1(T, M')^* \longrightarrow \mathrm{III}^2(M) \longrightarrow 0,$$

where $\mathrm{III}^2(M) = \mathrm{Ker}[H^2(k, M) \to \prod_{v \in \Omega_k} H^2(k_v, M)]$.

Exercise 17.6 Let $k = \mathbf{Q}$ and let the set S be reduced to the infinite place. Set $M = \mathbf{Z}$ and $M' = \mathbf{G}_m$ (which plays the role of the dual of M, cf. Exercise 17.7 below). If K is a field with separable closure $\overline{K}$ and with absolute Galois group $\Gamma_K = \mathrm{Gal}(\overline{K}/K)$, we then have (cf. Remark 6.1) $H^i(K, \mathbf{G}_m) := H^i(\Gamma_K, \overline{K}^*)$. Using Corollary 12.30, show that the analogues of Propositions 17.25 and 17.26 are false in this context (when S contains almost all places of k, we will see in Exercise 17.8 that the situation is better).

Exercise 17.7 Let k be a number field with absolute Galois group $G_k = \mathrm{Gal}(\overline{k}/k)$. Let M be a G_k-module of finite type. For any integer $r \geqslant 3$, we consider the diagonal map

$$\theta^r(M) : H^r(k, M) \longrightarrow \bigoplus_{v \in \Omega_{\mathbf{R}}} H^r(k_v, M)$$

induced by the restrictions $H^r(k, M) \to H^r(k_v, M)$.

(a) Assume that the G_k-module M is torsion-free. Show that for any integer $i \geqslant 2$, we have an isomorphism

$$H^{i-1}(k, M \otimes_{\mathbf{Z}} \mathbf{Q}/\mathbf{Z}) \simeq H^i(k, M).$$

Show that for any integer $n \geqslant 1$, the G_k-module $M \otimes_{\mathbf{Z}} \mathbf{Z}/n\mathbf{Z}$ is finite, then that for any $r \geqslant 4$, the homomorphism $\theta^r(M)$ is an isomorphism.

(b) We no longer assume M to be torsion-free. Show that there exists an integer $s \geqslant 0$ and an exact sequence of G_k-modules

$$0 \longrightarrow N \longrightarrow P \longrightarrow M \longrightarrow 0$$

with P of the form $\mathbf{Z}[G_k/U]^s \simeq (I_{G_k}^U(\mathbf{Z}))^s$ for a certain open normal subgroup U of G_k, and N of finite type and torsion-free.

(c) Let $r \geqslant 4$. Show that the homomorphism $\theta^r(M)$ is an isomorphism.
(We will see in Exercise 18.1 that the result of (c) remains valid for $r = 3$).

Exercise 17.8 Let k be a number field. In this exercise we keep the notation used in the book: in particular S is a set of places containing all the archimedean places and for any finite place v, we respectively denote by G_v and $G(v)$ the groups $\mathrm{Gal}(\overline{k}_v/k_v)$ and $\mathrm{Gal}(k_v^{\mathrm{nr}}/k_v)$, this last being the absolute Galois group of the residue field $k(v)$ at v. We denote by M a G_S-module of finite type, such that $n := \#M_{\mathrm{tors}}$ is invertible in $\mathcal{O}_{k,S}$. We denote by N the G_S-module $\mathrm{Hom}(M, E_S)$, and for v such that M is not ramified at v (For example $v \notin S$), we denote by $N(v)$ the $G(v)$-module $\mathrm{Hom}(M, \mathcal{O}_v^{\mathrm{nr}*})$. Lastly, for any place v of k, we denote by N_v the G_v-module $\mathrm{Hom}(M, \overline{k}_v^*)$.

(A) (a) Show that there exists an exact sequence of $G(v)$-modules:

$$0 \longrightarrow F \longrightarrow N(v) \longrightarrow N_1 \longrightarrow 0,$$

with F cohomologically trivial and N_1 finite of n-torsion (use Exercise 3.5 and its generalisation to profinite groups).

(b) Deduce that $H^2(k(v), N(v)) = 0$ and if M is torsion-free, we have furthermore $H^1(k(v), N(v)) = 0$.

(c) For $r \geqslant 1$, we call $\mathbf{P}_S^r(k, N)$ the restricted product of the $H^r(k_v, N_v)$ for $v \in S$, with respect to the images of the $H^r(k(v), N(v))$ in $H^r(k_v, N_v)$. Show that

$$\mathbf{P}_S^2(k, N) = \bigoplus_{v \in S} H^2(k_v, N_v)$$

and that if M is torsion-free, we have furthermore

$$\mathbf{P}_S^1(k, N) = \bigoplus_{v \in S} H^1(k_v, N_v).$$

(d) Show that for any $r \geqslant 0$, the analogue of the Lemma 17.21 remains valid, which is:

(i) we have isomorphisms:

$$\mathrm{Ext}^r_{G_S}(M, E_S) \simeq H^r(G_S, N);$$

(ii) for any place v of k, we have:

$$\mathrm{Ext}^r_{G_v}(M, \overline{k}_v^*) \simeq H^r(G_v, N_v);$$

(iii) for any place v not in S, we have:

$$\mathrm{Ext}^r_{G(v)}(M, \mathcal{O}_v^{\mathrm{nr}*}) \simeq H^r(k(v), N(v)),$$

with furthermore $H^r(k(v), N(v)) = 0$ if $r \geqslant 2$.

You can use Exercise 17.3.

(B) We assume until the end of this Exercise 17.8 that S contains almost all places of k. In particular (Exercise 17.1), the set P associated to G_S as in Remark 17.1 contains all prime numbers.

(a) Show that if $v \in S$, we have

$$\varinjlim_{F \subset k_S} F_v^* = \overline{k}_v^*,$$

the limit taken over all the finite Galois extensions F of k contained in k_S (if L is a finite extension of k_v, write $L = k_v[T]/Q(T)$, where Q is a separable polynomial in $k_v[T]$; then apply the weak approximation theorem and Krasner's lemma, drawing inspiration from the proof of Proposition 13.6).

(b) Drawing inspiration from the proof of Proposition 17.25, show that for any $r \geqslant 2$, we have

$$\operatorname{Ext}^r_{G_S}(M, I_S) \simeq \mathbf{P}^r_S(k, N) = \bigoplus_{v \in S} H^r(k_v, N_v).$$

(c) Drawing inspiration from the proof of Proposition 17.26, show that we have

$$\operatorname{Ext}^1_{G_S}(M, I_S) \simeq \mathbf{P}^1_S(k, N).$$

(d) Deduce that we still have a Poitou–Tate type exact sequence

$$H^1(G_S, N) \longrightarrow \mathbf{P}^1_S(k, N) \longrightarrow H^1(G_S, M)^*$$
$$\longleftrightarrow H^2(G_S, N) \longrightarrow \mathbf{P}^2_S(k, N) \longrightarrow H^0(G_S, M)^* \longrightarrow 0.$$

(To prove the surjectivity of the last map, proceed by duality using Exercise 16.3, (b).)

(d) Show that the groups $\text{III}^2_S(N)$ and $\text{III}^1_S(M)$ (with similar notations to those used in the book) are finite and dual to each other (use Exercise 16.3, c).

(e) Assume furthermore that M is torsion-free. For any abelian group B, denote by $B^\wedge$ its profinite completion. Show that we have another exact sequence of Poitou–Tate type

$$0 \longrightarrow H^0(G_S, M)^\wedge \longrightarrow \prod_{v \in S} H^0(k_v, M)^\wedge \longrightarrow H^2(G_S, N)^*$$
$$\longleftrightarrow H^1(G_S, M) \longrightarrow \prod_{v \in S} H^1(k_v, M) \longrightarrow H^1(G_S, N)^*.$$

(C) How can one extend the previous results to the case of a function field?

Remark. We can view N as a *group scheme of multiplicative type* over $\operatorname{Spec}(\mathcal{O}_{k,S})$, and the pair $(N_v, N(v))$ as a group scheme of multiplicative type over $\operatorname{Spec} \mathcal{O}_v$ when M is not ramified at v. The case where M is torsion-free corresponds to the case where N is a *torus*. In this language, it is possible to extend (by putting in a little bit more effort, a good context being that of the étale cohomology) the above sequence to a 9-term sequence, and thus obtain a duality between $\text{III}^2_S(M)$ and $\text{III}^1_S(N)$; cf. for example [20]. The method of this exercise does not apply directly to this extension, because of the problem with applying Theorem 17.18 with $r = 0$. Another approach consists of using a version of the duality theorem for class formations which uses Tate modified cohomology of profinite groups; this requires to properly define the cup-products in this context; see [42], Theorem 3.1.11, 8.4.4. and 8.6.7.

Exercise 17.9 Let k be a number field. Let K be a finite extension (not necessarily Galois) of k. We set $G_k = \operatorname{Gal}(\bar{k}/k)$ and $U = \operatorname{Gal}(\bar{k}/K)$. We define a G_k-module of finite type M by the exact sequence:

$$0 \longrightarrow \mathbf{Z} \longrightarrow I_{G_k}^U(\mathbf{Z}) \longrightarrow M \longrightarrow 0,$$

where the map $\mathbf{Z} \to I_{G_k}^U(\mathbf{Z})$ sends $a \in \mathbf{Z}$ to the function $f \in I_{G_k}^U(\mathbf{Z})$ defined by $f(x) = a$ for any x of G_k.

(a) Show that M is torsion-free.
(b) Let $T := \mathrm{Hom}(M, \overline{k}^*)$ and $R := \mathrm{Hom}(I_{G_k}^U(\mathbf{Z}), \overline{k}^*)$. Compare R and $I_{G_k}^U(\overline{k}^*)$. Show that we have an exact sequence of G_k-modules

$$0 \longrightarrow T \longrightarrow R \xrightarrow{N} \overline{k}^* \longrightarrow 0,$$

where N is induced by the norm $(K \otimes_k \overline{k})^* \to \overline{k}^*$ (cf. Remark 1.47).
(c) Show that $H^1(k, R) = 0$, and deduce that

$$H^1(k, T) = k^*/N_{K/k}K^*.$$

Show that if v is a place of k and K_v denotes the k_v-algebra $K \otimes_k k_v$, then

$$H^1(k_v, T_v) = k_v^*/N_{K_v/k_v}K_v^*,$$

where $N_{K_v/k_v} : K_v^* \to k_v^*$ is the norm and $T_v := \mathrm{Hom}(M, \overline{k}_v^*)$.
(d) Using Exercises 4.10 and 17.8, show that if there are infinitely many places v of k such that the group $k_v^*/N_{K_v/k_v}K_v^*$ is non-trivial, then the group $k^*/N_{K/k}K^*$ is infinite.

We will see in Exercise 18.8 that this condition is in fact always satisfied if $[K : k] \geqslant 2$, which generalises Exercise 15.1 to the non-Galois case.

Chapter 18
Some Applications

We keep the notation of the last chapter. In particular, k denotes a global field, S a non-empty set of places of k (containing all the archimedean places if k is a number field) and $G_S = \mathrm{Gal}(k_S/k)$ is the Galois group of the maximal extension of k unramified outside S. We also denote by $\mathcal{O}_S = \mathcal{O}_{k,S}$ the ring of S-integers, Ω_k (or simply by Ω) the set of places of k, and Ω_f the set of finite places.

18.1 Triviality of Some of the III^i

The aim of this paragraph is to present some vanishing results for the groups $\mathrm{III}^i(G_S, M)$ when M is a finite G_S-module and the set S is "large". We will in particular see that if S contains almost all places of k, then $\mathrm{III}^1_S(M) = 0$ if the action of G_S on M is trivial, and the same result is valid if the action of G_S on the Cartier dual M' is trivial provided we avoid one special case. We begin by recalling the following important number theoretic result.

Theorem 18.1 (Čebotarev) *Let L be a finite Galois extension of k, whose Galois group is denoted by G. Let C be a conjugacy class of G. For any finite place v of k which is not ramified in the extension L/k, we denote by Frob_v the Frobenius at v (it is an element of G, well defined up to conjugation). Then the Dirichlet density $\delta_{L/k}(C)$ of the set of places v such that $\mathrm{Frob}_v \in C$ is*

$$\delta_{L/k}(C) = \#C/\#G.$$

For a proof of this result (which is analytic in nature), see for example [15], Theorem 5.6. or Theorem B.32 in the appendix. Recall that the Dirichlet density of $S \subset \Omega_k$ is the limit (if it exists)

$$\delta(S) := \lim_{s \to 1} \frac{\sum_{\mathfrak{p} \in S \cap \Omega_f} (N\mathfrak{p})^{-s}}{\sum_{\mathfrak{p} \in \Omega_f} (N\mathfrak{p})^{-s}},$$

© Springer Nature Switzerland AG 2020
D. Harari, *Galois Cohomology and Class Field Theory*, Universitext,
https://doi.org/10.1007/978-3-030-43901-9_18

where the *absolute norm* $N\mathfrak{p}$ is the cardinality of the residue field of $\mathfrak{p}$.

Čebotarev theorem implies that the "proportion" of places v totally split in the extension L/k is $1/[L : k]$. We can also restrict ourselves to places v of *absolute degree* 1 of k (i.e., such that k_v is of degree 1 over $\mathbf{Q}_p$, where p is the prime number dividing v) as the other places form a set of density zero among places of k.

Proposition 18.2 *Let S be a set of places of k containing all the archimedean places. Let A be a finite abelian group equipped with the trivial action of G_S. Let $T \subset S$ be a set of places such that $\delta(T) > 1/p$, where p is the smallest prime divisor of $\#A$. Then the natural map (induced by the restrictions $H^1(G_S, A) \longrightarrow H^1(k_v, A)$)*

$$H^1(G_S, A) \longrightarrow \prod_{v \in T} H^1(k_v, A)$$

is injective. In particular, if $\delta(S) > 1/p$, we have $\text{III}^1_S(A) = 0$.

Proof We are immediately reduced to the case where $A = \mathbf{Z}/\ell^r\mathbf{Z}$ with $r \in \mathbf{N}^*$ and ℓ prime. The kernel of the map under consideration corresponds to continuous homomorphisms

$$\varphi : G_S \longrightarrow \mathbf{Z}/\ell^r\mathbf{Z}$$

whose restriction has a trivial decomposition subgroup at v for every place v of T. The kernel of φ is thus of the form $\text{Gal}(k_S/L)$, where L is a finite extension of k which is of degree ℓ^s (with $s \leqslant r$ and $\ell \geqslant p$) and totally split at places of T. Let X be the set of places of k where L/k splits totally. Then $T \subset X$. Čebotarev theorem implies that $\delta(X) = 1/\ell^s$, hence

$$\frac{1}{\ell} < \delta(T) \leqslant \delta(X) = \frac{1}{\ell^s}.$$

Thus $s = 0$, which means that φ is the trivial homomorphism. $\square$

Remark 18.3 In the case where T contains almost all places of k, Čebotarev theorem is not necessary, it is enough to apply Corollary 13.13 instead, which is its weak version. From this fact, Theorems 18.9 and 18.15 below, as well as their corollaries, only require Corollary 13.13, and not the full Čebotarev theorem.

Corollary 18.4 *Let T be a set of places of k of density $> 1/2$ (for example containing almost all places of k). Let A be a finite abelian group endowed with the trivial action of $G_k = \text{Gal}(\bar{k}/k)$. Then the natural map*

$$H^1(k, A) \longrightarrow \prod_{v \in T} H^1(k_v, A)$$

is injective.

It is the special case $S = \Omega_k$ in Proposition 18.2.

Corollary 18.5 *Let S be a set of places of k containing all archimedean places. Let A be a finite G_S-module of order invertible in $\mathcal{O}_S$, and such that the action of G_S on the Cartier dual A' is trivial. We further assume that $\delta(S) > 1/p$, where p is the smallest prime divisor of $\#A$. Then $\text{III}^2_S(A) = 0$.*

Proof This follows from previous proposition and from the Poitou–Tate duality (Theorem 17.13, b) between $\text{III}^2_S(A)$ and $\text{III}^1_S(A')$. $\qquad\qquad\square$

The study of $\text{III}^1_S(A)$ when the action of G_S on A' is non-trivial (for example $A = \mu_n$) is more complicated. We will need the following lemma:

Lemma 18.6 *Let p be a prime number. Let r be a positive integer and let G be a subgroup of $(\mathbf{Z}/p^r\mathbf{Z})^*$ that we can also view as a subgroup of $\text{Aut}(\mathbf{Z}/p^r\mathbf{Z})$. Let A be the G-module isomorphic to $\mathbf{Z}/p^r\mathbf{Z}$ as abelian group, with the natural action of G. Then*

$$\widehat{H}^i(G, A) = 0, \quad \forall i \in \mathbf{Z}$$

except in the case $p = 2$, $r \geqslant 2$, and $-1 \in G$, in which case $\widehat{H}^i(G, A) = \mathbf{Z}/2$ for any $i \in \mathbf{Z}$.

Naturally, the fact that $p = 2$ is an exceptional case is due to the non-cyclicity of the group $(\mathbf{Z}/2^r\mathbf{Z})^*$ for $r \geqslant 3$.

Proof We denote by v_p the p-adic valuation on $\mathbf{Z}$. Assume first $p \neq 2$. Then the group $(\mathbf{Z}/p^r\mathbf{Z})^*$ is cyclic of order $p^{r-1}(p-1)$. Let $\pi : (\mathbf{Z}/p^r\mathbf{Z})^* \to (\mathbf{Z}/p\mathbf{Z})^*$ be the canonical surjection, then the p-Sylow of G is the group $G_1 = \ker[G \xrightarrow{\pi} (\mathbf{Z}/p\mathbf{Z})^*]$. As the restriction $\widehat{H}^i(G, A) \to \widehat{H}^i(G_1, A)$ is injective by Lemma 3.2, we are reduced to the case where $G \subset G_1$. Let us then fix $\alpha \in \mathbf{Z}$ whose class $\bar{\alpha}$ modulo $p^r\mathbf{Z}$ generates G and denote by p^{r-s} the order of $\bar{\alpha}$ (we have $s \geqslant 1$). We can write $\alpha = 1 + p^s u$ with $v_p(u) = 0$. Then A^G is the kernel of $\bar{\alpha} - 1 : A \to A$, which gives $A^G = p^{r-s}A$. On the other hand we have

$$\sum_{i=0}^{p^{r-s}-1} \alpha^i = \frac{\alpha^{p^{r-s}} - 1}{\alpha - 1} = p^{r-s}v$$

with $v_p(v) = 0$: indeed, we have

$$\alpha^{p^{r-s}} = 1 + p^r u + \frac{p^{r+s}(p^{r-s} - 1)}{2}u^2 + \cdots, \qquad (18.1)$$

which shows that $v_p(\alpha^{p^{r-s}} - 1) = p^r$ as $p \geqslant 3$ and $s \geqslant 1$. We deduce $N_G(A) = p^{r-s}A = A^G$. Finally $\widehat{H}^0(G, A) = 0$. As A is finite, its Herbrand quotient is 1 (Theorem 2.20, b) hence $H^1(G, A) = 0$ and we conclude using the 2-periodicity of the cohomology of a cyclic group (Theorem 2.16).

Assume now $p = 2$ (and $r \geqslant 2$). Then the group $(\mathbf{Z}/2^r\mathbf{Z})^*$ is a direct product of the subgroup $\{\pm 1\}$ and a cyclic subgroup C of order 2^{r-2}. Three cases present themselves. If $G \subset C$, then G is cyclic and generated by the class of an $\alpha \in \mathbf{Z}$ such that $v_2(\alpha - 1) = s \geqslant 2$. This case is dealt with in exactly the same way as the case $p \neq 2$ since we again have $v_2(\alpha^{2^{r-s}} - 1) = 2^r$ by formula (18.1).

The second case is when G is cyclic and generated by an element which is not in C, i.e., by the class of an $\alpha \in \mathbf{Z}$ congruent to 3 modulo 4 and such that $\bar{\alpha} \neq -1$. Let us set $\beta = -\alpha$, then $\bar{\beta} \in C$ and $v_2(\beta - 1) = s$ with $2 \leqslant s \leqslant r - 1$. As $v_2(\alpha - 1) = 1$, we have $A^G = 2^{r-1}A$. Furthermore, the order of G is twice the order of $\bar{\alpha}^2 = \bar{\beta}^2$, that is, 2^{r-s}. Then

$$\sum_{i=0}^{2^{r-s}-1} \alpha^i = \frac{-\beta^{2^{r-s}} + 1}{\beta + 1}$$

is of 2-adic valuation $r - 1$ (always by the formula (18.1), applied to β). We deduce that $N_G(A) = 2^{r-1}A$, and we conclude in the same way as above.

There remains the case where G is of the form $G = \{\pm 1\} \times \langle \bar{\alpha} \rangle$ with α congruent to 1 modulo 4, which is the only case where $p = 2$ and $-1 \in G$. Let then H be the subgroup generated by $\bar{\alpha}$. It is of order 2^{r-s} with $s \geqslant 2$, and s is the largest integer $\leqslant r$ such that 2^s divides $\alpha - 1$. In this case, we have

$$\sum_{i=0}^{2^{r-s}-1} \alpha^i + \sum_{i=0}^{2^{r-s}-1} -\alpha^i = 0,$$

which implies $N_G A = 0$. Furthermore, $A^G = 2^{r-1}A$ and $I_G A = 2A$, hence $\widehat{H}^i(G, A) = \mathbf{Z}/2$ for $i = -1, 0$. As by what precedes we have $\widehat{H}^i(H, A) = 0$ for any i, Corollary 1.45 implies that for any $i \geqslant 1$,

$$H^i(G, A) = H^i(\{\pm 1\}, A^H) = H^i(\{\pm 1\}, 2^{r-s}A)$$

whose cardinality is that of $\widehat{H}^0(\{\pm 1\}, 2^{r-s}A)$ (as $\{\pm 1\}$ is cyclic), which is 2. The same argument using homology shows that the cardinality of $\widehat{H}^i(G, A)$ is 2 for $i < -1$. $\square$

We will apply the above lemma to cyclotomic extensions of a field F. Let $m \geqslant 2$ be an integer. Let $F(\mu_m)$ be the field obtained by adjoining the mth roots of unity to F. The choice of a primitive mth root of unity allows to identify the group of roots of unity μ_m to the additive group $\mathbf{Z}/m\mathbf{Z}$. The group $G_m := \mathrm{Gal}(F(\mu_m)/F)$ then embeds into $\mathrm{Aut}(\mathbf{Z}/m\mathbf{Z})$ and the action of G_m on μ_m corresponds to its action by automorphisms of $\mathbf{Z}/m\mathbf{Z}$, as in Lemma 18.6.

Remark 18.7 Let F be a field and $r \geqslant 1$. If $\mathrm{Car}\, F = 0$, we say that the field $\mathbf{Q}(\mu_{2^r}) \cap F$ is real if it has a real place, which is equivalent to saying that all its archimedean places are real, as it is a finite (abelian) Galois extension of $\mathbf{Q}$. The group $G_{2^r} = \mathrm{Gal}(F(\mu_{2^r})/F)$ is a subgroup of $\mathrm{Gal}(\mathbf{Q}(\mu_{2^r})/\mathbf{Q})$. It can happen that it is not cyclic but isomorphic to the product of $\mathbf{Z}/2\mathbf{Z}$ by a cyclic group if $r \geqslant 3$, while G_{p^r} is always cyclic for p prime $\geqslant 3$.

If F is a field of positive characteristic $\ell \neq p$, The extension $F(\mu_{p^r})/F$ is always cyclic as its Galois group G_{p^r} is isomorphic to that of the extension of finite fields $\mathbf{F}_\ell(\mu_{p^r})/(\mathbf{F}_\ell(\mu_{p^r}) \cap F)$.

Corollary 18.8 *Let p be a prime number and $r \in \mathbf{N}$. Let F be a field of characteristic $\neq p$. We set $G_{p^r} := \mathrm{Gal}(F(\mu_{p^r})/F)$. Then we have*

$$\widehat{H}^i(G_{p^r}, \mu_{p^r}) = 0$$

for any $i \in \mathbf{Z}$, except in the cases, that we call exceptional, *where $p = 2, r \geqslant 2$, and:*

- *Either (case "Ex_0") F is of characteristic zero and $\mathbf{Q}(\mu_{2^r}) \cap F$ is real;*
- *or (case "Ex_ℓ") F is of characteristic $\ell \geqslant 3$, we have $\ell \equiv -1 \bmod 2^r$, and $\mathbf{F}_\ell(\mu_{2^r}) \cap F = \mathbf{F}_\ell$.*

Furthermore, in the exceptional cases, we have

$$\widehat{H}^i(G_{p^r}, \mu_{p^r}) = \mathbf{Z}/2\mathbf{Z}$$

for any $i \in \mathbf{Z}$.

Proof First assume F is of characteristic zero and set $F_0 = F \cap \mathbf{Q}(\mu_{p^r})$. The group G_{p^r} is isomorphic to $\mathrm{Gal}(\mathbf{Q}(\mu_{p^r})/F_0)$, and we can view it as a subgroup of $\mathrm{Aut}(\mathbf{Z}/p^r\mathbf{Z}) \simeq (\mathbf{Z}/p^r\mathbf{Z})^*$. We then apply Lemma 18.6. The exceptional case occurs when $p = 2$ and $r \geqslant 2$, when the element $-1 \in (\mathbf{Z}/2^r\mathbf{Z})^*$ is in G_{2^r}. This last condition is equivalent to the condition that the complex conjugation (that sends a root of unity ζ to ζ^{-1}) over $\mathbf{Q}(\mu_{p^r})$ induces the identity on F_0, in other words, that F_0 is real.

Assume now that F is of characteristic $\ell \geqslant 3$. Let $\mathbf{F}_q = F \cap \mathbf{F}_\ell(\mu_{p^r})$, then the finite field $\mathbf{F}_q \subset \mathbf{F}_\ell(\mu_{p^r})$ is an extension of $\mathbf{F}_\ell$. The group G_{p^r} is now isomorphic to $\mathrm{Gal}(\mathbf{F}_\ell(\mu_{p^r})/\mathbf{F}_q)$, which is a subgroup of $(\mathbf{Z}/p^r\mathbf{Z})^*$. Lemma 18.6 again implies that the exceptional case corresponds to $p = 2$ and $r \geqslant 2$, if in addition $-1 \in G_{2^r}$. As G_{2^r} is generated by the Frobenius $x \mapsto x^q$, the condition can be expressed as follows: the multiplicative subgroup generated by q in $(\mathbf{Z}/2^r\mathbf{Z})^*$ contains -1. But the only cyclic subgroup of $(\mathbf{Z}/2^r\mathbf{Z})^* \simeq (\mathbf{Z}/2\mathbf{Z}) \times (\mathbf{Z}/2^{r-2}\mathbf{Z})$ containing -1 is $\{\pm 1\}$ (Note that $\{\pm 1\}$ correspond to a subgroup $(\mathbf{Z}/2\mathbf{Z}) \times \{0\} \subset (\mathbf{Z}/2\mathbf{Z}) \times (\mathbf{Z}/2^{r-2}\mathbf{Z})$). If we set $n = [\mathbf{F}_q : \mathbf{F}_\ell]$, the condition is $q = \ell^n \equiv -1 \bmod 2^r$. As n (which is a power of 2) is equal to 1 or is even, the only possibility is $n = 1$ and $\ell \equiv -1 \bmod 2^r$, since $r \geqslant 2$ and -1 is not a square modulo 4. $\qquad\square$

Theorem 18.9 *Let k be a global field. Let S be a set of places of k containing all the archimedean places. Let $p_1, \ldots, p_n$ be pairwise distinct prime numbers in $\mathcal{O}_S^*$ and m an integer of the form $m = p_1^{r_1} \cdots p_n^{r_n}$. Let T be a subset of S, we assume that T contains almost all the places of k. We finally assume that if k is a number field, we are not in the following exceptional case:*

($Ex_{0,T}$) One of the p_i is 2 with $r_i \geqslant 3$, the exceptional case (Ex_0) of Corollary 18.8 occurs for the extension $k(\mu_{2^{r_i}})/k$, and no place of T is inert in $k(\mu_{2^{r_i}})/k$.

Then the homomorphism

$$\beta_{S,T,m} : H^1(G_S, \mu_m) \longrightarrow \prod_{v \in T} H^1(k_v, \mu_m)$$

(induced by restrictions $H^1(G_S, \mu_m) \longrightarrow H^1(k_v, \mu_m)$) is injective. In particular,
$\text{III}^1_S(k, \mu_m) = 0$.

Remark 18.10 We can weaken the assumption on T (by stating it in terms of Dirichlet density. In particular the proof works if we assume T to be of Dirichlet density 1, provided we know the Čebotarev theorem). We can also describe more precisely the exceptional case for a number field, and in particular show that in this case we have $\text{III}^1_S(k, \mu_m) = \mathbf{Z}/2\mathbf{Z}$, cf. [42], Theorem 9.1.9.

Proof As the G_S-module μ_m is isomorphic to the product of the $\mu_{p_i^{r_i}}$, we are reduced to the case where $m = p^r$ with p prime. Let $K = k(\mu_m)$. Then K is a subextension of k_S as m is invertible in $\mathcal{O}_S$. We set $G = \text{Gal}(K/k)$, and we denote by G_v the decomposition subgroup of K/k at v, which is well defined as G is abelian. Denote by $\text{III}^1_S(T, m)$ the kernel of $\beta_{S,T,m}$ and by $\text{III}^1(K, T, m)$ that of the homomorphism $H^1(G, \mu_m) \to \prod_{v \in T} H^1(G_v, \mu_m)$ induced by the restrictions. Let T_K be the set of places of K above a place of T. We have (via the restriction-inflation exact sequence) a commutative diagram with exact rows and columns, whose maps i_1, i_2 and i_3 are injective:

$$
\begin{array}{ccccc}
0 \longrightarrow \text{III}^1(K,T,m) \longrightarrow & H^1(G,\mu_m) & \longrightarrow & \prod_{v\in T} H^1(G_v,\mu_m) \\
\quad\quad\quad i_1 \downarrow & i_2 \downarrow & & i_3 \downarrow \\
0 \longrightarrow \text{III}^1_S(T,m) \longrightarrow & H^1(G_S,\mu_m) & \longrightarrow & \prod_{v\in T} H^1(k_v,\mu_m) \\
\downarrow & \downarrow & & \downarrow \\
& H^1(G_{K,S},\mu_m) & \xrightarrow{\ i_4\ } & \prod_{w\in T_K} H^1(K_w,\mu_m)
\end{array}
$$

where $G_{K,S} := \text{Gal}(k_S/K)$ and the bottom right vertical map is induced by the maps $H^1(k_v, \mu_m) \to \prod_{w|v} H^1(K_w, \mu_m)$ for $v \in T$. As the action of $G_{K,S}$ on μ_m is trivial, Proposition 18.2 implies that the map i_4 is injective. By diagram chasing, we get that $\text{III}^1_S(T, m) = \text{III}^1(K, T, m)$.

On the other hand, if we are not in one of the exceptional cases of Corollary 18.8, then $H^1(G, \mu_m) = 0$ and we immediately obtain $\text{III}^1(K, T, m) = 0$, hence $\text{III}^1_S(T, m) = 0$. If we are in one of the exceptional cases, then $p = 2$ and $r \geqslant 2$. In the case of a number field, the assumption made if $r \geqslant 3$ ensures that there exists a place $v \in T$ which is inert in the extension $k(\mu_{2^r})/k$, and if $r = 2$ this remains true by Corollary 13.13 as this extension is then quadratic or trivial. In the case of a function field, this extension is always cyclic (cf. Remark 18.7), and the assumption that T contains almost all places of k implies again (with Corollary 13.13) that one of the places v of T is inert for $k(\mu_{2^r})/k$. We then have $\text{III}^1(K, T, m) = 0$ as the groups G and G_v are equal. This finishes the proof. $\qquad\square$

Remark 18.11 The exceptional case of Theorem 18.9 can effectively happen if m is divisible by 8, even if $T = S$ is the set of all places, see Exercise 18.6 (d). However it does not exist when $\sqrt{-1} \in k$, as the field $k(\sqrt{-1})$ is then not real.

Corollary 18.12 *Let S be a set of places of k containing all the archimedean places and such that $\Omega_k - S$ is finite. Let $m \in \mathcal{O}_S^*$ be an integer. We furthermore assume that if k is a number field, we are not in the exceptional case $(Ex_{0,S})$ of Theorem 18.9. Then $\text{III}_S^2(k, \mathbf{Z}/m) = 0$.*

Proof This follows directly from Poitou–Tate duality between the groups $\text{III}_S^2(k, \mathbf{Z}/m)$ and $\text{III}_S^1(k, \mu_m)$, and from Theorem 18.9 applied to $T = S$. ☐

Corollary 18.13 (Grunwald–Wang) *Let $m \geqslant 1$ be an integer. Let T be a subset of places of k with $\Omega_k - T$ finite. We furthermore assume that if k is a number field, we are not in the exceptional case $Ex_{0,T}$ of Theorem 18.9. Then the map*

$$k^*/k^{*^m} \to \prod_{v \in T} k_v^*/k_v^{*^m}$$

(induced by the restrictions $k^* \to k_v^*$) is injective.

Proof We have $H^1(k, \mu_m) = k^*/k^{*^m}$ and $H^1(k_v, \mu_m) = k_v^*/k_v^{*^m}$. We then apply Theorem 18.9 with $S = \Omega_k$. ☐

Remark 18.14 The same result is valid (with the same proof, provided we know Čebotarev theorem) if T is only assumed to have Dirichlet density 1. Similarly, in Corollary 18.12, it is enough to assume S to be of Dirichlet density 1. Historically, the first example of the exceptional case of Corollary 18.13 is due to Wang (1948): it happens for $k = \mathbf{Q}, T = \Omega_\mathbf{Q} - \{2\}, m = 8$ and is related to the fact that the restriction $H^1(\mathbf{Q}, \mathbf{Z}/8) \to H^1(\mathbf{Q}_2, \mathbf{Z}/8)$ is not surjective; see Exercise 15.4 and Exercise 18.6, (b) and (c).

18.2 The Strict Cohomological Dimension of a Number Field

Let S be a set of places of k such that S contains all the archimedean places. As we have already pointed out (Remark 17.15) to determine $scd_p(G_S)$ for p invertible in $\mathcal{O}_S$ is an open problem in general (in the case of a number field). Nevertheless, there is an answer when S contains almost all places of k.

Theorem 18.15 *Let k be a global field. Let S be a set of places of k containing all the archimedean places and such that $\Omega_k - S$ is finite. Let p be a prime number invertible in $\mathcal{O}_S$. Furthermore we assume that k is totally imaginary if k is a number field and $p = 2$. Then $scd_p(G_S) = 2$.*

In particular, we have $scd(k) = 2$ for totally imaginary number fields.

Proof By Corollary 17.14, we have $\mathrm{cd}_p(G_S) \leqslant 2$. On the other hand p (and even p^∞) divides the order of G_S by Remark 17.1, which implies that there exist cyclic extensions of degree p of k contained in k_S, and hence in particular that $H^2(G_S, \mathbf{Z})[p] = H^1(G_S, \mathbf{Q}/\mathbf{Z})[p]$ is nonzero. We deduce that $\mathrm{scd}_p(G_S) \geqslant 2$. By Proposition 5.14, it is enough to check that $H^2(U, \mathbf{Q}_p/\mathbf{Z}_p) = 0$ for any open subgroup U of G_S. As each finite extension k' of k contained in k_S satisfies the assumptions of the theorem (replacing S by the set of places of k' dividing a place of S), we are reduced to showing that $H^2(G_S, \mathbf{Q}_p/\mathbf{Z}_p) = 0$. If p is odd or $\sqrt{-1} \in k$, or if k is a function field, Corollary 18.12 implies, by passing to the limit, that the map

$$H^2(G_S, \mathbf{Q}_p/\mathbf{Z}_p) \longrightarrow \prod_{v \in S} H^2(k_v, \mathbf{Q}_p/\mathbf{Z}_p)$$

is injective. We conclude using the fact that (Theorem 10.6, Remark 10.7 and Proposition 6.15) the absolute Galois group Γ_v of k_v verifies $\mathrm{scd}_p(\Gamma_v) = 2$ for v finite, along with the observation that for an archimedean v, the assumptions we have made imply $\mathrm{scd}_p(k_v) = 0$.

Assume now that k is a totally imaginary number field with $p = 2$ and $\sqrt{-1} \notin k$. By what we have seen, if we set $K = k(\sqrt{-1})$, we have $H^2(G_{K,S}, \mathbf{Q}_2/\mathbf{Z}_2) = 0$, where $G_{K,S} = \mathrm{Gal}(k_S/K)$. But we know that the corestriction

$$H^2(G_{K,S}, \mathbf{Q}_2/\mathbf{Z}_2) \longrightarrow H^2(G_S, \mathbf{Q}_2/\mathbf{Z}_2)$$

is surjective (Lemma 5.9) as $\mathrm{cd}_2(G_S) = 2$. The result follows. $\square$

Remark 18.16 Here again, it is enough to assume that S is of Dirichlet density 1.

Corollary 18.17 *Let k be a global field. We have $H^3(k, \mathbf{Z}) = 0$ and for $r \geqslant 4$, the natural map*

$$H^r(k, \mathbf{Z}) \longrightarrow \bigoplus_{v \in \Omega_\mathbf{R}} H^r(k_v, \mathbf{Z})$$

is an isomorphism. In particular, for $r \geqslant 3$, the group $H^r(k, \mathbf{Z})$ is zero if r is odd, and isomorphic to $(\mathbf{Z}/2)^t$ if r is even, where t is the number of real places of k.

Proof Assume first $r \geqslant 4$ (cf. also Exercise 17.7 for this case, in a slightly more general context). Then

$$H^r(k, \mathbf{Z}) \simeq H^{r-1}(k, \mathbf{Q}/\mathbf{Z}) \simeq \varinjlim_n H^{r-1}(k, \mathbf{Z}/n)$$

is also isomorphic to

$$\varinjlim_n \bigoplus_{v \in \Omega_\mathbf{R}} H^{r-1}(k_v, \mathbf{Z}/n).$$

Indeed, this follows from Theorem 17.13 applied to the $\mathbf{Z}/n$ (since we have $(r-1) \geqslant 3$). But $\varinjlim_n \bigoplus_{v \in \Omega_\mathbf{R}} H^{r-1}(k_v, \mathbf{Z}/n)$ is also $\bigoplus_{v \in \Omega_\mathbf{R}} H^{r-1}(k_v, \mathbf{Q}/\mathbf{Z})$. This follows from the fact that inductive limits commute to direct sums. The result follows (for the last assertion, observe that for v real, the group $H^r(k_v, \mathbf{Z})$ is isomorphic to $H^1(\mathbf{R}, \mathbf{Z})$ if r is odd and to $\widehat{H}^0(\mathbf{R}, \mathbf{Z})$ if r is even by Theorem 2.16).

It remains to show that $H^3(k, \mathbf{Z}) = 0$. If $\sqrt{-1} \in k$, then k is totally imaginary and the result follows from Theorem 18.15. Otherwise, let $G_k = \mathrm{Gal}(\bar{k}/k)$, let us set $L = k(\sqrt{-1})$ and $U = \mathrm{Gal}(\bar{k}/L)$. The field L (and thus the profinite group U) is of strict cohomological dimension 2, which implies, for any $r \geqslant 3$, $H^r(G_k, \mathbf{Z}[G_k/U]) = H^r(U, \mathbf{Z}) = 0$. Let σ be the generator of the group G_k/U (which is of order 2), we have an exact sequence of G_k-modules

$$0 \longrightarrow \mathbf{Z} \xrightarrow{1+\sigma} \mathbf{Z}[G_k/U] \xrightarrow{1-\sigma} \mathbf{Z}[G_k/U] \xrightarrow{\sigma \mapsto 1} \mathbf{Z} \longrightarrow 0$$

which (by cutting this exact sequence into two short exact sequences) yields isomorphisms $H^r(G_k, \mathbf{Z}) \simeq H^{r+2}(G_k, \mathbf{Z})$ for any integer $r \geqslant 3$. In particular, we have $H^3(G_k, \mathbf{Z}) = H^5(G_k, \mathbf{Z}) = 0$. $\qquad\square$

18.3 Exercises

Exercise 18.1 Show that the result of Exercise 17.7, (c), still holds true for $r = 3$ (use Exercise 17.7, b).

Exercise 18.2 (*after [45], Sect. 1*) Let k be a global field with absolute Galois group G_k. Let A be a finite G_k-module. We denote by $\mathrm{III}^1_\omega(A)$ the subgroup of $H^1(k, A)$ consisting of those a whose image $a_v \in H^1(k_v, A)$ is zero for almost all places v of k.

(a) Show that if the action of G_k on A is trivial, then $\mathrm{III}^1_\omega(A) = 0$.
(b) Fix a finite Galois extension K of k such that the action of $G_K = \mathrm{Gal}(\bar{k}/K)$ on A is trivial. Show that $\mathrm{III}^1_\omega(A)$ is a subgroup of $H^1(G, A)$, where $G := \mathrm{Gal}(K/k)$, and deduce that $\mathrm{III}^1_\omega(A)$ is finite.
(c) Show that $\mathrm{III}^1_\omega(A)$ is the subgroup of $H^1(G, A)$ consisting of those b such that the restriction of b to $H^1(H, A)$ is zero for every cyclic subgroup H of G.
(d) Show that $\mathrm{III}^1_\omega(A)$ is also the subgroup of $H^1(k, A)$ consisting of those a whose restriction to $H^1(C, A)$ is zero for every procyclic (i.e., topologically generated by one element) closed subgroup C of G_k. Deduce that if $b \in \mathrm{III}^1_\omega(A)$, then for any real completion k_v of k, the restriction $b_v \in H^1(k_v, A)$ is zero.
(e) Can previous results be generalised to the case where A is only assumed to be of finite type?
(f) In an analogous manner, we define $\mathrm{III}^2_\omega(A)$ as the subgroup of $H^2(k, A)$ consisting of elements whose image in $H^2(k_v, A)$ is zero for almost all places v of k. Show that for a finite A, we have $\mathrm{III}^2_\omega(A) = H^2(k, A)$ (this group is infinite if $A \neq 0$, see next exercise), but if A is a free $\mathbf{Z}$-module of finite type, the results analogous to (a), (b), (c), and (d) still hold true.

Exercise 18.3 Let k be a number field with absolute Galois group G_k. Let A be a finite nonzero G_k-module.

(a) Let n be an integer $\geqslant 2$. Show that if v is a finite place of k, the groups $H^1(k_v, \mu_n)$, $H^1(k_v, \mathbf{Z}/n)$ and $H^2(k_v, \mu_n)$ are nonzero. Can we say the same thing about $H^2(k_v, \mathbf{Z}/n)$?

(b) Using the Čebotarev theorem, show that there exist infinitely many places v of k such that the group $H^1(k_v, A)$ is nonzero. Same question with $H^2(k_v, A)$.

(c) Show that the group $H^2(k, A)$ is infinite.

Exercise 18.4 *(sequel to the previous exercise)* Let k be a number field with absolute Galois group G_k and A be a finite nonzero G_k-module. We endow $\prod_{v \in \Omega_k} H^1(k_v, A)$ with the product topology (each finite group $H^1(k_v, A)$ is endowed with discrete topology) and we denote by $\overline{H^1(k, A)}$ the closure of the image of $H^1(k, A)$ in $\prod_{v \in \Omega_k} H^1(k_v, A)$.

(a) With the notation of Exercise 18.2, show that we have an exact sequence

$$0 \longrightarrow \overline{H^1(k, A)} \longrightarrow \prod_{v \in \Omega_k} H^1(k_v, A) \longrightarrow \text{III}^1_\omega(A)^* \longrightarrow \text{III}^2(A) \longrightarrow 0,$$

where $*$ as usual denotes the dual (use the Exercise 17.5).

(b) Using Exercise 18.3, show that $\overline{H^1(k, A)}$ is infinite.

Exercise 18.5 *(around Grunwald–Wang, cf. [42], Theorem 9.2.3)* Let k be a number field. Let S be a set of places of k containing all the archimedean places and let T be a finite subset of S. Let A be a finite G_S-module whose order is invertible in $\mathcal{O}_{k,S}$. We denote by A' the Cartier dual of A.

(a) We denote by N the kernel of the natural map

$$H^1(G_S, A') \longrightarrow \prod_{v \in S-T} H^1(k_v, A')$$

and by C the cokernel of the map $\beta^1_{S,T} : H^1(G_S, A) \longrightarrow \bigoplus_{v \in T} H^1(k_v, A)$. Show that we have an exact sequence

$$0 \longrightarrow \text{III}^1_S(k, A') \longrightarrow N \longrightarrow C^* \longrightarrow 0.$$

(see also Exercise 17.5).

(b) Assume that the action of G_S on A' is trivial and that the density $\delta(S)$ is $> 1/p$, where p is the smallest prime divisor of $\#A$. Show that $\beta^1_{S,T}$ is surjective.

(c) Assume that $A = \mathbf{Z}/m\mathbf{Z}$ (with the trivial action of G_S) with m invertible in $\mathcal{O}_{k,S}$. We make an additional assumption that S contains almost all the places of k. Show that if m is odd or $\sqrt{-1} \in k$, then $\beta^1_{S,T}$ is surjective.

Exercise 18.6 (a) Show that -1 is not a square in $\mathbf{Q}_2$, but that $\mathbf{Q}_2(\sqrt{7})$ is a field extension of $\mathbf{Q}_2$ in which -1 becomes a square (observe that $-1 = 7(1 - 8/7)$).

(b) Let a be the class of 16 in $\mathbf{Q}^*/\mathbf{Q}^{*8} \simeq H^1(\mathbf{Q}, \mu_8)$. Show that the restriction a_p of a to $H^1(\mathbf{Q}_p, \mu_8)$ is zero for p an odd prime and $p = \infty$, but that it is nonzero for $p = 2$ (observe that $16 = 2^4 = (-2)^4 = (1 + \sqrt{-1})^8$).

(c) Using Exercise 17.5, recover the result of the Exercise 15.4, that is, that the restriction $H^1(\mathbf{Q}, \mathbf{Z}/8) \to H^1(\mathbf{Q}_2, \mathbf{Z}/8)$ is not surjective.

(d) Let $k = \mathbf{Q}(\sqrt{7})$. Show that the class of 16 in k^*/k^{*8} is in the kernel of the natural homomorphism

$$k^*/k^{*8} \longrightarrow \prod_{v \in \Omega_k} k_v^*/k_v^{*8}.$$

Exercise 18.7 This exercise shows some applications of the Čebotarev theorem.

(a) Let a and m be two strictly positive coprime integers. Considering the cyclotomic extension $\mathbf{Q}(\zeta)/\mathbf{Q}$, where ζ is a primitive mth root of unity, recover Dirichlet theorem saying that there exists a strictly positive density of prime numbers congruent to a modulo m.

(b) Let G be a finite group. Let H be a subgroup of G with $H \neq G$. Show that

$$\bigcup_{g \in G} gHg^{-1} \neq G$$

(observe that gHg^{-1} only depends on the left H-coset of g).

(c) Let k be a number field. Let K be a finite extension (not necessarily Galois) of k with $[K : k] \geqslant 2$. Show that there is a strictly positive density of places v such that no place w of K above v satisfies $K_w = k_v$ (use Exercise 12.2).

(d) Let L/k be a finite extension (not necessarily Galois) of number fields and L' a Galois cyclic extension of L. Show that there exists a strictly positive density of places v of k such that there exists a place w_L of L above v satisfying: w_L is split in L/k, and inert for the extension L'/L.

Exercise 18.8 Let G be a finite group. Let H be a subgroup of G with $H \neq G$. We admit the following result from the theory of finite groups[1] [14], that uses the classification of simple finite groups, and extends Exercise 18.7 (b): there exists a element of G whose order is a power of a prime and is not in $\bigcup_{g \in G} gHg^{-1}$.

Let k be a number field and K be a finite non-trivial extension of k.

(a) Show that there exists a prime number p such that for infinitely many places v of k, all the degrees $[K_w : k_v]$ (for w place of K above v) are divisible by p (use Exercise 12.2, d).

(b) Using Exercise 17.9, show that $k^*/N_{K/k}K^*$ is infinite.

(c) Show that the kernel of the restriction $\mathrm{Br}\, k \to \mathrm{Br}\, K$ is infinite.

---

[1] Thanks to J.-L. Colliot-Thélène and O. Wittenberg for providing a reference.

Appendix A
Some Results from Homological Algebra

In this appendix, we recall (without proof, except when a complete reference could not be found) some basic results from homological algebra that we have used in this book. One may consult [35] for generalities on categories and [54] for more details on homological algebra.

A.1 Generalities on Categories

Definition A.1 A *category* $\mathcal{C}$ is the following data:

- a class $\mathrm{Ob}(\mathcal{C})$ of *objects*;
- for any pair (A, B) of objects in $\mathcal{C}$ a set $\mathrm{Hom}_{\mathcal{C}}(A, B)$ (or simply $\mathrm{Hom}(A, B)$ when it is clear what $\mathcal{C}$ is) of *morphisms* from A to B; an element f of $\mathrm{Hom}(A, B)$ will be denoted $f : A \to B$;
- for any object $A \in \mathrm{Ob}(\mathcal{C})$, an identity morphism $\mathrm{Id}_A : A \to A$;
- for any triple (A, B, C) of objects, a composition function

$$\mathrm{Hom}(A, B) \times \mathrm{Hom}(B, C) \longrightarrow \mathrm{Hom}(A, C)$$

denoted $(f, g) \mapsto g \circ f$ (we will often shorten $g \circ f$ to gf), verifying the two following axioms:

- $(hg)f = h(gf)$ for all morphisms $f : A \to B, g : B \to C, h : C \to D$;
- $f = \mathrm{Id}_B \circ f = f \circ \mathrm{Id}_A$ for any morphism $f : A \to B$.

Example A.2

(a) The sets form a category (that we denote by $\mathcal{E}ns$), taking as morphisms the usual functions.

(b) The groups form a category (denoted $\mathcal{G}r$), the morphisms being the group morphisms. The same is true for abelian groups, or for rings (the morphisms are then the ring morphisms). We denote by $\mathcal{A}b$ and $\mathcal{A}nn$ these categories.

© Springer Nature Switzerland AG 2020
D. Harari, *Galois Cohomology and Class Field Theory*, Universitext,
https://doi.org/10.1007/978-3-030-43901-9

(c) If R is a ring, left R-modules form a category that we denote by $\mathcal{M}od_R$, the morphisms are the morphisms of R-modules. Naturally, we have an analogous result for the right R-modules.

(d) Topological spaces form a category $\mathcal{T}op$, the morphisms being the continuous maps.

Note that in general $\mathrm{Ob}(\mathcal{C})$ is not a set (otherwise we very quickly run into paradoxes of the type "the set of all sets"). A category whose objects form a set is called *small*.

Definition A.3 An *isomorphism* in $\mathcal{C}$ is a morphism $f : A \to B$ such that there exists a morphism $g : B \to A$ (necessarily unique, denoted by $g = f^{-1}$) satisfying $fg = \mathrm{Id}_B$ and $gf = \mathrm{Id}_A$.

A *monomorphism* in $\mathcal{C}$ is a morphism $f : B \to C$ such that for every pair e_1, e_2 of morphisms $A \to B$, the equality $fe_1 = fe_2$ implies $e_1 = e_2$.

An *epimorphism* in $\mathcal{C}$ is a morphism $f : B \to C$ such that for every pair e_1, e_2 of morphisms $A \to B$, the equality $e_1 f = e_2 f$ implies $e_1 = e_2$.

For example, in the category $\mathcal{A}b$ (resp. in $\mathcal{M}od_R$, in $\mathcal{T}op$), the monomorphisms are the injective morphisms. The epimorphisms in $\mathcal{A}b$ and $\mathcal{M}od_R$ are the surjective morphisms, but this is no longer true for example in the category of rings (for example take the inclusion $\mathbf{Z} \to \mathbf{Q}$) or Hausdorff topological spaces (take for example the inclusion $\mathbf{Q} \to \mathbf{R}$).

Definition A.4 Let B be an object in a category $\mathcal{C}$. A subobject A of B is an object endowed with a monomorphism $i : A \to B$ (we will identify $i : A \to B$ and $i' : A' \to B$ if i' factors through i and i factors through i').

Similarly, we define the notion of quotient object by replacing a monomorphism by an epimorphism and reversing the arrows.

Definition A.5 An object A in a category $\mathcal{C}$ is *noetherian* if any increasing sequence (for the ordering "being a subobject") of subobjects of A is stationary. A category $\mathcal{C}$ is *noetherian* if it is equivalent (cf. Definition A.15) to a small category and all its objects are noetherian.

For example, the category of modules of finite type over a noetherian commutative ring is a noetherian category. Similarly we have a definition of an *artinian* category by working with the quotient objects and reversing the arrows.

Definition A.6 We say that an object A of $\mathcal{C}$ is *initial* (resp. *final*) if for any object B of $\mathcal{C}$, there exists one and only one morphism $A \to B$ (resp. one and only one morphism $B \to A$). An object which is both initial and final is called a *zero-object* (or simply *zero*, or *null object*), that we generally denote by 0.

Note that two initial (resp. final) objects are isomorphic in $\mathcal{C}$ and the isomorphism between them is unique. We can then talk of "the initial (resp. final) object" of $\mathcal{C}$ when it exists.

Example A.7

(a) In $\mathcal{E}ns$, the only initial object is $\varnothing$, the final objects are the singletons (hence there is no zero).

(b) In $\mathcal{A}b$ or $\mathcal{M}od_R$, the trivial group (resp. the trivial R-module) 0 is a zero-object.

Definition A.8 Let $\mathcal{C}$ be a category which has a zero 0. If A and B are two objects of $\mathcal{C}$, we denote again by 0 the morphism $A \to B$ which is the composite of $A \to 0$ and $0 \to B$.

(a) A *kernel* of a morphism $f : B \to C$ is a morphism $i : A \to B$ satisfying $fi = 0$, and which is universal for this property (meaning: for any morphism $i' : A' \to B$ with $fi' = 0$, there exists a unique morphism $g : A' \to A$ such that $i' = ig$).

(b) A *cokernel* of a morphism $f : B \to C$ is a morphism $p : C \to D$ such that $pf = 0$, and universal for that property (meaning: for any morphism $p' : C \to D'$ with $p'f = 0$, there exists a unique morphism $g : D \to D'$ such that $p' = gp$).

Note that each kernel is a monomorphism, and two kernels of the same morphism are isomorphic (in an obvious way). Similarly, a cokernel is an epimorphism and two cokernels of the same morphism are isomorphic. Naturally, the notions of kernel and cokernel in the category $\mathcal{A}b$ or $\mathcal{M}od_R$ coincide with the usual notions. We can also define the notion of co-image of f as an epimorphism $p : B \to D$ such that there exists $\tilde{f} : D \to C$ verifying $f = \tilde{f} \circ p$, with p universal for this property.

Definition A.9 Let $\mathcal{C}$ be a category. The *opposite category* $\mathcal{C}^{op}$ of $\mathcal{C}$ is the category whose objects are the same as those of $\mathcal{C}$, but the morphisms and their composites are reversed.

For any morphism $f : A \to B$ in $\mathcal{C}$, we thus have a morphism $f^{op} : B \to A$ in $\mathcal{C}^{op}$. The morphism f is a monomorphism if and only if f^{op} is an epimorphism (and vice versa). Similarly, to pass to the opposite morphism transforms the kernels into cokernels (and vice versa), and the passage to the opposite category swaps the notions of initial and final objects.

Definition A.10 A *subcategory* of a category $\mathcal{C}$ is a category $\mathcal{C}'$ whose objects are objects of $\mathcal{C}$, and such that for all objects A, B of $\mathcal{C}'$, the set $\mathrm{Hom}_{\mathcal{C}'}(A, B)$ is a subset of $\mathrm{Hom}_{\mathcal{C}}(A, B)$. We say that such a subcategory is *full* if we have $\mathrm{Hom}_{\mathcal{C}'}(A, B) = \mathrm{Hom}_{\mathcal{C}}(A, B)$ for all objects A, B of $\mathcal{C}'$.

For example, abelian groups form a full subcategory of the category of groups.

Definition A.11 Let $\mathcal{C}$ be a category. Let $(A_i)_{i \in I}$ be a family of objects of $\mathcal{C}$. A *product* $A = \prod_{i \in I} A_i$ is an object in $\mathcal{C}$, endowed with morphisms $p_i : A \to A_i$, such that for any object B of $\mathcal{C}$ and any family of morphisms $f_i : B \to A_i$, there exists a unique morphism $f : B \to A$ such that $p_i \circ f = f_i$ for any $i \in I$.

By definition, if a product $\prod_{i \in I} A_i$ exists, it is unique up to an isomorphism. The products exist for example in the category of sets, groups, abelian groups, R-modules (if R is a ring), but not in the category of fields. The product of two objects $A \times B$ exists in the category of $\mathbf{Z}$-modules of finite type, but a product of an infinite family does not.

Definition A.12 Let C be a category. Let $(A_i)_{i \in I}$ be a family of objects of C. A *coproduct* $A = \coprod_{i \in I} A_i$ is an object of C, equipped with morphisms $u_i : A_i \to A$, such that for any object B of C and any family of morphisms $g_i : A_i \to B$, there exists a unique morphism $g : A \to B$ such that $g \circ u_i = g_i$ for any $i \in I$.

A coproduct in C is thus a product in C^{op} (and vice versa). The coproduct in the category of sets is the disjoint union, in $\mathcal{M}od_R$ it is the direct sum.

A.2 Functors

Definition A.13 A *functor* (sometimes called "covariant functor") F from a category C to a category $\mathcal{D}$ is the data for any object A of C of an object $F(A)$ (that we will also denote by FA) of $\mathcal{D}$, and for any morphism $f : A \to B$ in C of a morphism $F(f) : F(A) \to F(B)$, satisfying:

$F(\mathrm{Id}_A) = \mathrm{Id}_{F(A)}$ for any object A of C.
$F(gf) = F(g)F(f)$ for all morphisms f, g in C.
A *contravariant functor* from C to $\mathcal{D}$ is a functor from C^{op} to $\mathcal{D}$.

We define in an obvious way the composite GF of two functors $F : C \to \mathcal{D}$ and $G : \mathcal{D} \to \mathcal{E}$, and the identity functor $\mathrm{Id} : C \to C$. It will often happen that to define a functor, we only define $F(A)$ for objects A of C, the definition of morphisms $F(f)$ when f is a morphism in C being completely obvious.

Example A.14

(a) We define a functor from $\mathcal{G}r$ to $\mathcal{A}b$ by associating to every group G its abelianisation G^{ab} (i.e., the quotient of G by its derived subgroup).

(b) By associating to every abelian group G the product group $G \times G$, we obtain a functor from $\mathcal{A}b$ to itself.

(c) We can construct functors called *forgetful* by retaining only some structures on the objects of a category (and forgetting others). For example, we obtain such a functor from $\mathcal{A}b$ to $\mathcal{G}r$ (resp. from $\mathcal{G}r$ to $\mathcal{E}ns$) by only considering the underlying group of an abelian group (resp. the underlying set of a group).

(d) Let k be a field. We obtain a contravariant functor from the category of k-vector spaces to itself by associating to every k-vector space E its dual E^*.

Definition A.15 Let C and $\mathcal{D}$ be two categories. Let F and G be two functors from C to $\mathcal{D}$. A *natural transformation* (or a morphism of functors) η from F to G is the assignment for any object A of C of a morphism $\eta_A : F(A) \to G(A)$ in $\mathcal{D}$, so that for any morphism $F : A \to A'$ in C, the following diagram is commutative:

$$
\begin{array}{ccc}
F(A) & \xrightarrow{\ F(f)\ } & F(A') \\
\eta_A \downarrow & & \downarrow \eta_{A'} \\
G(A) & \xrightarrow{\ G(f)\ } & G(A')
\end{array}
$$

If furthermore all of the η_A are isomorphisms, we say that η is a *natural isomorphism*. An *equivalence of categories* is a functor $F : C \to D$ such that there exists a functor $G : D \to C$ and natural isomorphisms $\mathrm{Id}_C \simeq GF$ and $\mathrm{Id}_D \simeq FG$.

Example A.16

(a) Let T be the functor from Ab to itself that associates to every abelian group A its torsion subgroup $T(A)$. Then the inclusion $T(A) \to A$ induces a natural transformation from T to Id_{Ab}.

(b) Let k be a field. Let C be the category of finite-dimensional k-vector spaces. The contravariant functor $E \mapsto E^*$ is an equivalence of categories from C to its opposite category (one sometimes speaks of an *anti-equivalence of categories* from C to C).

(c) The forgetful functor which to a $\mathbf{Z}$-module associates the underlining abelian group is an equivalence of categories between $Mod_{\mathbf{Z}}$ and Ab.

Definition A.17 A functor $F : C \to D$ is *faithful* if for all morphisms f_1, f_2 in C with $f_1 \neq f_2$, the morphisms $F(f_1)$ and $F(f_2)$ are distinct. The functor F is called *full* if every morphism $g : F(A) \to F(B)$ in D is of the form $F(f)$ with $f : A \to B$ morphism in C. A functor which is both full and faithful will be called *fully faithful*. We say that the functor F is *essentially surjective* if each object in D is isomorphic to $F(A)$ for a certain object A in C.

The following theorem ([35], Sect. IV.4, Th. 1) gives a useful criterion to obtain an equivalence of categories.

Theorem A.18 *Let $F : C \to D$ be a functor. Then F is an equivalence of categories if and only if it is fully faithful and essentially surjective.*

Lastly, the following notion often plays an important role, namely in the abelian categories (which figure in the next paragraph).

Definition A.19 Let C and D be two categories. Two functors $L : C \to D$ and $R : D \to C$ are called *adjoint* if for all objects A in C and B in D, we have a natural bijection

$$\tau = \tau_{AB} : \mathrm{Hom}_D(L(A), B) \simeq \mathrm{Hom}_C(A, R(B))$$

between A and B, meaning that for all morphisms $f : A \to A'$ in C and $g : B \to B'$ in D, the following diagram commutes:

$$
\begin{array}{ccccc}
\mathrm{Hom}_D(L(A'), B) & \xrightarrow{Lf^*} & \mathrm{Hom}_D(L(A), B) & \xrightarrow{g_*} & \mathrm{Hom}_D(L(A), B') \\
\downarrow{\tau} & & \downarrow{\tau} & & \downarrow{\tau} \\
\mathrm{Hom}_C(A', R(B)) & \xrightarrow{f^*} & \mathrm{Hom}_C(A, R(B)) & \xrightarrow{Rg_*} & \mathrm{Hom}_C(A, R(B')).
\end{array}
$$

In these cases we also say that L is a *left adjoint* of R, or that R is a *right adjoint* of L.

Appendix A: Some Results from Homological Algebra

For example, if R is a ring and B is a left R-module, we can for any abelian group C consider the group $\mathrm{Hom}_{\mathbf{Z}}(B, C)$ as a right R-module via the left action of R on the first factor. The left adjoint functor of $\mathrm{Hom}_{\mathbf{Z}}(B, .)$ (from abelian groups to right R-modules) is $A \mapsto A \otimes_R B$ (cf. [4], Sect. 4.1).

A.3 Abelian Categories

Definition A.20 An *additive category* $\mathcal{C}$ is a category whose sets $\mathrm{Hom}_{\mathcal{C}}(A, B)$ (for A, B objects of $\mathcal{C}$) are endowed with a structure of abelian group, satisfying the following properties:

(i) The composition of morphisms is distributive with respect to the addition: if A, B, C, D are objects of $\mathcal{C}$ and $f : A \to B, g : B \to C, g' : B \to C, h : C \to D$ are morphisms, we have

$$h(g + g')f = hgf + hg'f \in \mathrm{Hom}(A, D).$$

(ii) $\mathcal{C}$ has a zero-object 0.

(iii) If A and B are objects of $\mathcal{C}$, the product $A \times B$ exists.

In particular, in an additive category, for any object A, $\mathrm{Hom}(A, A)$ is endowed with the structure of a ring (the unit being the identity morphism). Furthermore, it is easy to see that the product $A \times B$ is also the coproduct of A and B; it will often be denoted by $A \oplus B$ (in $\mathcal{M}od_R$, it corresponds to the usual direct sum).

Definition A.21 An *abelian category* (denoted by $\mathcal{C}$) is an additive category additionally satisfying:

(i) Every morphism in $\mathcal{C}$ has a kernel and a cokernel.

(ii) Every monomorphism is the kernel of its cokernel.

(iii) Every epimorphism is the cokernel of its kernel.

We easily see that in an abelian category, every morphism $f : A \to B$ factors uniquely as $f = i \circ p$, where $p : A \to B'$ is an epimorphism and $i : B' \to B$ is a monomorphism, with i defined as the kernel of the cokernel of f. The subobject $B' := \mathrm{Im}\, f$ of B is called the *image* of f. In particular, the notion of *exact sequence* (finite or infinite)

$$\cdots \longrightarrow A_{n-1} \longrightarrow A_n \longrightarrow A_{n+1} \longrightarrow \cdots$$

is given by the usual property that the image of a map is the kernel of the next one. Injective morphisms (= with zero kernel) coincide with monomorphisms and surjective morphisms (= with zero cokernel) with epimorphisms. We can also synthesise axioms (a) and (b) by the statement that the morphism that f induces between its co-image and its image is an isomorphism.

Example A.22

(a) The category of abelian groups $\mathcal{A}b$ is abelian.

(b) If R is a ring, the category $\mathcal{M}od_R$ is abelian.

(c) The opposite category of an abelian category is abelian.

(d) Let R be a commutative ring. The R-modules of finite type form an additive subcategory of $\mathcal{M}od_R$. This subcategory is abelian if and only if R is noetherian.

(e) A full subcategory of an abelian category containing 0 and stable by direct sums is additive. If furthermore it is stable by kernel and cokernel, it is an abelian category.

An *additive functor* F between two abelian categories $\mathcal{C}$ and $\mathcal{D}$ is a functor such that if A and B are objects in $\mathcal{C}$, the map induced between abelian groups $\mathrm{Hom}_{\mathcal{C}}(A, B)$ and $\mathrm{Hom}_{\mathcal{D}}(F(A), F(B))$ is a morphism of abelian groups. When we speak of a functor between two abelian categories, it will be understood (except an explicit mention to the contrary) that this functor is additive.

Definition A.23 Let $F : \mathcal{C} \to \mathcal{D}$ be a functor defined between two abelian categories. We say that F is left *exact* (resp. right exact) if for every exact sequence

$$0 \longrightarrow A_1 \longrightarrow A_2 \longrightarrow A_3 \longrightarrow 0,$$

the sequence

$$0 \longrightarrow F(A_1) \longrightarrow F(A_2) \longrightarrow F(A_3)$$

(resp. the sequence $F(A_1) \to F(A_2) \to F(A_3) \to 0$) is exact. We will say that F is exact if it is both left and right exact. We similarly define the notion of contravariant left exact (resp. right exact, resp. exact) functor F by that the corresponding covariant functor $F' : \mathcal{C}^{op} \to \mathcal{D}$ has the same property.

We easily see that the left exactness of a functor F is equivalent to the property that for every exact sequence

$$0 \longrightarrow A_1 \longrightarrow A_2 \longrightarrow A_3,$$

the sequence

$$0 \longrightarrow F(A_1) \longrightarrow F(A_2) \longrightarrow F(A_3)$$

is exact. A similar statement holds true for right exactness.

Example A.24

(a) If $\mathcal{C}$ is an abelian category and A is an object of $\mathcal{C}$, the covariant functor $\mathrm{Hom}_{\mathcal{C}}(A, .)$ from $\mathcal{C}$ to $\mathcal{A}b$ is left exact. Same is true for the contravariant functor $\mathrm{Hom}_{\mathcal{C}}(., A)$. Si D is a divisible abelian group, the functor $\mathrm{Hom}(., D)$ from $\mathcal{A}b$ to itself is exact (consequence of Zorn theorem, see also [54], Cor. 2.3.2).

(b) Let R be a commutative ring. Let A be an R-module. The functor $. \otimes A$ from $\mathcal{M}od_R$ to itself is right exact. We say that A is *flat* if this functor is exact. For example it is the case if R is a principal ideal domain and A is torsion-free.

(c) If G is a finite group, the functor $A \mapsto A^G$ (from the category of G-modules to the category of abelian groups) is left exact.

Definition A.25 Let $\mathcal{C}$ be an abelian category. A functor $F : \mathcal{C} \to \mathcal{A}b$ is called *representable* by an object A of $\mathcal{C}$ if there exists a natural isomorphism from F to the functor $\mathrm{Hom}_{\mathcal{C}}(A, .)$. A contravariant functor from $\mathcal{C}$ to $\mathcal{A}b$ is called representable by an object I from $\mathcal{C}$ if it is naturally isomorphic to $\mathrm{Hom}_{\mathcal{C}}(., I)$.

The following theorem (essentially "formal") (cf. [18], Sect. 3: here we consider the case of a noetherian category, which is the opposite category of an artinian category) was used in the book to prove the existence of the dualising module (paragraph Sect. 10.1).

Theorem A.26 *Let $\mathcal{C}$ be a noetherian abelian category. Let F be a contravariant functor which is right exact from $\mathcal{C}$ to $\mathcal{A}b$. Then the functor F is* Ind-representable, *meaning the following: there exists a filtered inductive system (I_i) of objects of $\mathcal{C}$ such that the functor F is naturally isomorphic to the functor*

$$A \longmapsto \varinjlim_{i} \mathrm{Hom}(A, I_i).$$

Proof Consider the pairs (A, x) with $A \in \mathcal{C}$ and $x \in F(A)$. We will say that such a pair is *minimal* if for every epimorphism $A \to B$ in $\mathcal{C}$ which is not an isomorphism, we have $x \notin F(B)$ (recall that as F is contravariant and right exact, we can view $F(B)$ as a subgroup of $F(A)$). If now (A, x) and (A', x') are pairs as above (not necessarily minimal), we will say that (A', x') *dominates* (A, x) if there exists a morphism $u : A \to A'$ such that $x = F(u)(x')$. Observe that as the category $\mathcal{C}$ is noetherian, each pair (A, x) is dominated by a minimal pair: indeed, to obtain such a pair, it is enough to consider a subobject A_0 of A, such that A/A_0 contains an element y satisfying $F(p)(y) = x$ (where $p : A \to A/A_0$ is the projection), with A_0 maximal for this property.

Let us now prove a lemma:

Lemma A.27 *A morphism $u : A \to A'$ as above is unique if we further assume that (A', x') is minimal.*

If $v : A \to A'$ satisfies $x = F(v)(x')$, then $(F(u - v))(x') = 0$. As F is left exact, the epimorphism $(u - v) : A \to \mathrm{Im}(u - v)$ induces an injection $F(u - v) : F(\mathrm{Im}(u - v)) \to F(A)$, which shows that x' is in the kernel of $F(i) : F(A') \to F(\mathrm{Im}(u - v))$, where $i : \mathrm{Im}(u - v) \to A'$ is the canonical injection. Let us set $B = \mathrm{Coker}(u - v)$. The exact sequence

$$0 \longrightarrow \mathrm{Im}(u - v) \overset{i}{\longrightarrow} A' \longrightarrow B \longrightarrow 0$$

implies, applying F, that $F(B)$ is the kernel of $F(i)$, and hence finally that $x' \in F(B)$. The minimality of (A', x') imposes then $B = A'$, i.e., $u = v$. This finishes the proof of the lemma.

We thus have an order relation on the set of minimal pairs by setting $(A, x) \leqslant (A', x')$ if (A', x') dominates (A, x). This order relation is filtering since if (A_1, x_1) and (A_2, x_2) are two minimal pairs, then they are dominated by $(A_1 \oplus A_2, (x_1, x_2))$, which is itself dominated by a minimal pair as we have seen above. Let then $(I_i, x_i)_{i \in I}$ be the associated inductive filtered system. Set $F(I) := \varprojlim_i F(I_i)$. The x_i then define a canonical element $\underline{x} = (x_i)$ of $F(I)$. If A is an object of $\mathcal{C}$ and $f = (f_i)$ is in $\mathrm{Hom}(A, I) := \varinjlim_i \mathrm{Hom}(A, I_i)$, we have associated to f the element $F(f)(\underline{x})$ of $F(A)$. We thus define a homomorphism Φ of $\mathrm{Hom}(A, I)$ in $F(A)$, which is clearly functorial in A.

It remains to prove that Φ is an isomorphism. It is injective since as $f_i : A \to I_i$ is a morphism with $F(f_i)(x_i) = 0$, the fact that the pair (I_i, x_i) is minimal implies $f_i = 0$ by Lemma A.27 (indeed, the zero morphism A in I_i induces the zero morphism from $F(I_i)$ to $F(A)$). Let us prove that Φ is surjective. Fix $x \in F(A)$. We know that the pair (A, x) is dominated by a certain minimal pair (I_i, x_i). Then, the relation $(A, x) \leqslant (I_i, x_i)$ provides a morphism $f_i : A \to I_i$ such that $F(f_i)(x_i) = x$, which means that $x = \Phi(f)$, where $f \in \mathrm{Hom}(A, I)$ is induced by f_i. $\square$

A.4 Categories of Modules

Abelian categories that we meet most often are subcategories from $\mathcal{M}od_R$, where R is a ring. The following theorem (see [54], Th. 1.6.1) in practice allows, even when we are in an abelian category which is not small, to reduce to a subcategory of a category of modules. For example, we can prove many properties by "diagram chasing" by considering the abelian category generated (in an obvious way) by the objects and morphisms of the diagram in question.

Theorem A.28 (Freyd–Mitchell) *Let $\mathcal{C}$ be a small abelian category. Then, there exists a ring R and a fully faithful exact functor from $\mathcal{C}$ to $\mathcal{M}od_R$, that identifies $\mathcal{C}$ with a full subcategory of $\mathcal{M}od_R$.*

This allows in particular to extend the following lemma to every abelian category, that we can easily prove by diagram chasing in the category of modules:

Lemma A.29 (Snake lemma) *Let $\mathcal{C}$ be an abelian category. Consider a commutative diagram in $\mathcal{C}$ with exact rows:*

$$\begin{array}{ccccccc} A' & \longrightarrow & B' & \overset{p}{\longrightarrow} & C' & \longrightarrow & 0 \\ \downarrow f & & \downarrow g & & \downarrow h & & \\ 0 & \longrightarrow & A & \underset{i}{\longrightarrow} & B & \longrightarrow & C. \end{array}$$

Then there is an exact sequence

$$\text{Ker } f \longrightarrow \text{Ker } g \longrightarrow \text{Ker } h \xrightarrow{\partial} \text{Coker } f \longrightarrow \text{Coker } g \longrightarrow \text{Coker } h,$$

where ∂ is defined by the formula:

$$\partial(c') = i^{-1} g p^{-1}(c')$$

for any $c' \in C'$ (by abuse of notation, we denote $p^{-1}(c')$ any lift of c' in B' by p, and i^{-1} the converse morphism from $i : A \to \text{Im } i \subset B$).

Furthermore, if $A' \to B'$ is a monomorphism, the same is true of $\text{Ker } f \to \text{Ker } g$. If $B \to C$ is an epimorphism, the same is true of $\text{Coker } f \to \text{Coker } g$.

Let R be a ring. In the category $\mathcal{M}od_R$, the *inductive limit* (or *direct limit*, or *colimit*) $\varinjlim_{i \in I} A_i$ of an inductive system $(A_i)_{i \in I}$ indexed by a filtered ordered set I is well defined, the inductive limit of the sets A_i being naturally endowed with a structure of an R-module by the assumption that I is filtered. Furthermore, by [54], Th. 2.6.15 we have (a) below, and we get (b) by the universal property of tensor product:

Proposition A.30 *(a) The functor $\varinjlim_{i \in I}$ (from the inductive systems of R-modules indexed by I to $\mathcal{M}od_R$) is exact.*

(b) If R is a commutative ring, $(A_i)_{i \in I}$ is an inductive system of R-modules, and B is an R-module, we have:

$$(\varinjlim_{i \in I} A_i) \otimes_R B = \varinjlim_{i \in I}(A_i \otimes B).$$

In particular (b) applies to the direct sum of a family $(A_i)_{i \in I}$ of R-modules, which is a special case of the inductive limit.

The situation is more complicated for projective limits. Let $(A_n)_{n \in \mathbf{N}}$ be a projective system of R-modules indexed by the integers, i.e., a family

$$\cdots \longrightarrow A_{n+1} \longrightarrow A_n \longrightarrow \cdots \longrightarrow A_1 \longrightarrow A_0$$

of R-modules equipped with morphisms $d_n : A_{n+1} \to A_n$ for any $n \in \mathbf{N}$. The *projective limit* (or simply the *limit*) $\varprojlim_n A_n$ is well defined in $\mathcal{M}od_R$, but it is generally only a left exact functor (from the abelian category $\mathcal{M}od_R^{\mathbf{N}}$ of projective systems of R-modules to $\mathcal{M}od_R$).

Definition A.31 We say that a projective system (A_n, d_n) as above satisfies the *Mittag–Leffler condition* (abbreviated ML) if for every $n \in \mathbf{N}$, the image of the transition maps $A_{n+m} \to A_n$ is the same for any m sufficiently large.

Note that the condition (ML) is automatically satisfied if the modules A_n are all finite, or if all the transition maps d_n are surjective.

Proposition A.32 (cf. [42], Prop. 2.7.4) *Let*

$$0 \longrightarrow (A_n) \longrightarrow (B_n) \longrightarrow (C_n) \longrightarrow 0$$

be a short exact sequence in $\mathcal{M}od_R^{\mathbf{N}}$. *Assume that the projective system* (A_n) *satisfies (ML). Then the sequence*

$$0 \longrightarrow \varprojlim_n A_n \longrightarrow \varprojlim_n B_n \longrightarrow \varprojlim_n C_n \longrightarrow 0$$

is exact.

A.5 Derived Functors

In this paragraph, we denote by $\mathcal{C}$ an abelian category.

Definition A.33 An object I in $\mathcal{C}$ is called *injective* if the contravariant functor $\mathrm{Hom}_{\mathcal{C}}(., I)$ (from $\mathcal{C}$ to $\mathcal{A}b$) is exact. An object P in $\mathcal{C}$ is called *projective* if the covariant functor $\mathrm{Hom}_{\mathcal{C}}(P, .)$ is exact. We say that $\mathcal{C}$ has *enough injectives* (resp. enough projectives) if any object is isomorphic to a subobject of an injective object (resp. for any object A, there exists a surjection $P \to A$ with P projective).

The next proposition ([54], Prop. 2.3.10) follows easily from the definitions.

Proposition A.34 *Let* $\mathcal{C}$ *and* $\mathcal{D}$ *be two abelian categories. Let* $F : \mathcal{D} \to \mathcal{C}$ *be a functor. If* F *has a left adjoint functor which is exact, then for any injective object* I *of* $\mathcal{D}$, *the object* $F(I)$ *is injective in* $\mathcal{C}$ *(we say that* F preserves the injectives). *Similarly, a functor which has an exact right adjoint preserves the projectives.*

For example, if I is an injective abelian group, the R-module (right or left) $\mathrm{Hom}_{\mathbf{Z}}(R, I)$ is injective in $\mathcal{M}od_R$, since the forgetful functor from $\mathcal{M}od_R$ to $\mathcal{A}b$ is the left adjoint of $\mathrm{Hom}_{\mathbf{Z}}(R, .)$.

Example A.35 (a) The injectives in the category $\mathcal{A}b$ are the divisible abelian groups ([54], Cor. 2.3.2). The projectives are the free abelian groups. The only projective in the category of finite abelian groups is the trivial group 0.

(b) More generally, the projectives in the category $\mathcal{M}od_R$ are projective modules in the usual sense, i.e., direct factors of a free module. Thus the category $\mathcal{M}od_R$ has enough projectives. It also has enough injectives ([54], Cor. 2.3.11 and Exercise 2.3.5): let indeed I be a divisible abelian group, for example $I = \mathbf{Q}/\mathbf{Z}$. Then every right R-module A is a quotient of a free R-module, which implies (applying $\mathrm{Hom}_{\mathbf{Z}}(., I)$) that every left R-module embeds into a module M of the form $\mathrm{Hom}_{\mathbf{Z}}(R, J)$, where J is an injective abelian group (as a product of injectives). Such an M is injective in $\mathcal{M}od_R$ by Proposition A.34.

Besides, Baer criterion ([54], Crit. 2.3.1, p. 39) says that a left R-module A is injective as soon as for every left ideal J in R, the induced homomorphism $\mathrm{Hom}_R(R, A) \to \mathrm{Hom}_R(J, A)$ is surjective, in other words as soon as every morphism of R-modules from J to A extends to a morphism of R-modules from R to A.

(c) If G is a finite group, the category of discrete G-modules (which is equivalent to the category of left modules over the ring $\mathbf{Z}[G]$) has enough injectives and also enough projectives. If G is only profinite, this category C_G still has enough injectives ([54], Lem. 6.11.10): the point is that if A is a discrete G-module, it embeds into a G-module I which is injective in the category of G-modules. It is easy to see that then A also embeds into $\bigcup_U I^U$ (where U ranges through open subgroups of G) and that this last one is injective in C_G. On the other hand for G profinite and infinite, the category C_G does not have enough projectives (Exercise 4.2).

(d) The category Sh_X of sheaves of abelian groups on a topological space X has enough injectives ([54], Ex. 2.3.12).

(e) An object is injective in C if and only if it is projective in C^{op}, and vice versa.

f) If R is a left noetherian ring (for example the ring $\mathbf{Z}[G]$, where G is a finite group), an inductive limit of injective left R-modules is an injective left R-module. This can be easily seen using the Baer criterion and the fact that every left ideal J in R is generated by a finite number of elements: indeed, this last property implies that a morphism from J to an inductive limit $\varinjlim A_i$ of left R-modules comes from a morphism $J \to A_i$ for a certain i.

In the book, we have used (to prove Proposition 4.25) the following characterisation of the injectives of C_G:

Proposition A.36 *Let G be a profinite group. Let A be a discrete G-module. Then A is injective in C_G if and only if G has a basis $\mathcal{B}$ of neighbourhoods of 1 consisting of normal open subgroups U satisfying the following : the G/U-module A^U is injective.*

Proof *(by J. Riou)*. If A is injective in C^G, then it is straightforward that for any open subgroup U of G, the G/U-module A^U is injective. Assume conversely that for every U in $\mathcal{B}$, the G/U-module A^U is injective. We use the following lemma.

Lemma A.37 *Assume that for every U in $\mathcal{B}$ and for every injective morphism of G-modules $M \hookrightarrow \mathbf{Z}[G/U]$, the induced homomorphism*

$$\mathrm{Hom}_G(\mathbf{Z}[G/U], A) \longrightarrow \mathrm{Hom}_G(M, A)$$

is surjective. Then the G-module A is injective.

Assume for now that the lemma is proved. Let $B \hookrightarrow \mathbf{Z}[G/U]$ be an injective morphism of G-modules. Let $B \to A$ be a morphism of G-modules. As B is a sub-G-module of $\mathbf{Z}[G/U]$, we have $B = B^U$, and the morphism $B \to A$ factors through a morphism $B \to A^U$, that we can view as a morphism of G/U-modules.

As A^U is injective in $C_{G/U}$, this morphism extends to a morphism of G/U-modules $\mathbf{Z}[G/U] \to A^U$, that we can also view as a morphism of G-modules $\mathbf{Z}[G/U] \to A$. Thus the criterion of the lemma is satisfied.

It remains to prove the lemma. The argument is the same as the one used to prove the Baer criterion. Let $i : B \hookrightarrow C$ be an injective morphism of G-modules. Let $f : B \to A$ be a morphism of G-modules, that we would like to extend to a morphism $C \to A$. Zorn's lemma implies that there exists a sub-G-module B' of C endowed with a morphism $f' : B' \to A$ extending f, and maximal with respect to this property (meaning that if $f_1 : B_1 \to A$ is a morphism of G-modules with $B' \subset B_1 \subset C$ and f_1 extends f', then $B_1 = B'$ and $f_1 = f'$). Assume for contradiction that $B' \neq C$. We thus have an element $c \in C$ which is not in B', and as C is a discrete G-module, there exists an open subgroup U of G (that can be assumed to be in $\mathcal{B}$, by shrinking it if necessary) fixing c. This implies that we have a morphism of G-modules $j :$ $\mathbf{Z}[G/U] \to C$ whose image is not contained in B'. Let $M \subset \mathbf{Z}[G/U]$ be the inverse image of B' by j. The assumption then says that the morphism $f' \circ j : M \to A$ extends to a morphism $g : \mathbf{Z}[G/U] \to A$. Let us set $B_1 = B' + \operatorname{Im} j$. We can then extend f' to a morphism of G-modules $f_1 : B_1 \to A$ by setting

$$f_1(x + j(y)) = f'(x) + g(y)$$

for any $x \in B'$ and any $y \in \mathbf{Z}[G/U]$. This makes sense since if $x \in B'$ verifies $x = j(y)$, then $y \in M$ and $f'(x) = g(y)$ as g extends $f' \circ j$ on M. We obtain a contradiction since B_1 contains B' strictly. $\square$

Definition A.38 A *complex* $A^\bullet$ in $\mathcal{C}$ is a family of objects $A^i, i \in \mathbf{Z}$ and morphisms (called *coboundaries*, one also sometimes calls them *(co)differentials*) $d^i : A^i \to A^{i+1}$ such that $d^{i+1} \circ d^i = 0$ for every i (when the objects are only defined on a certain interval, ex. $i \geqslant 0$, we set $A^i = 0$ for other superscripts i). A *morphism of complexes* $f : A^\bullet \to B^\bullet$ is a family of morphisms $f^i : A^i \to B^i$ that commute with the coboundaries d^i.

We see immediately that the complexes in $\mathcal{C}$ form an abelian category. We call a *morphism of degree d* between two complexes $A^\bullet$ and $B^\bullet$ a morphism from $A^\bullet$ to the shifted complex $B^\bullet[d]$ (defined by $(B^\bullet[d])^i = B^{i+d}$).

Definition A.39 The ith *cohomology object* of a complex $A^\bullet$ is

$$h^i(A^\bullet) := \operatorname{Ker} d^i / \operatorname{Im} d^{i-1}.$$

If $f : A^\bullet \to B^\bullet$ is a morphism of complexes, it induces a natural map $h^i(f) :$ $h^i(A^\bullet) \to h^i(B^\bullet)$.

Remark A.40 The above definitions correspond to *cochain complexes* and to their cohomology, which will be those we are mostly concerned with in this book. We can also define *chain complexes* $(A_i)_{i \in \mathbf{Z}}$ for which we have boundaries (instead of coboundaries) $d_i : A_i \to A_{i-1}$ satisfying $d_i \circ d_{i+1} = 0$, and for which we can consider the *homology objects* $h_i(A^\bullet) := \operatorname{Ker} d_i / \operatorname{Im} d^{i+1}$. One passes from one notion to the other via the identification $A^i = A_{-i}$, which amounts to passing to the opposite category.

An application of the snake lemma is ([54], Th. 1.3.1):

Theorem A.41 *Let*

$$0 \longrightarrow A^\bullet \overset{f}{\longrightarrow} B^\bullet \overset{g}{\longrightarrow} C^\bullet \longrightarrow 0$$

be a short exact sequence of complexes. Then we have natural maps $\delta^i : h^i(C^\bullet) \to h^{i+1}(A^\bullet)$ *that give rise to a long exact sequence*

$$\cdots \longrightarrow h^i(A^\bullet) \overset{h^i(f)}{\longrightarrow} h^i(B^\bullet) \overset{h^i(g)}{\longrightarrow} h^i(C^\bullet) \overset{\delta^i}{\longrightarrow} h^{i+1}(A^\bullet) \longrightarrow \cdots$$

Definition A.42 We say that two morphisms of complexes f, g are *homotopic* (we will write $f \sim g$) if there exists a family of morphisms $k^i : A^i \to B^{i-1}$ (not necessarily commuting with the d^i) such that $f - g = dk + kd$. If $f \sim g$, the morphisms $h^i(f)$ and $h^i(g)$ induced on the cohomology are the same. Two complexes $A^\bullet$ and $B^\bullet$ are *homotopic* if there exist morphisms $f : A^\bullet \to B^\bullet$ and $g : B^\bullet \to A^\bullet$ such that $f \circ g$ and $g \circ f$ are homotopic to the identity (in this case the cohomology of the two complexes is the same, cf. [54], Lem. 1.4.5).

The theory of derived functors is based on the following proposition ([54], Lem. 2.2.5, Th. 2.2.6, Lem. 2.3.6 and Th. 2.3.7).

Proposition A.43 *Let* C *be an abelian category with enough injectives. Then every object* A *of* C *admits an* injective resolution; *i.e., there exists a complex*

$$I^\bullet = 0 \longrightarrow I^0 \longrightarrow I^1 \longrightarrow \cdots$$

(defined in degrees $i \geqslant 0$*) and a morphism* $A \to I^0$ *such that all the objects of* $I^\bullet$ *are injective and we have an exact sequence*

$$0 \longrightarrow A \longrightarrow I^0 \longrightarrow I^1 \longrightarrow \cdots$$

Furthermore, two injective resolutions are homotopic. We have an analogous result if C *has enough projectives, replacing injective resolutions by* projective resolutions, *i.e., exact sequence*

$$\cdots \longrightarrow P_1 \longrightarrow P_0 \longrightarrow A \longrightarrow 0,$$

where each P_i *is projective.*

We deduce the following definition.

Definition A.44 Let C be an abelian category with enough injectives. Let $F : C \to D$ be a covariant left exact functor. The (right) *derived functors* $R^i F, i \geqslant 0$ are defined as follows: for any object A of C, fix an injective resolution $I^\bullet$ of A and set

$$R^i F(A) = h^i(F(I^\bullet)).$$

This definition is justified by the fact that the derived functors are independent of the chosen realisation up to an isomorphism of additive functors ([54], Lem. 2.4.1 applied to $F^{op} : C^{op} \to D^{op}$), which follows from Proposition A.43. Note that the functor F is isomorphic to $R^0 F$ as

$$(R^0 F)(A) = \mathrm{Ker}[F(I^0) \longrightarrow F(I^1)] = F(A)$$

since F is left exact. On the other hand, we have $R^i F(I) = 0$ if I is injective and $i > 0$, an injective resolution of I being then simply

$$0 \longrightarrow I \longrightarrow I \longrightarrow 0.$$

In an analogous way, we define left derived functors of a right exact covariant functor in a category with enough projectives by working with the homology of a projective resolution. We also have the same notions for a contravariant functor (if it is left exact, we obtain right derived functors by working with projective resolutions. If it is right exact we obtain left derived functors by working with injective resolutions).

Example A.45

(a) Let G be a finite group. For any G-module A, the cohomology groups $H^i(G, A)$ are derived functors of the functor $A \mapsto A^G$ (which is right exact) from the category of G-modules to that of abelian groups. Same holds true if G is profinite.

(b) If X is a topological space and $\mathcal{F}$ is a sheaf of abelian groups on X, then the groups $H^i(X, \mathcal{F})$ are derived functors of the functor "global section" $\mathcal{F} \mapsto \Gamma(X, \mathcal{F})$ from the category of sheaves of abelian groups on X to $\mathcal{A}b$.

(c) Let R be a ring. If A and B are left R-modules, the groups $\mathrm{Ext}^i_R(A, B)$ are defined as right derived functors of the functor $\mathrm{Hom}_R(A, .)$ (applied to B) from $\mathcal{M}od_R$ to $\mathcal{A}b$. In particular this applies to $R = \mathbf{Z}$, or to $R = \mathbf{Z}[G]$ for any finite group G. More generally, we can define the groups $\mathrm{Ext}^i_C(A, B)$ if A and B are two objects of an abelian category with enough injectives, considering derived functors from $\mathrm{Hom}_C(A, .)$. The $\mathrm{Ext}^i_C(A, .)$ then form a *cohomological functor* (cf. Theorem A.46 below), and the $\mathrm{Ext}^i_C(., B)$ form a *homological functor* (the argument is the same as in Remark 1.37).

(d) Let R be a ring. The category $\mathcal{M}od_R^{\mathbf{N}}$ of projective systems of R-modules (indexed by $\mathbf{N}$) has enough injectives. The first derived functor $\varprojlim^1$ of functor $\varprojlim_n$ (from $\mathcal{M}od_R^{\mathbf{N}}$ to $\mathcal{M}od_R$) may be nonzero; however, all the higher derived functors are always zero ([42], Prop. 2.7.4). If (A_n) is a projective system of R-modules satisfying (ML), then its $\varprojlim^1$ is zero.

The following theorem ([54], Th. 2.4.6, Theorem 2.4.7, and Exercise 2.4.3 applied
to $F^{\mathrm{op}} : \mathcal{C}^{\mathrm{op}} \to \mathcal{D}^{\mathrm{op}}$) summarise the main properties of derived functors.

Theorem A.46 *Let $\mathcal{C}$ be an abelian category with enough injectives. Let $F : \mathcal{C} \to \mathcal{D}$
be a left exact covariant functor.*

(a) The family of the $(R^i F)_{i \geqslant 0}$ forms a cohomological functor *(also called
δ-functor), which is: for any exact sequence*

$$0 \longrightarrow A' \longrightarrow A \longrightarrow A'' \longrightarrow 0 \tag{A.1}$$

*in $\mathcal{C}$, we have coboundary morphisms $\delta^i : R^i F(A'') \to R^{i+1} F(A')$ inducing a long
exact sequence*

$$\cdots \longrightarrow R^i F(A') \longrightarrow R^i F(A) \longrightarrow R^i F(A'') \overset{\delta^i}{\longrightarrow} R^{i+1} F(A') \longrightarrow \cdots$$

*and such that if we have a morphism from a short exact sequence (A.1) to another
short exact sequence $0 \to B' \to B \to B'' \to 0$, then the δ^i induce a commutative
diagram:*

$$\begin{array}{ccc} R^i F(A'') & \overset{\delta^i}{\longrightarrow} & R^{i+1} F(A') \\ \downarrow & & \downarrow \\ R^i F(B'') & \overset{\delta^i}{\longrightarrow} & R^{i+1} F(B'). \end{array}$$

(b) The family of the $(R^i F)_{i \geqslant 0}$ forms a universal δ-functor, *meaning the following:
if $S = (S^i)_{i \geqslant 0}$ is another δ-functor and $f^0 : R^0 F = F \to S^0$ is a natural transformation, then there exists a unique morphism of δ-functors $(f^i)_{i \geqslant 0}$ from $(R^i F)_{i \geqslant 0}$ to
S which extends f^0 ("morphism of δ-functors" means a family of natural transformations $f^i : R^i F \to S^i$ commuting with the δ^i).*

(c) If $(J^j)_{j \geqslant 0}$ is a family of acyclic *objects for the functor F (i.e., such that
$R^i F(J^j) = 0$ for every $i > 0$) inducing a resolution*

$$0 \longrightarrow A \longrightarrow J^0 \longrightarrow J^1 \longrightarrow \cdots$$

*of an object A, then for any $i \geqslant 0$, we have $R^i F(A) \simeq h^i(F(J^\bullet))$ (thus we can
compute the derived functors using acyclic resolutions, not necessarily injective).*

We have analogous results with other derived functors (starting with a contravariant or right exact functor F), which give cohomological or homological derived
functors.

Remark A.47 We have in fact a more general result than (b): each cohomological
δ-functor T such that the T^i for $i > 0$ are *effaceable* (i.e., such that for every object
A, there exists a monomorphism $u : A \to I$ such that $T^i(u) = 0$) is universal, [54],

Exer. 2.4.5. The proof of this fact is similar to that of the universality of derived functors.

We sometimes need (in particular to define the tensor product of two complexes, or to define spectral sequences) the following notion:

Definition A.48 Let C be an abelian category. A *double complex* $A^{\bullet\bullet}$ is a family of objects $(A^{pq})_{p,q \in \mathbb{Z}}$ equipped with morphisms (in general called *differentials*) $d'^{pq} : A^{pq} \to A^{p+1,q}$ and $d''^{pq} : A^{pq} \to A^{p,q+1}$, satisfying (with obvious notation): $d' \circ d' = 0$, $d'' \circ d'' = 0$, and $d' \circ d'' + d'' \circ d' = 0$ (the last equality is often called the *rule of signs*).

Definition A.49 We furthermore assume that C is a category $\mathcal{M}od_R$ of modules over a ring R. The *total complex* $T^{\bullet} = \mathrm{Tot}(A^{\bullet\bullet})$ associated to a double complex $A^{\bullet\bullet}$ is defined by $T^n = \bigoplus_{p+q=n} A^{pq}$, the differentials $d : A^n \to A^{n+1}$ are given by the sum of maps

$$d = d' + d'' : A^{pq} \longrightarrow A^{p+1,q} \oplus A^{p,q+1}.$$

The fact that $T^{\bullet}$ is complex follows from the rule of signs of Definition A.48 ([54], alinéa 1.2.6).

Remark A.50 We have here given the definition of $\mathrm{Tot}(A^{\bullet\bullet})$ which uses the direct sum (and not the product) of the A^{pq} since it is that definition that allows to define the tensor product of two complexes (cf. definition A.51 below). The definition with the product (denoted Tot^Π or simply Tot in [54]) would be indispensable if we wanted for example to consider the internal Hom between the complexes. See also the footnote in the proof of Theorem 2.25.

Definition A.51 Let R be a commutative ring. Let $C^{\bullet}$ and $D^{\bullet}$ be two complexes in the abelian category $\mathcal{M}od_R$. The *tensor product*

$$A^{\bullet\bullet} = C^{\bullet} \otimes_R D^{\bullet}$$

is the double complex defined as $A^{pq} := C^p \otimes_R D^q$, with differentials

$$d'^{pq} = d_C^p \otimes_R \mathrm{Id}_{D^q}; \quad d''^{pq} = (-1)^p \, \mathrm{Id}_{C^p} \otimes_R d_D^q.$$

If there is no risk of confusion, we will also denote by $C^{\bullet} \otimes_R D^{\bullet}$ the total complex associated to $A^{\bullet\bullet}$.

A.6 Ext and Tor

Let R be a ring. Let A be a left R-module. Recall that in the category $\mathcal{M}od_R$, the functors $\mathrm{Ext}_R^i(A, .)$ are defined as the right derived functors of the functor $\mathrm{Hom}_R(A, .)$

from $\mathcal{M}od_R$ to $\mathcal{A}b$. We can in fact compute $\text{Ext}^i_R(A, B)$ by deriving "on the other side" using the following result ([54], Th. 2.7.6):

Theorem A.52 *Let A and B be two R-modules. Then, for any $i \geqslant 0$, we have:*

$$\text{Ext}^i_R(A, B) = (R^i \operatorname{Hom}_R(A, .))(B) = (R^i \operatorname{Hom}_R(., B))(A),$$

where the $R^i \operatorname{Hom}_R(., B)$ are the right derived functors of the contravariant functor (which is left exact) $\operatorname{Hom}_R(., B)$. In particular, if A or B is m-torsion (for a certain integer $m > 0$), then the groups $\text{Ext}^i_R(A, B)$ are m-torsion.

The Ext behave well with respect to products and direct sums ([54], Prop. 3.3.4):

Proposition A.53 *Let (A_i) and (B_j) be two families of R-modules. Let A and B be two R-modules. Then, for any $n \geqslant 0$, we have*

$$\text{Ext}^n_R\left(\bigoplus_i A_i, B\right) = \prod_i \text{Ext}^n_R(A_i, B),$$

$$\text{Ext}^n_R\left(A, \prod_j B_j\right) = \prod_j \text{Ext}^n_R(A, B_j).$$

In the special case $R = \mathbf{Z}$, the category $\mathcal{M}od_{\mathbf{Z}}$ is simply that of abelian groups. We have in this case the following vanishing result ([54], Lem. 3.3.1), which easily follows from the fact that the injectives of $\mathcal{A}b$ are the abelian divisible groups and that a quotient of a divisible abelian group is divisible:

Proposition A.54 *Let A and B be abelian groups. Then $\text{Ext}^i_{\mathbf{Z}}(A, B) = 0$ for $i \geqslant 2$.*

Corollary A.55 *If B is an abelian group, we have $\text{Ext}^1(\mathbf{Z}/p\mathbf{Z}, B) = B/pB$.*

If R is a commutative ring, we can also define the $\text{Tor}^i_R(A, B)$ as the left derived functors of $. \otimes_R B$ as well as those of $A \otimes_R .$ ([54], Th. 2.7.2). These derived functors are thus trivial if A or B is a flat R-module. We have in particular the following proposition, when $R = \mathbf{Z}$ ([54], Prop. 3.1.2, 3.1.3. and 3.1.4):

Proposition A.56 *Let A and B be two abelian groups. Then the group $\text{Tor}_{\mathbf{Z}}(A, B) := \text{Tor}^1_{\mathbf{Z}}(A, B)$ is a torsion abelian group, and $\text{Tor}^i_{\mathbf{Z}}(A, B) = 0$ if $i \geqslant 2$. The group $\text{Tor}_{\mathbf{Z}}(\mathbf{Q}/\mathbf{Z}, B)$ is the torsion subgroup B_{tors} of B. If A or B is torsion-free, we also have $\text{Tor}_{\mathbf{Z}}(A, B) = 0$.*

It is on the other hand possible to define groups $\text{Ext}^i_{\mathcal{C}}(A, B)$ when A and B are two objects of an arbitrary abelian category $\mathcal{C}$, in terms of *Yoneda r-extensions* ([54], Vista 3.4.6). When $\mathcal{C}$ has enough injectives (or enough projectives), they coincide with the groups Ext^i defined using derived functors.

Definition A.57 Let $\mathcal{C}$ be an abelian category, that we assume, for simplicity, to have enough injectives. Let M, N, P be objects of $\mathcal{C}$. We have for all $r, s \geqslant 0$ a bilinear map (compatible with the structure of cohomological functor of the "right" Ext and the structure of homological functor of the "left" Ext):

$$\text{Ext}^r_C(N, P) \times \text{Ext}^s_C(M, N) \longrightarrow \text{Ext}^{r+s}_C(M, P), \quad (f, g) \longmapsto f \cdot g$$

which can for example be defined via the interpretation of the groups Ext^r_C in terms of Yoneda r-extensions: for $f \in \text{Ext}^r_C(N, P)$ and $g \in \text{Ext}^s_C(M, N)$, we obtain $f \cdot g \in \text{Ext}^{r+s}_C(M, P)$ by concatenating the r-extension of N by P corresponding to f and the s-extension of M by N corresponding to g.

Remark A.58 The bilinear map $(f, g) \mapsto f \cdot g$ is easily understood in the language of derived categories. The group $\text{Ext}^r_C(M, N)$ can then be seen as $\text{Hom}(M, N[r])$ (in the derived category of C) where $N[r]$ is the complex with one term obtained by placing N in degree $-r$, cf. [54], Sect. 10.7. It is equivalent to view $\text{Ext}^r_C(M, N)$ as the set of homotopy classes of morphisms $M^\bullet \to N^\bullet$ of degree r between complexes, where $M^\bullet$ and $N^\bullet$ are injective resolutions of M and N, respectively.

It turns out that the interpretation of spectral sequences in the language of derived categories (loc. cit., Sect. 10.8) is very useful to check some compatibilities, like for example the one in Proposition 16.17.

A.7 Spectral Sequences

Spectral sequences (in particular the spectral sequence of composed functors of Grothendieck, that we will see below) are a powerful tool allowing to obtain information about cohomology groups, that they endow with filtrations. They are notably used to obtain vanishing or finiteness theorems about these groups. Here we will limit ourselves to spectral sequences *concentrated in the first quadrant*, which are the only ones we use in this book, and for which the results are slightly easier to state than for example for bounded spectral sequences in the sense of [54], Déf. 5.2.5. (and a fortiori for the not necessarily bounded spectral sequences).

In all this paragraph, C is an abelian category.

Definition A.59 Let $a \in \mathbf{N}$. A *cohomological spectral sequence concentrated in the first quadrant* (in the sequel we will simply say *spectral sequence*), *starting with* E_a, is the data:

(i) of a family (E^{pq}_r) of objects of C defined for $r \geqslant a, p \geqslant 0, q \geqslant 0$ (by convention, for the other indices, $E^{pq}_r = 0$).

(ii) of morphisms

$$d^{pq}_r : E^{pq}_r \longrightarrow E^{p+r,q-r+1}_r$$

such that the composites $d^{pq}_r \circ d^{p-r,q+r-1}_r$ are zero. In other words, if for r fixed we place the object E^{pq}_r to the point with coordinates (p, q) of the lattice $\mathbf{Z}^2$, the objects on the right with slope $(1 - r)/r$ form a cochain complex, the maps going from left to right. We sometimes say that for a fixed r, the E_r^{pq} form a *page* E_r of the spectral sequence.

(iii) of an isomorphism between E_{r+1}^{pq} and the cohomology at E_r^{pq} of the complex defined above, that is,

$$E_{r+1}^{pq} = \operatorname{Ker} d_r^{pq} / \operatorname{Im} d_r^{p-r,q+r-1}.$$

Such a spectral sequence will be denoted by $(E_r^{pq})_{r \geqslant a}$ or simply (E_a^{pq}). To avoid mixing up of indices, we will also denote E_r^{pq} by $E_r^{p,q}$.

For example, for $r = 1$, we obtain a complex on each horizontal line (= line with slope 0) of the lattice. The objects of the "page" E_{r+1} are deduced from the preceding page E_r by taking cohomology of different complexes of E_r. Note also that the slope of differentials d_r changes at each page, and that each differential d_r^{pq} increases the total degree $p + q$ of the object E_r^{pq} by 1. Besides, the fact that we have imposed $E_r^{pq} = 0$ if p or q is < 0 implies immediately that for fixed p, q, there exists $r_0 \geqslant a$ such that $E_r^{pq} = E_{r_0}^{pq}$ for $r \geqslant r_0$.

Definition A.60 We denote by E_∞^{pq} the "stable" value of E_r^{pq}, i.e., the object E_r^{pq} for r large enough.

Remark A.61 For any p, the object $E_\infty^{0,p}$ is a subgroup of the object $E_a^{0,p}$ of the initial page, and for any q, the object $E_\infty^{q,0}$ is a quotient of the object $E_a^{q,0}$ of the initial page. More generally $E_\infty^{p,q}$ is always a subquotient of the object $E_a^{p,q}$ by point (iii) of the Definition A.59.

Definition A.62 Let $H^* = (H^n)_{n \geqslant 1}$ be a family of objects of $\mathcal{C}$. We say that a spectral sequence $(E_r^{pq})_{r \geqslant a}$ *converges to* H^* if each H^n admits a finite filtration

$$0 = F^{n+1} H^n \subset F^n H^n \subset \cdots \subset F^1 H^n \subset F^0 H^n = H^n$$

such that for each p, q, we have

$$E_\infty^{pq} \simeq F^p H^{p+q} / F^{p+1} H^{p+q}.$$

We will then often write

$$(E_a^{pq}) \implies H^{p+q}.$$

The maps $H^n \to E_\infty^{0,n} \subset E_a^{0,n}$ and $E_a^{n,0} \to E_\infty^{n,0} \subset H^n$ are called the *boundary maps* of the spectral sequence converging to H^n.

Example A.63 Let $A^{\cdot\cdot}$ be a double complex in the category of modules $\mathcal{M}od_R$, with $A^{pq} = 0$ if $p < 0$ or $q < 0$. Let $T^{\cdot}$ be the total complex associated to $A^{\cdot\cdot}$. Let

$$E_1^{pq} = H^q(A^{p\cdot}, d'')$$

be the cohomology groups of $A^{\cdot\cdot}$ "in the direction q". Then we have (cf. [54], Sect. 5.6 for the homology version) a spectral sequence

$$E_1^{pq} \implies H^{p+q}(T^{\bullet}).$$

We can also start this spectral sequence with

$$E_2^{pq} = H^p(H^q(A^{\bullet\bullet})).$$

The next proposition is an immediate consequence of the fact that if a spectral sequence converges to H^*, then each object H^n is filtered by the objects whose successive quotients are the E_∞^{pq} with total degree $p + q = n$.

Proposition A.64 *Assume that the spectral sequence $(E_r^{pq})_{r \geqslant a}$ converges to $H^* = (H^n)_{n \geqslant 1}$. Let $n \geqslant 1$ be fixed. If all the initial E_a^{pq} of total degree n is zero (resp. finite), then H^n is zero (resp. finite).*

We will often apply the above proposition to Grothendieck spectral sequences (that start at E_2), obtained using the spectral sequence of composed functors (see below), as for example in the Hochschild–Serre spectral sequence (Theorem 1.44) where the H^n are cohomology groups of degree n.

Spectral sequences that converge to a family H^* are mainly interesting when families E_r^{pq} have many zero terms. Here is a proposition that we easily deduce from the definitions and which illustrates the most common situations in which we can compute the H^n.

Proposition A.65 *Let $(E_r^{pq})_{r \geqslant a}$ be a spectral sequence converging to $H^* = (H^n)_{n \geqslant 1}$.*

(a) (collapse). Let $r \geqslant a$ be fixed. Assume that there exists an index q_0 such that we have $E_r^{pq} = 0$ if $q \neq q_0$ (resp. there exists an index p_0 such that $E_r^{pq} = 0$ if $p \neq p_0$). Then H^n is the term $E_r^{n-q_0,q_0}$ (resp. $E_r^{p_0,n-p_0}$). One says in this case that the spectral sequence collapses at the page E_r.

(b) (spectral sequence with two columns). Assume that $E_2^{pq} = 0$ if $p > 1$. Then we have, for any $n \geqslant 1$, an exact sequence

$$0 \longrightarrow E_2^{1,n-1} \longrightarrow H^n \longrightarrow E_2^{0,n} \longrightarrow 0.$$

(c) (spectral sequence with two lines). Assume that $E_2^{pq} = 0$ if $q > 1$. Then we have a long exact sequence

$$0 \longrightarrow E_2^{1,0} \longrightarrow H^1 \longrightarrow E_2^{0,1} \longrightarrow E_2^{2,0} \longrightarrow H^2 \longrightarrow E_2^{0,2} \longrightarrow \cdots$$

It is also possible, by writing out definitions of filtrations on H^1 and H^2, to obtain the *exact sequence of lower degrees* associated to a spectral sequence beginning at the page E_2.

Proposition A.66 *Let $(E_r^{pq})_{r \geqslant 2}$ be a spectral sequence converging to $H^* = (H^n)_{n \geqslant 1}$. Then we have an exact sequence*

$$0 \longrightarrow E_2^{1,0} \longrightarrow H^1 \longrightarrow E_2^{0,1} \longrightarrow E_2^{2,0} \longrightarrow \mathrm{Ker}[H^2 \longrightarrow E_2^{0,2}] \longrightarrow E_2^{3,0}$$

and the sequence

$$\mathrm{Ker}[H^2 \longrightarrow E_2^{0,2}] \longrightarrow E_2^{3,0} \longrightarrow H^3$$

is a complex.

All spectral sequences used in this book are obtained from the following general theorem ([54], Th. 5.8.3), which is the *spectral sequence of composed functors of Grothendieck*.

Theorem A.67 *Let $\mathcal{A}$, $\mathcal{B}$, $\mathcal{C}$ be abelian categories such that $\mathcal{A}$ and $\mathcal{B}$ have enough injectives. Let $G : \mathcal{A} \to \mathcal{B}$ and $F : \mathcal{B} \to \mathcal{C}$ be left exact covariant functors. Assume that G sends each injective object of $\mathcal{A}$ to an F-acyclic object $\mathcal{B}$ (i.e., an object B of $\mathcal{B}$ such that $R^i F(B) = 0$ for any $i > 0$, for example an injective object in $\mathcal{B}$). Then, if A is an object in $\mathcal{A}$, we have a convergent spectral sequence (starting at the page E_2):*

$$E_2^{pq} = (R^p F)((R^q G)(A)) \implies R^{p+q}(FG)(A).$$

Some examples of application of this theorem are:

(a) the Hochschild–Serre spectral sequence (Theorem 1.44).

(b) the spectral sequence of the Ext (Theorem 16.14).

In these two cases, we often use Propositions A.64 and A.65 to obtain interesting information about cohomology groups or the groups Ext, when we have vanishing or finiteness theorems for the terms E_2^{pq}.

Appendix B
A Survey of Analytic Methods

This appendix gives an introduction to analytic methods, including Dirichlet series and the Čebotarev density theorem.

In this appendix, we outline how analytic methods are used in class field theory. In addition to their historic importance, these methods provide for example a direct proof of the first inequality (that we have proved in this book using the second inequality and Kummer extensions, see Theorem 13.21). These methods also prove the very important Čebotarev theorem, that we have already encountered (without proof) in Chap. 18. These two results are the main objectives of this appendix. For notational convenience, we will restrict ourselves to the case of a number field.

B.1 Dirichlet Series

In this paragraph, we collect some classical results from complex analysis, which will be used in what follows. For any complex number $z \neq 0$, we denote by $\arg z \in [-\pi, \pi[$ its argument. For all strictly positive real numbers b, c, ε, we set

$$D(b, c, \varepsilon) := \{s \in \mathbf{C}, \operatorname{Re}(s) \geqslant b + c, \, |\arg(s - b)| \leqslant \pi/2 - \varepsilon\}.$$

Definition B.1 A *Dirichlet series* is a series (of complex variable s) of the form

$$f(s) = \sum_{n=1}^{\infty} \frac{a(n)}{n^s},$$

where $a(n) \in \mathbf{C}$ and $n^s := \exp(s \log n)$, the symbol log denoting the Neperian logarithm.

Example B.2 The *Riemann ζ function* corresponds to the case where $a(n) = 1$ for any n. That is, it is defined by

© Springer Nature Switzerland AG 2020
D. Harari, *Galois Cohomology and Class Field Theory*, Universitext,
https://doi.org/10.1007/978-3-030-43901-9

$$\zeta(s) = \sum_{n=1}^{\infty} \frac{1}{n^s}.$$

Theorem B.3 *Let $f(s) = \sum_{n=1}^{\infty} a(n)/n^s$ be a Dirichlet series. For any positive real number x, let us set*

$$S(x) := \sum_{n \leqslant x} a(n).$$

Suppose that there exists a positive real constant b such that $|S(x)| = O(x^b)$ when x tends to $+\infty$. Then, for all $c, \varepsilon > 0$, the series converges uniformly on $D(b, c, \varepsilon)$, and the function $f(s)$ is holomorphic on the half plane $\mathrm{Re}(s) > b$.

For a proof, see for example [26], Prop. IV.2.1.

Example B.4 The Riemann ζ function is holomorphic on the half plane $\mathrm{Re}(s) > 1$, since in this case the associated function S sends any $x \in \mathbf{R}_+^*$ to its integral part. Furthermore, ζ extends to a meromorphic function on the half plane $\mathrm{Re}(s) > 0$, admitting a unique pole (which is simple) at $s = 1$ with residue 1 (see for example [40], Th. V.1.3).

We immediately deduce from the previous example and from Theorem B.3 (applied to the series $f(s) - \alpha\zeta(s)$) the following corollary:

Corollary B.5 *Let $f(s) = \sum_{n=1}^{\infty} a(n)/n^s$ be a Dirichlet series. Assume that there exists $\alpha \in \mathbf{C}$ and a real number b with $0 < b < 1$, such that*

$$|S(x) - \alpha x| = O(x^b)$$

when x tends to $+\infty$, where $S(x) = \sum_{n \leqslant x} a(n)$. Then the function f is holomorphic on the half plane $\mathrm{Re}(s) > 1$ and extends to a meromorphic function on the half plane $\mathrm{Re}(s) > b$, admitting a unique possible pole at $s = 1$ with residue α.

In what follows, if z is a complex number which is not a $\leqslant 0$ real (hence with argument $\arg z \in] -\pi, \pi[$), we set $\log z := \log |z| + i \arg z$. Recall that for a z with modulus < 1, we have a power series expansion

$$-\log(1 - z) = \sum_{n=1}^{\infty} \frac{z^n}{n}. \tag{B.1}$$

Recall also that if $(u_n)_{n \geqslant 1}$ is a sequence of nonzero complex numbers, we say that the infinite product $\prod u_n$ is *absolutely convergent* if the series $\sum |u_n - 1|$ converges. This implies in particular that $\prod u_n$ converges (to a nonzero complex number). If u_n is of the form $u_n = 1/(1 - v_n)$ with $|v_n| < 1$, then $\log u_n$ is well defined and the absolute convergence of $\prod u_n$ is equivalent to that of the series $\sum -\log(1 - v_n)$. We have an analogous statement for the absolute uniform convergence if $(u_n(s))$ is a sequence of functions of complex variable s.

B.2   Dedekind ζ Function; Dirichlet L-Functions

Let k be a number field, whose group of fractional ideals we denote $\mathcal{I}_k$, and Specm$(\mathcal{O}_k)$ the set of nonzero prime ideals (we will identify Specm$(\mathbf{Z})$ with the set of prime numbers). Let $\mathfrak{a}$ be a nonzero ideal of the ring of integers $\mathcal{O}_k$. Denote by $N(\mathfrak{a})$ the *absolute norm* of $\mathfrak{a}$, which is the cardinality of $(\mathcal{O}_k/\mathfrak{a})$.

Definition B.6   The *Dedekind ζ function* of k is the Dirichlet series

$$\zeta_k(s) = \sum_{\mathfrak{a}} \frac{1}{N(\mathfrak{a})^s},$$

the sum being taken over all nonzero ideals $\mathfrak{a}$ of $\mathcal{O}_k$.

Remark B.7   Note that we can identify the expression of $\zeta_k(s)$ to that of a Dirichlet series, by collecting for any $n \in \mathbf{N}^*$ the ideals $\mathfrak{a}$ of norm n in the above sum.

Proposition B.8   *The function ζ_k is holomorphic on the half plane* Re$(s) > 1$, *with:*

$$\zeta_k(s) = \prod_{\mathfrak{p}\in\mathrm{Specm}(\mathcal{O}_k)} \frac{1}{1 - N(\mathfrak{p})^{-s}}.$$

Proof   Let $c > 1$. We will show that the infinite product

$$\prod_{\mathfrak{p}\in\mathrm{Specm}(\mathcal{O}_k)} \frac{1}{1 - N(\mathfrak{p})^{-s}}$$

converges absolutely and uniformly on the half plane Re$(s) \geqslant c$ (which will give the desired result by [46], Chap. VI, Sect. 2, Lemma 1). This is equivalent to showing this property for the series

$$\sum_{\mathfrak{p}\in\mathrm{Specm}(\mathcal{O}_k)} -\log(1 - N(\mathfrak{p})^{-s}).$$

But we have $|-\log(1 - N(\mathfrak{p})^{-s})| \leqslant -\log(1 - N(\mathfrak{p})^{-c})$ as soon as Re$(s) \geqslant c$ (use for example the series expansion (B.1)) and it is thus enough to show that the series with positive terms

$$\sum_{\mathfrak{p}\in\mathrm{Specm}(\mathcal{O}_k)} -\log(1 - N(\mathfrak{p})^{-c})$$

converges. As $-\log(1 - N(\mathfrak{p})^{-c}) \sim N(\mathfrak{p})^{-c}$ when $N(\mathfrak{p})$ tends to $+\infty$, is it equivalent to saying that

$$\sum_{\mathfrak{p}\in\mathrm{Specm}(\mathcal{O}_k)} N(\mathfrak{p})^{-c}$$

converges. Above each prime number p, there are at most $d := [k : \mathbf{Q}]$ prime ideals of $\mathcal{O}_k$. Each such prime ideal $\mathfrak{p}$ verifies $N(\mathfrak{p})^{-c} \leqslant p^{-c}$. We deduce that

$$\sum_{\mathfrak{p} \in \mathrm{Specm}(\mathcal{O}_k)} N(\mathfrak{p})^{-c} \leqslant \sum_{p \in \mathrm{Specm}(\mathbf{Z})} d \cdot p^{-c} \leqslant d \cdot \sum_{n \geqslant 1} n^{-c},$$

which proves the desired convergence with the Example B.4.

We then obtain the equality of $\zeta_k(s)$ with the desired infinite product for $\mathrm{Re}(s) > 1$, using the identity

$$\prod_{\mathfrak{p}} \frac{1}{1 - N(\mathfrak{p})^s} = \prod_{\mathfrak{p}} (1 + \frac{1}{N(\mathfrak{p})^s} + \frac{1}{N(\mathfrak{p})^{2s}} + \cdots) = \sum_{\mathfrak{a}} \frac{1}{N(\mathfrak{a})^s}.$$

The last equality follows from the existence and the uniqueness of the decomposition of any nonzero ideal of $\mathcal{O}_k$ as a product of prime ideals. □

We will now extend the definition of these functions ζ_k to a more general context. Let Ω_k be the set of all places of k and Ω_f be the set of finite places (they correspond to nonzero prime ideals of $\mathcal{O}_k$). Recall (paragraph Sect. 15.4) that a cycle $\mathcal{M}$ of k is a formal product $\mathcal{M} = \prod_{v \in \Omega_k} v^{n_v}$, where $n_v \in \mathbf{N}$, n_v is zero for almost every v, and $n_v \in \{0, 1\}$ if v is archimedean. If v_0 is a place of k, the notation $(v_0, \mathcal{M}) = 1$ means that v_0 does not divide $\mathcal{M}$, in other words $n_{v_0} = 0$.

Let $\mathcal{I}_k^{\mathcal{M}} \subset \mathcal{I}_k$ be the subgroup of $\mathcal{I}_k$ consisting of fractional ideals I of k which are prime to $\mathcal{M}$ (that is: such that no prime ideal $\mathfrak{p}$ dividing $\mathcal{M}$ appears in the decomposition of I). We denote by $\mathcal{P}_k^{\mathcal{M}}$ the group of principal fractional ideals (a) verifying $a \equiv 1 \bmod v^{n_v}$ for every place v dividing $\mathcal{M}$ (with the notation of paragraph Sect. 15.4). We have defined (Definition 15.20) the group of ray class field $H_k^{\mathcal{M}} := C_k/C_k^{\mathcal{M}} = I_k/I_k^{\mathcal{M}} \cdot k^*$ in terms of idèles. We will often shorten the notations $\mathcal{I}_k^{\mathcal{M}}, \mathcal{P}_k^{\mathcal{M}}, I_k^{\mathcal{M}}, C_k^{\mathcal{M}} \ldots$ to $\mathcal{I}^{\mathcal{M}}, \mathcal{P}^{\mathcal{M}}, I^{\mathcal{M}}, C^{\mathcal{M}} \ldots$ when there is no ambiguity on the field k. Lastly, denote by $I_1^{\mathcal{M}}$ the group of idèles $\alpha = (\alpha_v)$ of k satisfying $\alpha_v = 1$ for every place v (including the archimedean places) dividing $\mathcal{M}$.

Proposition B.9 *Let $I := I_k$ be the group of idèles of k and $C := C_k$ its group of idèle classes. Let $\mathcal{M} = \prod_{v \in \Omega_k} v^{n_v}$ be a cycle of k. Then we have an isomorphism $\bar{i}$ between $H^{\mathcal{M}} = C/C^{\mathcal{M}}$ and the finite group $\mathcal{I}^{\mathcal{M}}/\mathcal{P}^{\mathcal{M}}$. Furthermore, if K/k is a finite Galois extension, this isomorphism is compatible with the norm maps $N_{K/k} : C_K \to C_k$ and $N_{K/k} : \mathcal{I}_K \to \mathcal{I}_k$ (in an obvious sense). If $c \in C/C^{\mathcal{M}}$ is the class of an idèle $\alpha = (\alpha_v)$ of $I_1^{\mathcal{M}}$, we have*

$$\bar{i}(c) = i(\alpha) := \prod_{(v, \mathcal{M})=1} v^{\mathrm{val}(x_v)}. \tag{B.2}$$

Note that we have already encountered the special case $\mathcal{M} = 1$ of this statement in Sect. 15.4. The finiteness of $\mathcal{I}^{\mathcal{M}}/\mathcal{P}^{\mathcal{M}}$ is classical, see for example [26], Cor. IV.1.6. Recall also that $N_{K/k} : \mathcal{I}_K \to \mathcal{I}_k$ is the homomorphism sending each nonzero prime ideal $\mathfrak{q}$ of $\mathcal{O}_K$ to $\mathfrak{p}^f$, where f is the residual degree of the extension K/k and $\mathfrak{p} := \mathfrak{q} \cap \mathcal{O}_k$.

Proof First observe that $I = I_1^{\mathcal{M}} \cdot I^{\mathcal{M}} \cdot k^*$. Indeed, if $\alpha \in I$, we can find (by approximation Theorem 12.18) an $a \in k^*$ such that $\alpha_v a \equiv 1 \bmod v^{n_v}$ for any place v dividing $\mathcal{M}$. Thus we can write $\alpha_v a = \beta_v \gamma_v$, where β_v and γ_v are defined by: $\beta_v = 1$ if v divides $\mathcal{M}$, $\beta_v = \alpha_v a$ otherwise; $\gamma_v = \alpha_v a$ if v divides $\mathcal{M}$, $\gamma_v = 1$ otherwise. We obtain that the idèle $\beta := (\beta_v)$ is in $I_1^{\mathcal{M}}$ and the idèle $\gamma := (\gamma_v)$ is in $I^{\mathcal{M}}$. Thus we can write $\alpha = \beta \cdot \gamma \cdot a^{-1} \in I_1^{\mathcal{M}} \cdot I^{\mathcal{M}} \cdot k^*$.

We deduce

$$H^{\mathcal{M}} = I/I^{\mathcal{M}} k^* = I_1^{\mathcal{M}} \cdot I^{\mathcal{M}} \cdot k^*/I^{\mathcal{M}} k^* \cong I_1^{\mathcal{M}}/(I^{\mathcal{M}} \cdot k^*) \cap I_1^{\mathcal{M}}.$$

The homomorphism $i : I_1^{\mathcal{M}} \to \mathcal{I}^{\mathcal{M}}/\mathcal{P}^{\mathcal{M}}$ defined by the formula (B.2) is clearly surjective. Its kernel N consists of the $(\alpha_v) \in I_1^{\mathcal{M}}$ such that there exists $b \in k^*$ verifying

$$(b) = \prod_{(v,\mathcal{M})=1} v^{\mathrm{val}(\alpha_v)} \in \mathcal{P}^{\mathcal{M}},$$

that is: such that there exists $b \in k^*$ such that $\alpha_v b^{-1} \equiv 1 \bmod v^{n_v}$ for every place v. Thus $N = (I^{\mathcal{M}} \cdot k^*) \cap I_1^{\mathcal{M}}$, and i induces an isomorphism $\bar{i}$ between $H^{\mathcal{M}}$ and $\mathcal{I}^{\mathcal{M}}/\mathcal{P}^{\mathcal{M}}$. The compatibility of $\bar{i}$ with the maps $N_{K/k}$ when K is a finite Galois extension of k is straightforward.   □

Definition B.10 A *Dirichlet character* is a homomorphism χ from the group of ray class field $H^{\mathcal{M}}$ to $\mathbf{C}^*$. In other words χ is a character of $\mathcal{I}^{\mathcal{M}}$ whose kernel contains $\mathcal{P}^{\mathcal{M}}$, for a certain cycle $\mathcal{M}$. The *Dirichlet L-function* associated to χ is the series

$$L(s, \chi) := \sum_{\mathfrak{a}} \frac{\chi(\mathfrak{a})}{N(\mathfrak{a})^s},$$

the sum being taken over the nonzero ideals $\mathfrak{a}$ of $\mathcal{O}_k$ (we make a convention that $\chi(\mathfrak{a}) = 0$ if $\mathfrak{a} \notin \mathcal{I}^{\mathcal{M}}$, which is if $\mathfrak{a}$ is not prime to $\mathcal{M}$).

Example B.11 For $\mathcal{M} = 1$ and $\chi = 1$, we recover the Dedekind ζ_k function of the number field k. For $k = \mathbf{Q}$ and $\mathcal{M} = m \cdot p_\infty$ (where $m \in \mathbf{N}^*$ and p_∞ is the infinite place), we have $H^{\mathcal{M}} \cong (\mathbf{Z}/m\mathbf{Z})^*$ and we recover the classical Dirichlet series ([46], Chap. VI, Sect. 2)

$$\sum_{n \geqslant 1} \frac{\chi(n)}{n^s},$$

where χ is a character of $(\mathbf{Z}/m\mathbf{Z})^*$ (extended by 0 to the integers non-prime to m).

In a way identical to Proposition B.8, we show that the function $L(s, \chi)$ is holomorphic on the half plane $\mathrm{Re}(s) > 1$, and in this region can be written as an infinite product

$$L(s, \chi) = \prod_{(\mathfrak{p}, \mathcal{M})=1} \frac{1}{1 - N(\mathfrak{p})^{-s} \chi(\mathfrak{p})},$$

the product taken over all the nonzero prime ideals of $\mathcal{O}_k$ not dividing $\mathcal{M}$. We will see that unlike in the case of ζ_k, the function $L(s, \chi)$ does not have a pole at 1 if χ is different from the trivial character χ_0.

Definition B.12 Let $\mathcal{M}$ be a cycle of k. Let $\mathfrak{k}$ an element of $H_\mathcal{M} = \mathcal{I}^\mathcal{M}/\mathcal{P}^\mathcal{M}$. We define the *partial ζ function* associated to $\mathfrak{k}$ by

$$\zeta(s, \mathfrak{k}) := \sum_{\mathfrak{a} \in \mathfrak{k}} \frac{1}{N(\mathfrak{a})^s}.$$

The sum is taken over the ideals $\mathfrak{a} \subset \mathcal{O}_k$ of $\mathcal{I}^\mathcal{M}$ which are in the class $\mathfrak{k}$ modulo $\mathcal{P}^\mathcal{M}$.

We thus have for any Dirichlet character χ of $H_\mathcal{M}$:

$$L(s, \chi) = \sum_{\mathfrak{k} \in H_\mathcal{M}} \chi(\mathfrak{k})\zeta(s, \mathfrak{k}).$$

In particular, we obtain

$$\zeta_k(s) = \sum_{\mathfrak{k} \in H_\mathcal{M}} \zeta(s, \mathfrak{k}).$$

Note that $\zeta(s, \mathfrak{k})$ is a Dirichlet series (cf. Remark B.7) whose associated function S (as in Theorem B.3) is

$$S(x, \mathfrak{k}) := \#\{\mathfrak{a} \in \mathfrak{k}, \ \mathfrak{a} \subset \mathcal{O}_k, \ N(\mathfrak{a}) \leqslant x\}.$$

Theorem B.13 *Let $d := [k : \mathbf{Q}]$. Let $\mathcal{M}$ be a cycle of k. There exists $g_\mathcal{M} \in \mathbf{R}_+^*$ (depending on $\mathcal{M}$ but not on $\mathfrak{k}$) such that for any class $\mathfrak{k} \in H_\mathcal{M}$:*

$$|S(x, \mathfrak{k}) - g_\mathcal{M} x| = O(x^{1-\frac{1}{d}})$$

when x tends to $+\infty$.

The proof of this statement (which is certainly the most technical of this appendix) uses methods of geometry of numbers (notably Minkowski theorem); see [33], Chap. VI, Theorem 3. The value of $g_\mathcal{M}$ (which depends on $[k : \mathbf{Q}]$, on the discriminant of k, its class number and its unit group) is related to the *analytic class number formula*, cf. [40], remark after Th 2.2.

Corollary B.14 *The function $\zeta(s, \mathfrak{k})$ is holomorphic for $\mathrm{Re}(s) > 1$, admits a meromorphic extension to the half plane $\mathrm{Re}(s) > 1 - 1/d$, with a unique simple pole at $s = 1$, the residue being the strictly positive real number $g_\mathcal{M}$.*

Proof This follows from Theorem B.13 and Corollary B.5.   □

Corollary B.15 *The Dedekind ζ_k function admits a meromorphic extension to the half plane* $\mathrm{Re}(s) > 1 - 1/d$, *with a unique simple pole at* $s = 1$.

Proof Indeed, we have $\zeta_k(s) = \sum_{\mathfrak{k} \in H_{\mathcal{M}}} \zeta(s, \mathfrak{k})$.   □

Corollary B.16 *Let χ be a non-trivial character of $H_{\mathcal{M}}$. Then the function $L(s, \chi)$ extends to a holomorphic function on the half plane* $\mathrm{Re}(s) > 1 - 1/d$.

Proof We have

$$L(s, \chi) = \sum_{\mathfrak{k} \in H_{\mathcal{M}}} \chi(\mathfrak{k}) \zeta(s, \mathfrak{k}).$$

By Corollary B.14, it is enough to show that

$$\sum_{\mathfrak{k} \in H_{\mathcal{M}}} \chi(\mathfrak{k}) \cdot g_{\mathcal{M}} = 0.$$

But as $g_{\mathcal{M}}$ does not depend on $\mathfrak{k}$, this follows from the orthogonality relation (cf. [46], Chap. VI, Prop. 4) :

$$\sum_{\mathfrak{k} \in H_{\mathcal{M}}} \chi(\mathfrak{k}) = 0,$$

which holds for any non-trivial character of the finite group $H_{\mathcal{M}}$.   □

B.3 Complements on the Dirichlet Density

Let $T \subset \mathrm{Specm}(\mathcal{O}_k)$ be a set of nonzero prime ideals of $\mathcal{O}_k$. Recall that the Dirichlet density (that we will simply call *density* in this appendix) of T is the limit (if it exists)

$$\delta(T) := \lim_{s \to 1^+} \frac{\sum_{\mathfrak{p} \in T} N(\mathfrak{p})^{-s}}{\sum_{\mathfrak{p} \in \mathrm{Specm}(\mathcal{O}_k)} N(\mathfrak{p})^{-s}}.$$

The notation $\lim_{s \to 1^+}$ means that s tend to 1 from the right as a real variable. Similarly, a notation such as $f(s) \sim_{1^+} g(s)$ will in what follows mean that the function $f(s)/g(s)$ tends to 1 when s tends to 1 from the right as a real variable.

Note that, by definition, we have $\delta(T_1 \cup T_2) = \delta(T_1) + \delta(T_2)$ if T_1 and T_2 are disjoint (when at least two of the densities $\delta(T_1)$, $\delta(T_2)$, and $\delta(T_1 \cup T_2)$ are defined).

Proposition B.17 *Let $T \subset \mathrm{Specm}(\mathcal{O}_k)$ be a set of nonzero prime ideals of $\mathcal{O}_k$. Set*

$$\zeta_{k,T}(s) = \prod_{\mathfrak{p} \in T} \frac{1}{1 - N(\mathfrak{p})^{-s}}.$$

(a) The function

$$g(s) := \log(\zeta_{k,T}(s)) - \sum_{\mathfrak{p} \in T} N(\mathfrak{p})^{-s}$$

is holomorphic in the region defined by $\mathrm{Re}(s) > 1/2$.

(b) We have

$$\sum_{\mathfrak{p} \in \mathrm{Specm}(\mathcal{O}_k)} (N\mathfrak{p})^{-s} \sim_{1^+} -\log(s-1)$$

(c) The density of T is the limit (if it exists)

$$\lim_{s \to 1^+} \frac{\sum_{\mathfrak{p} \in T}(N\mathfrak{p})^{-s}}{-\log(s-1)}.$$

In particular, if T is finite, we have $\delta(T) = 0$.

(d) Let m be a strictly positive integer. Assume that $\zeta_{k,T}(s)^m$ *extends to a meromorphic function in a neighbourhood of* 1 *with a simple pole at* $s = 1$. *Then T has density* $1/m$.

(e) Assume that $\zeta_{k,T}(s)$ *extends to a holomorphic function in a neighbourhood of* 1. *Then T is of density* 0.

Proof (a) Using the series development (B.1), we obtain

$$\log(\zeta_{k,T}(s)) = \sum_{\mathfrak{p} \in T} \sum_{n \geqslant 1} \frac{1}{nN(\mathfrak{p})^{sn}},$$

hence

$$g(s) = \sum_{\mathfrak{p} \in T} \sum_{n \geqslant 2} \frac{1}{nN(\mathfrak{p})^{sn}}.$$

Let $c > 1/2$. If $\mathrm{Re}(s) \geqslant c$, we have the inequality

$$\left| \frac{1}{nN(\mathfrak{p})^{sn}} \right| \leqslant \frac{1}{N(\mathfrak{p})^{cn}}$$

for any $n \geqslant 2$ and any $\mathfrak{p} \in \mathrm{Specm}(\mathcal{O}_k)$. We deduce, using the formula for the sum of a geometric series

$$\sum_{\mathfrak{p} \in T} \sum_{n \geqslant 2} \left| \frac{1}{nN(\mathfrak{p})^{sn}} \right| \leqslant \sum_{\mathfrak{p} \in \mathrm{Specm}(\mathcal{O}_k)} \frac{1}{(1 - N(\mathfrak{p})^{-c})N(\mathfrak{p})^{2c}}.$$

As $c > 1/2$, this last series is convergent since $N(\mathfrak{p}) \geqslant 2$, hence $(1 - N(\mathfrak{p})^{-c}) \geqslant 1 - 1/\sqrt{2} > 0$ and we know that $\sum_{\mathfrak{p} \in \mathrm{Specm}(\mathcal{O}_k)} N(\mathfrak{p})^{-2c}$ is convergent by Proposition B.8 since $2c > 1$. We finally deduce that g is holomorphic in the region $\text{Re}(s) > 1/2$.

(b) Let us apply (a) with $T = \text{Specm}(\mathcal{O}_k)$. We then know (Corollary B.15) that $\zeta_{k,T} = \zeta_k$ has a simple pole at $s = 1$. Thus we can write

$$\zeta_k(s) = \frac{a}{s-1} + h(s)$$

with $a \neq 0$ and h a holomorphic function in a neighbourhood of 1. We necessarily have $a > 0$ since for $s > 1$ real, we have $\zeta_k(s) > 1$. We deduce

$$\log(\zeta_k(s)) \sim_{1^+} -\log(s-1). \tag{B.3}$$

as in the right neighbourhood of 1, we know that $\log a$ is negligible compared to $-\log(s-1)$. On the other hand (a) gives

$$\sum_{\mathfrak{p}\in\text{Specm}(\mathcal{O}_k)} N(\mathfrak{p})^{-s} = \log(\zeta_k(s)) - g(s)$$

with g holomorphic in a neighbourhood of 1, which along with (B.3) implies

$$\sum_{\mathfrak{p}\in\text{Specm}(\mathcal{O}_k)} N(\mathfrak{p})^{-s} \sim_{1^+} -\log(s-1).$$

(c) follows immediately from (b).

(d) We have by assumption

$$\zeta_{k,T}(s)^m = \frac{a}{s-1} + g(s),$$

with g holomorphic in a neighbourhood of 1 and $a \neq 0$ (hence $a \in \mathbf{R}_+^*$). We deduce that

$$m\log(\zeta_{k,T}(s)) \sim_{1^+} -\log(s-1).$$

The assertion (a) then implies that

$$m\sum_{\mathfrak{p}\in T} N(\mathfrak{p})^{-s} \sim_{1^+} -\log(s-1),$$

which by (c) means that $\delta(T) = 1/m$.

(e) The assumption is now that $\zeta_{k,T}$ is holomorphic in a neighbourhood of 1. As for $s > 1$ real, we clearly have $\zeta_{k,T}(s) \geq 1$, we also have $\log(\zeta_{k,T})$ holomorphic in a neighbourhood of 1, which by (a) means that

$$\sum_{\mathfrak{p} \in T} N(\mathfrak{p})^{-s}$$

is holomorphic in a neighbourhood of 1, hence by (c) that $\delta(T) = 0$.   □

Remark B.18 The same proof shows that if $\zeta_{k,T}(s)^m$ extends to a meromorphic function in a neighbourhood of 1 with a pole of order $r > 0$ at 1, then $\delta(T) = r/m$. In such cases, we sometimes say that T is of *polar density* r/m.

Recall that the *absolute degree* of a nonzero prime ideal $\mathfrak{p}$ of $\mathcal{O}_k$ is the degree $[k_\mathfrak{p} : \mathbf{Q}_p]$, where p is the prime number that $\mathfrak{p}$ divides.

Corollary B.19 *(a) Assume that T contains no prime ideal of absolute degree 1. Then $\zeta_{k,T}$ is holomorphic in a neighbourhood of 1 and $\delta(T) = 0$.*

(b) If $S \subset \mathrm{Specm}(\mathcal{O}_k)$ contains all the prime ideals of absolute degree 1, then $\zeta_{k,S}$ extends to a meromorphic function in a neighbourhood of 1 admitting a simple pole at 1, and we have $\delta(S) = 1$.

Proof (a) By assumption, if $\mathfrak{p} \in T$, we have $N(\mathfrak{p}) = p^f$ with $f \geqslant 2$, where p is the prime number such that $\mathfrak{p}$ divides p. If $d = [k : \mathbf{Q}]$, there are at most d prime ideals above each prime number p. Thus, the infinite product

$$\zeta_{k,T}(s) = \prod_{\mathfrak{p} \in T} \frac{1}{1 - N(\mathfrak{p})^{-s}}$$

can be written as a product of d infinite products, each of those products being of the form

$$\prod_{p \in A} \frac{1}{1 - p^{-f_p s}},$$

where $f_p \geqslant 2$ and A is a subset of $\mathrm{Specm}(\mathbf{Z})$. Such a product converges absolutely and uniformly (by domination) on every region of $\mathbf{C}$ of the form $\mathrm{Re}(s) > c$ with $c > \frac{1}{2}$ because

$$\prod_{p \in \mathrm{Specm}(\mathbf{Z})} \frac{1}{1 - p^{-2c}} = \zeta(2c)$$

converges. Thus $\zeta_{k,T}$ is holomorphic at 1 (with $\zeta_{k,T}(1) \in \mathbf{R}_+^*$) and we conclude with Proposition B.17, (e).

(b) Let T be the complement of S in $\mathrm{Specm}(\mathcal{O}_k)$. Then by definition we have $\zeta_k(s) = \zeta_{k,T}(s) . \zeta_{k,S}(s)$. The result then follows from (a) and Proposition B.17, (b) along with Corollary B.15.   □

B.4   The First Inequality

We begin with a special case of the Čebotarev theorem, whose proof in the general case will be given in Sect. B.6.

Proposition B.20 *Let k be a number field. Let K be a finite Galois extension of k. Then the set of prime ideals of $\mathcal{O}_k$ totally split in the extension K/k has density $\frac{1}{[K:k]}$.*

Proof Let S be the set of prime ideals of $\mathcal{O}_k$ totally split in K/k. Let T be the set of prime ideals of $\mathcal{O}_K$ which are above a prime ideal of S. Let us set $d := [K:k]$. For any $\mathfrak{p} \in S$, there are exactly d prime ideals $\mathfrak{q}$ which are above $\mathfrak{p}$ and they verify by definition: $N_{K/k}(\mathfrak{q}) = \mathfrak{p}$, hence

$$N(\mathfrak{q}) = N(N_{K/k}(\mathfrak{q})) = N(\mathfrak{p}).$$

We deduce

$$\zeta_{K,T}(s) = \zeta_{k,S}(s)^d. \qquad (B.4)$$

Besides, T contains all the prime ideals of $\mathcal{O}_K$ unramified in K/k which are of absolute degree 1 (since a fortiori, such an ideal is above an ideal of $\mathcal{O}_k$ totally split in K/k). Furthermore, the prime ideals ramified in K/k are finite in number, hence their density is zero by Proposition B.17 (c). By Corollary B.19 (b) and equality (B.4), the function $\zeta_{k,S}(s)^d$ extends to a meromorphic function in a neighbourhood of 1 with a simple pole at 1. We conclude with Proposition B.17, (d).   □

We are now ready to prove a partial equidistribution result (which will be made precise in Theorem B.28) on prime ideals in different classes of $\mathcal{I}^{\mathcal{M}}/\mathcal{P}^{\mathcal{M}}$ (the notations are those of Proposition B.9):

Proposition B.21 *Let $\mathcal{M}$ be a cycle of k. Let H be a subgroup of $\mathcal{I}^{\mathcal{M}}$ containing $\mathcal{P}^{\mathcal{M}}$. Let δ be the density of the set $H \cap \mathrm{Specm}(\mathcal{O}_k)$ of prime ideals which are in H. Then:*

(i) if $L(1, \chi) \neq 0$ for every non-trivial character χ of $\mathcal{I}^{\mathcal{M}}/H$, then

$$\delta = 1/[\mathcal{I}^{\mathcal{M}} : H];$$

(ii) otherwise, $\delta = 0$. In this case, $L(1, \chi) = 0$ for one and only one non-trivial character $\chi = \chi_1$ of $\mathcal{I}^{\mathcal{M}}/H$, and the zero of $L(s, \chi_1)$ at $s = 1$ is simple.

We will see in paragraph Sect. B.5 (as a consequence of the translation of the existence Theorem 15.9 in terms of ideals) that we are in fact always in the case (i).

Proof Let us set $h = [\mathcal{I}^{\mathcal{M}} : H]$. Let χ be a character of $\mathcal{I}^{\mathcal{M}}/H$, which is a character of $\mathcal{I}^{\mathcal{M}}$ whose restriction to H is 1. The function

$$L(s, \chi) = \prod_{(\mathfrak{p}, \mathcal{M})=1} \frac{1}{1 - \chi(\mathfrak{p})N(p)^{-s}}$$

(the product being over the set of prime ideals of $\mathcal{O}_k$ not dividing $\mathcal{M}$) verifies

$$\log(L(s,\chi)) - \sum_{(\mathfrak{p},\mathcal{M})=1} \frac{\chi(\mathfrak{p})}{N(\mathfrak{p})^s} = g(s), \qquad (B.5)$$

with $g(s)$ holomorphic for $\mathrm{Re}(s) > 1/2$ (the proof is exactly the same as that of Proposition B.17, (a)).

In addition we know, via the orthogonality relation ([46], Chap. VI, Corollary of Prop. 4) that for any prime ideal $\mathfrak{p} \in \mathrm{Specm}(\mathcal{O}_k)$ not dividing $\mathcal{M}$, the expression $\sum_\chi \chi(\mathfrak{p})$ (the sum being over all the characters χ of $\mathcal{I}^{\mathcal{M}}/H$) is h or 0 depending on whether $\mathfrak{p}$ is in H or not. Summing (B.5) over all the characters of $\mathcal{I}^{\mathcal{M}}/H$, we thus obtain

$$\sum_\chi \log(L(s,\chi)) = h \sum_{\mathfrak{p}\in H} N(\mathfrak{p})^{-s} + g_1(s),$$

with g_1 holomorphic in a neighbourhood of 1.

If now χ is not the trivial character χ_0, then $L(s,\chi)$ is holomorphic in a neighbourhood of 1 (Corollary B.16), and we can write $L(s,\chi) = (s-1)^{m(\chi)} u(s)$ with $m(\chi) \in \mathbf{N}$ and $u(1) \neq 0$, hence we deduce

$$\log(L(s,\chi)) = m(\chi)\log(s-1) + g_\chi(s),$$

where the function g_χ is holomorphic in a neighbourhood of 1. On the other hand we have

$$\zeta_k(s) = L(s,\chi_0) \cdot \prod_{\mathfrak{p}|\mathcal{M}} \frac{1}{1 - N(\mathfrak{p})^{-s}},$$

which shows that (as there are only finitely many factors in the product) that $\log(L(s,\chi_0))$ is the sum of $\log(\zeta_k(s))$ and a function holomorphic in a neighbourhood of 1. We conclude using Corollary B.15 that

$$\log(L(s,\chi_0)) = -\log(s-1) + g_0(s),$$

where g_0 is holomorphic in a neighbourhood of 1. Putting it all together, we obtain

$$h \sum_{\mathfrak{p}\in H} N(\mathfrak{p})^{-s} = -(1 - \sum_{\chi\neq\chi_0} m(\chi))\log(s-1) + g_2(s),$$

with g_2 holomorphic in a neighbourhood of 1. In the case (i), this gives

$$\sum_{\mathfrak{p}\in H} N(\mathfrak{p})^{-s} \sim_{1^+} -\frac{1}{h}\log(s-1),$$

and hence $\delta = 1/h$ using Proposition B.17, (c). In the case (ii), as $\sum_{\mathfrak{p} \in H} N(\mathfrak{p})^{-s} \in \mathbf{R}_+^*$, the only possibility is that one and only one of the $m(\chi)$ is nonzero (and it is necessarily equal to 1), in which case $h \sum_{\mathfrak{p} \in H} N(\mathfrak{p})^{-s} = g_2(s)$ and $\delta = 0$ by loc. cit. $\qquad \square$

From this we will deduce the first inequality (cf. Theorem 13.21), initially in the language of ideals. If K is a finite extension of a number field k and $\mathcal{M}$ is a cycle of k, then the set S of prime ideals dividing $\mathcal{M}$ is finite. Let S_K be the set of prime ideals of K above a prime ideal in S. Denote by $\mathcal{I}_k^{\mathcal{M}}$ (resp. $\mathcal{I}_K^{\mathcal{M}}$) the group of those fractional ideals of k (resp. of K) that contain no element of S (resp. of S_K) in their decomposition (in Proposition B.21, we have shortened $\mathcal{I}_k^{\mathcal{M}}$ as $\mathcal{I}^{\mathcal{M}}$). Recall also that $\mathcal{P}_k^{\mathcal{M}}$ is the group of principal fractional ideals (a) of $\mathcal{I}_k$ verifying $a \equiv 1 \bmod v^{n_v}$ for every place v dividing $\mathcal{M}$.

Theorem B.22 (First inequality, "classical" form) *Let k be a number field. Let K be a finite Galois extension of k. Let $\mathcal{M}$ be a cycle of k. Then*

$$[\mathcal{I}_k^{\mathcal{M}} : \mathcal{P}_k^{\mathcal{M}}.N_{K/k}(\mathcal{I}_K^{\mathcal{M}})] \leqslant [K : k].$$

Furthermore, for any non-trivial character χ of $\mathcal{I}_k^{\mathcal{M}}/\mathcal{P}_k^{\mathcal{M}}.N_{K/k}(\mathcal{I}_K^{\mathcal{M}})$, we have $L(1, \chi) \neq 0$.

Proof Let $d = [K : k]$. Let us set $H = \mathcal{P}_k^{\mathcal{M}} \cdot N_{K/k}(\mathcal{I}_K^{\mathcal{M}})$, this is a subgroup of $\mathcal{I}_k^{\mathcal{M}}$ which contains $\mathcal{P}_k^{\mathcal{M}}$. Let δ be the density of the set of prime ideals of $\mathcal{O}_k$ which are in H. If a prime ideal $\mathfrak{p}$ of $\mathcal{O}_k$ is totally split in K/k, then $\mathfrak{p}$ is the norm of each ideal of $\mathcal{O}_K$ above it, therefore $\mathfrak{p} \in H$. By Proposition B.20, we have $\delta \geqslant d^{-1} > 0$.

We are therefore necessarily in the case (i) of Proposition B.21, which implies that $\delta = [\mathcal{I}_k^{\mathcal{M}} : H]^{-1}$. As we have seen that $\delta \geqslant d^{-1}$, we conclude that

$$[\mathcal{I}_k^{\mathcal{M}} : H] \leqslant d.$$

Proposition B.21 also gives $L(1, \chi) \neq 0$ for any non-trivial character χ of $\mathcal{I}_k^{\mathcal{M}}/H$. $\qquad \square$

Corollary B.23 (First inequality, "idelic" form) *Let k be a number field. Let K be a finite Galois extension of k. Let $C_k = I_k/k^*$ and $C_K = I_K/K^*$ be the ideal class groups of k an K respectively. Then*

$$[C_k : N_{K/k}C_K] \leqslant [K : k].$$

Proof Let S be a finite set of places of k containing all archimedean places and all those ramified in K/k. Let us set $n_v = 1$ if v is archimedean and choose, for v finite in S, a sufficiently large $n_v > 0$ so that each element x_v of k_v^* verifying $\mathrm{val}(x_v - 1) \geqslant n_v$ is a norm of the local extension K_v/k_v (such an n_v exists by Proposition 9.7 (b) and Lemma 9.9 (c)). If v is a finite place of k not in S, each element of $\mathcal{O}_v^*$ is a norm of K_v/k_v by Lemma 9.8 (a).

For $v \notin S$, let us set $n_v = 0$ and consider the cycle $\mathcal{M} := \prod_{v \in \Omega_k} v^{n_v}$. With the notation of the paragraph Sect. 15.4 (in particular Definition 15.20), we then have $I_k^{\mathcal{M}} \subset N_{K/k} I_K$, hence we deduce that $C_k^{\mathcal{M}} \subset N_{K/k} C_K$. Let $H_k^{\mathcal{M}} = C_k / C_k^{\mathcal{M}}$ be the ray class group modulo $\mathcal{M}$, and $H_K^{\mathcal{M}} := C_K / C_K^{\mathcal{M}_K}$, where $\mathcal{M}_K$ is the cycle of K obtained by associating to each place w of K the integer n_v with $w|v$ (thus $H_K^{\mathcal{M}}$ is the ray class group modulo $\mathcal{M}_K$). We obtain

$$C_k / N_{K/k} C_K \cong H_k^{\mathcal{M}} / N_{K/k} H_K^{\mathcal{M}} \cong \mathcal{I}_k^{\mathcal{M}} / \mathcal{P}_k^{\mathcal{M}} \cdot N_{K/k}(\mathcal{I}_K^{\mathcal{M}})$$

by Proposition B.9. It is then enough to apply Theorem B.22. $\qquad\square$

Remark B.24 We have obtained the first inequality totally independently of the global class field theory developed in part III of this book. We will now use this theory (in particular the existence theorem) to make precise certain analytic results encountered above.

B.5 Class Field Theory in Terms of Ideals

As observed in Remark 15.5, a disadvantage of the classical definition of the Artin reciprocity map $\psi_{K/k}$ for an abelian extension K/k is that we cannot a priori define it on the whole group of fractional ideals $\mathcal{I}_k$, but only on the subgroup generated by the prime ideals which are unramified in K/k. Furthermore, it is necessary (in particular to obtain the desired kernel of $\psi_{K/k}$) to work with a cycle $\mathcal{M}$ such that K is contained in the ray class field $k^{\mathcal{M}}$ modulo $\mathcal{M}$. Such a cycle is called a *declaration cycle* for the extension K/k. As we have seen (Theorem 15.21 and Definition 15.22), such a cycle always exists (but note that to obtain this result, we had to use the fact that $N_{K/k} C_K$ is a subgroup of finite index of C_k, which follows from cohomological theory of global class field theory). Once we have the existence Theorem 11.20, we have the following statement:

Proposition B.25 *Let $\mathcal{M}$ be a cycle of k. Let χ be a character of the ray class group $H_k^{\mathcal{M}} \cong \mathcal{I}_k^{\mathcal{M}} / \mathcal{P}_k^{\mathcal{M}}$ (cf. Proposition B.9). Then, there exists a finite abelian extension K of k such that the kernel of χ is $N_{K/k} H_K^{\mathcal{M}}$.*

Proof By definition, we have $H_k^{\mathcal{M}} = C_k / C_k^{\mathcal{M}}$ and the kernel of χ is thus of the form $U / C_k^{\mathcal{M}}$, where U is a subgroup of C_k containing $C_k^{\mathcal{M}}$. Theorem 15.21 (the part using the existence Theorem 11.20) implies that there exists a finite abelian extension K of k such that $U = N_{K/k} C_K$, hence $\text{Ker}\, \chi = N_{K/k} H_K^{\mathcal{M}}$. $\qquad\square$

Corollary B.26 *For any cycle $\mathcal{M}$ of k and any non-trivial Dirichlet character χ : $H_k^{\mathcal{M}} \to \mathbf{C}^*$, we have $L(1, \chi) \neq 0$.*

Proof This follows immediately from the second assertion of Theorem B.22 and of the Proposition B.25. $\qquad\square$

Remark B.27 It is possible to obtain Corollary B.26 by purely analytic methods, without using class field theory: see [9], Exp. VIII, theorem 2.

We can now state a theorem about the equidistribution of prime ideals in a more precise form than Proposition B.21:

Theorem B.28 *Let $\mathcal{M}$ be a cycle of k. Let $\mathfrak{k}_0$ be an element of $H_k^{\mathcal{M}} = \mathcal{I}_k^{\mathcal{M}}/\mathcal{P}_k^{\mathcal{M}}$. Then, the set of prime ideals of $\mathcal{O}_k$ which are in the class $\mathfrak{k}_0$ has density $1/\#H_k^{\mathcal{M}}$.*

Proof Let χ be a character of $H_k^{\mathcal{M}}$. The formula (B.5) gives:

$$\log(L(s,\chi)) - \sum_{(\mathfrak{p},\mathcal{M})=1} \frac{\chi(\mathfrak{p})}{N(\mathfrak{p})^s} = g(s),$$

with $g(s)$ holomorphic near 1, that we can rewrite

$$\log(L(s,\chi)) = \sum_{\mathfrak{k}\in H_k^{\mathcal{M}}} \chi(\mathfrak{k}) \sum_{\mathfrak{p}\in\mathfrak{k}} \frac{1}{N(\mathfrak{p})^s} + g(s).$$

If χ is not the trivial character χ_0, observe that $\chi(\mathfrak{k}_0^{-1}) \cdot \log(L(s,\chi))$ is holomorphic near 1 by Corollary B.26, while for χ_0 we have with the formula (B.3):

$$\chi_0(\mathfrak{k}_0^{-1}) \cdot \log(L(s,\chi_0)) \sim \log \zeta_k(s) \sim -\log(s-1).$$

Summing over different characters, we obtain

$$\sum_{\mathfrak{k}\in H_k^{\mathcal{M}}} \sum_{\chi} \chi(\mathfrak{k}\mathfrak{k}_0^{-1}) \sum_{\mathfrak{p}\in\mathfrak{k}} \frac{1}{N(\mathfrak{p})^s} \sim -\log(s-1).$$

As we have by orthogonality of characters ([46], Chap. VI, Corollary to Prop. 4) $\sum_{\chi} \chi(\mathfrak{k}\mathfrak{k}_0^{-1}) = 0$ except for $\mathfrak{k}\mathfrak{k}_0^{-1} = 1$, we obtain

$$\#H_k^{\mathcal{M}} \sum_{\mathfrak{p}\in\mathfrak{k}_0} \frac{1}{N(\mathfrak{p})^s} \sim -\log(s-1),$$

which means exactly (with Proposition B.17, (c)) that the desired density is $1/\#H_k^{\mathcal{M}}$.
□

We can also reformulate the reciprocity law in terms of the Artin map constructed in Remark 15.5:

Theorem B.29 (Reciprocity law in the langage of ideals)
Let K/k be an abelian extension of a number field. Let $\mathcal{M}$ be a declaration cycle for K/k. Then the Artin map (defined in the Remark 15.5)

$$\psi_{K/k} : \mathcal{I}_k^{\mathcal{M}} \longrightarrow \mathrm{Gal}(K/k)$$

induces an isomorphism of $\mathcal{I}_k^{\mathcal{M}} / N_{K/k} \mathcal{I}_K^{\mathcal{M}} \cdot \mathcal{P}_k^{\mathcal{M}}$ *with* $\mathrm{Gal}(K/k)$. *Furthermore,* $\psi_{K/k}$ *is compatible with the norm residue symbol, that is: the diagram*

$$
\begin{array}{ccc}
C_k & \xrightarrow{\;(.,\,K/k)\;} & \mathrm{Gal}(K/k) \\
\bar{\imath} \downarrow & & \downarrow \mathrm{Id} \\
\mathcal{I}_k^{\mathcal{M}} / \mathcal{P}_k^{\mathcal{M}} & \xrightarrow{\;\psi_{K/k}\;} & \mathrm{Gal}(K/k)
\end{array}
$$

is commutative.

To simplify the notation, we have again denoted by $\bar{\imath} : C_k \to \mathcal{I}_k^{\mathcal{M}} / \mathcal{P}_k^{\mathcal{M}}$ the homomorphism induced by the isomorphism $C_k / C_k^{\mathcal{M}} \to \mathcal{I}_k^{\mathcal{M}} / \mathcal{P}_k^{\mathcal{M}}$ of Proposition B.9.

Proof The assumption that $\mathcal{M}$ is a declaration cycle for K/k means that K is a subextension of the ray class field $k^{\mathcal{M}}$, or that the congruence subgroup $C_k^{\mathcal{M}}$ verifies $\mathrm{rec}_k(C_k^{\mathcal{M}}) \subset \mathrm{Gal}(k^{\mathrm{ab}}/K)$, where $\mathrm{rec}_k : C_k \to \mathrm{Gal}(k^{\mathrm{ab}}/k)$ is the reciprocity map (induced by passage to the limit of the norm residue symbols $(., k'/k)$ for k' finite abelian extension of k). Thus, the isomorphism $\bar{\imath} : C_k / C_k^{\mathcal{M}} \cong \mathcal{I}_k^{\mathcal{M}} / \mathcal{P}_k^{\mathcal{M}}$ induces a commutative diagram

$$
\begin{array}{ccc}
C_k / C_k^{\mathcal{M}} & \xrightarrow{\;(.,\,K/k)\;} & \mathrm{Gal}(K/k) \\
\bar{\imath} \downarrow & & \downarrow \mathrm{Id} \\
\mathcal{I}_k^{\mathcal{M}} / \mathcal{P}_k^{\mathcal{M}} & \xrightarrow{\quad u \quad} & \mathrm{Gal}(K/k)
\end{array}
$$

and we need to show that u is the Artin map $\psi_{K/k}$. For any place v of k and any $a_v \in k_v$, denote by $[a_v]$ the idèle $(1, 1, \dots, a_v, 1, \dots)$. We have seen in the proof of Proposition B.9 that every class of $C_k / C_k^{\mathcal{M}}$ is represented by an idèle (α_v) of $I_{k,1}^{\mathcal{M}}$, which is an idèle verifying $\alpha_v = 1$ for every place v dividing $\mathcal{M}$. This implies that $C_k / C_k^{\mathcal{M}}$ is generated by the classes of the idèles c of the form $[\pi_v]$, where v is a finite place of k (corresponding to a $\mathfrak{p} \in \mathrm{Specm}(\mathcal{O}_k)$) not dividing $\mathcal{M}$ (hence unramified) and π_v a uniformiser of k_v. For such a class c, we have $\bar{\imath}(c) = \mathfrak{p}$ in $\mathcal{I}_k^{\mathcal{M}} / \mathcal{P}_k^{\mathcal{M}}$ via the formula (B.2). Thus

$$u(\bar{\imath}(c)) = (c, K/k) = ([\pi_v], K/k) = \psi_{K/k}(\mathfrak{p}) \in \mathrm{Gal}(K/k)$$

is the Frobenius at v by definition of $\psi_{K/k}$ (Remark 15.5) and of the norm residue symbol $(., K/k)$ (cf. formula (14.1) and Proposition 9.8, (a)). We thus have $u = \psi_{K/k}$ and $\psi_{K/k}$ is surjective as $(., K/k)$ is.

It remains to show that the image of $N_{K/k} C_K$ by the map $\bar{\imath} : C_k \to \mathcal{I}_k^{\mathcal{M}} / \mathcal{P}_k^{\mathcal{M}}$ is exactly $N^{\mathcal{M}} / \mathcal{P}_k^{\mathcal{M}}$, where $N^{\mathcal{M}} := N_{K/k} \mathcal{I}_K^{\mathcal{M}} \cdot \mathcal{P}_k^{\mathcal{M}}$. We already have $C_k^{\mathcal{M}} \subset N_{K/k} C_K$ as $C_k^{\mathcal{M}}$ is in the kernel of the reciprocity map $(., K/k) : C_k \to \mathrm{Gal}(K/k)$. Let us denote by $I_{K,1}^{\mathcal{M}}$ the subgroup of I_K consisting of idèles (β_w) verifying $\beta_w = 1$ for every place w above a place of v dividing $\mathcal{M}$. As we have seen in the proof of the Proposition B.9, we have $I_K = I_{K,1}^{\mathcal{M}} \cdot I_K^{\mathcal{M}} \cdot K^*$, hence:

$$N_{K/k}C_K/C_k^{\mathcal{M}} = N_{K/k}I_K \cdot k^*/I_k^{\mathcal{M}} \cdot k^* = N_{K/k}I_{K,1}^{\mathcal{M}} \cdot I_k^{\mathcal{M}} \cdot k^*/I_k^{\mathcal{M}} \cdot k^*.$$

The elements of $N_{K/k}C_K/C_k^{\mathcal{M}}$ are the classes represented by the elements of $N_{K/k}I_{K,1}^{\mathcal{M}}$ in $I_{k,1}^{\mathcal{M}}$. By formula (B.2), they are exactly the classes sent by $\bar{i}$ to the ideals belonging to $N_{K/k}\mathcal{I}_K^{\mathcal{M}} \subset \mathcal{I}_k^{\mathcal{M}}$, hence

$$i(N_{K/k}C_K/C_k^{\mathcal{M}}) = N_{K/k}\mathcal{I}_K^{\mathcal{M}} \cdot \mathcal{P}_k^{\mathcal{M}}/\mathcal{P}_k^{\mathcal{M}} = N^{\mathcal{M}}/\mathcal{P}_k^{\mathcal{M}},$$

which finishes the proof. □

Remark B.30 The kernel $N^{\mathcal{M}} = N_{K/k}\mathcal{I}_K^{\mathcal{M}} \cdot \mathcal{P}_k^{\mathcal{M}}$ of the Artin map $\psi_{K/k} : \mathcal{I}_k^{\mathcal{M}} \to \mathrm{Gal}(K/k)$ is called the *group of declared ideals modulo* $\mathcal{M}$ for the extension K/k. The correspondence $K \mapsto N^{\mathcal{M}}$ is a bijection between the subextensions of the ray class field $k^{\mathcal{M}}$ and the subgroups of $\mathcal{I}^{\mathcal{M}}$ containing $\mathcal{P}^{\mathcal{M}}$.

We can now prove the "abelian" case of the Čebotarev theorem:

Corollary B.31 *Let K/k be a finite abelian extension of a number field with Galois group G. Let $\sigma \in G$. Then, the density of the set of prime ideals $\mathfrak{p} \in \mathrm{Specm}(\mathcal{O}_k)$ which are unramified in K/k and such that $\psi_{K/k}(\mathfrak{p}) = \sigma$ (in other words: such that the Frobenius at $\mathfrak{p}$ is σ) is $1/[K : k]$.*

Proof Choose a declaration cycle $\mathcal{M}$ of the extension K/k. By Theorem B.29, the Artin map $\psi_{K/k}$ induces an isomorphism of $\mathcal{I}_k^{\mathcal{M}}/N^{\mathcal{M}}$ with G for a certain subgroup $N^{\mathcal{M}}$ of $\mathcal{I}_k^{\mathcal{M}}$ containing $\mathcal{P}_k^{\mathcal{M}}$. Let $c \in \mathcal{I}_k^{\mathcal{M}}/N^{\mathcal{M}}$ be the class sent to σ by this isomorphism. Proposition B.21 implies that the density of the set of prime ideals whose class modulo $N^{\mathcal{M}}$ is c is

$$\frac{1}{[\mathcal{I}_k^{\mathcal{M}} : \mathcal{P}_k^{\mathcal{M}}]} \cdot [N^{\mathcal{M}} : \mathcal{P}_k^{\mathcal{M}}] = \frac{1}{[\mathcal{I}_k^{\mathcal{M}} : N^{\mathcal{M}}]} = \frac{1}{\#G} = \frac{1}{[K : k]}.$$

□

B.6   Proof of the Čebotarev Theorem

We have already encountered (Theorem 18.1) the Čebotarev theorem, but we have not provided a proof. To conclude this appendix, we are going to fill this gap here.

Theorem B.32 (Čebotarev theorem) *Let K/k be a finite Galois extension of a number field, whose Galois group we denote by G. Let C be a conjugacy class in G. Let δ be the density of the set T of prime ideals $\mathfrak{p} \in \mathrm{Specm}(\mathcal{O}_k)$, unramified in K/k and such that $\mathrm{Frob}_{\mathfrak{p}} \in C$. Then $\delta = \#C/\#G$.*

Recall that $\mathrm{Frob}_\mathfrak{p} \in G$ is the Frobenius at $\mathfrak{p}$, which is well defined up to conjugation so that the condition $\mathrm{Frob}_\mathfrak{p} \in C$ makes sense. In this statement, we can also restrict to prime ideals of absolute degree 1 by Corollary B.19.

Proof Let $\sigma \in C$, then C is the set of $\tau\sigma\tau^{-1}$ with $\tau \in G$. Denote by d the cardinality of G, m the order of σ in G and c the cardinality of C. Let $M \subset K$ be the field fixed by σ, then K is a cyclic extension of M of degree m. In the remainder of the proof, we only consider prime ideals that are unramified in the various finite extensions of number fields that appear, a finite set of prime ideals being of zero density.

Let T_M be the set of prime ideals $\mathfrak{q} \in \mathrm{Specm}(\mathcal{O}_M)$ which are split in M/k and whose Frobenius (relatively to the abelian extension K/M) is $\sigma \in \mathrm{Gal}(K/M)$. By Corollaries B.19 and B.31, the density of T_M is $\frac{1}{m}$. Denote by T_K the set of prime ideals $\mathfrak{k} \in \mathrm{Specm}(\mathcal{O}_K)$ whose Frobenius in $\mathrm{Gal}(K/k)$ (cf. Definition 12.13) is σ. We will need a lemma:

Lemma B.33 (a) The map $u_1 : \mathfrak{k} \to \mathfrak{k} \cap \mathcal{O}_M$ is bijective from T_K to T_M.

(b) The map $u_2 : \mathfrak{k} \to \mathfrak{k} \cap \mathcal{O}_k$ is surjective from T_K onto T, and the preimage of each element of T under u_2 has cardinality d/cm.

Assume that the lemma is proved. Then the map $\mathfrak{q} \to \mathfrak{q} \cap \mathcal{O}_k$ is surjective from T_M onto T, and there are d/cm prime ideals of T_M above each prime ideal of T. Furthermore, if $\mathfrak{q} \in T_M$, we have $N_{M/k}(\mathfrak{q}) = \mathfrak{p}$ hence $N(\mathfrak{q}) = N(\mathfrak{p})$. It follows that

$$\sum_{\mathfrak{p} \in T} N\mathfrak{p}^{-s} = \frac{cm}{d} \sum_{\mathfrak{q} \in T_M} N\mathfrak{q}^{-s} \sim_{1^+} -\frac{cm}{d} \cdot \frac{1}{m} \log(s-1) = -\frac{c}{d} \log(s-1),$$

which shows that the density δ of T is $\frac{c}{d}$.

It remains to prove the lemma. Let us first show that if $\mathfrak{k} \in T_K$, then $\mathfrak{q} := \mathfrak{k} \cap \mathcal{O}_M$ is in T_M. Let us set $\mathfrak{p} := \mathfrak{k} \cap \mathcal{O}_k$. The group $\mathrm{Gal}(K_\mathfrak{k}/k_\mathfrak{p})$ is generated by σ and as σ fixes $M_\mathfrak{q}$, we have $M_\mathfrak{q} = k_\mathfrak{p}$, which proves that $\mathfrak{q}$ is split in M/k and the Frobenius associated to $\mathfrak{q}$ in $\mathrm{Gal}(K/M)$ is σ, which implies $\mathfrak{q} \in T_M$. Furthermore, the residual degree of $K_\mathfrak{k}/M_\mathfrak{q}$ is the same as that of $K_\mathfrak{k}/k_\mathfrak{p}$, that is, m, which means that $\mathfrak{k}$ is the only prime ideal of K above $\mathfrak{q}$, hence the injectivity of u_1. Finally if $\mathfrak{q} \in T_M$, there exists a prime ideal $\mathfrak{k}$ of $\mathcal{O}_K$ above $\mathfrak{q}$. As by definition of T_M we have $M_\mathfrak{q} = k_\mathfrak{p}$, we obtain that the Frobenius at $\mathfrak{k}$ in $\mathrm{Gal}(K/k)$ is also its Frobenius in $\mathrm{Gal}(K/M)$, which is σ. Thus $\mathfrak{k} \in T_K$, which shows that u_1 is surjective and concludes the proof of (a).

Let us prove the item (b) of the lemma. Fix $\mathfrak{p}_0 \in T$ and choose a prime ideal $\mathfrak{k}_0 \in \mathrm{Specm}(\mathcal{O}_K)$ above $\mathfrak{p}_0$. By definition, we have $\mathfrak{k}_0 \in T_K$. For any $\tau \in G$, we have (cf. Definition 12.13)

$$\mathrm{Frob}_{\tau \cdot \mathfrak{k}_0} = \tau \, \mathrm{Frob}_{\mathfrak{k}_0} \, \tau^{-1} = \tau\sigma\tau^{-1}.$$

Thus $\mathrm{Frob}_{\tau \cdot \mathfrak{k}_0} = \sigma$ if and only if τ is in the centraliser Z of σ in G. The cardinality of Z is $\frac{d}{c}$ as G/Z is in bijection with C (G acts transitively by conjugation on C and

Z is the stabiliser of an element). If $D_0 \subset G$ is the decompositions subgroup of $\mathfrak{k}_0$, we have a bijection $\tau \mapsto \tau.\mathfrak{k}_0$ of Z/D_0 over the set of prime ideals of T_K above $\mathfrak{p}_0$. Thus, this last set has cardinality

$$\frac{\#Z}{\#D_0} = \frac{\#Z}{m} = \frac{d}{cm},$$

since D_0 is the cyclic group generated by σ. This proves (b) of the lemma and concludes the proof of the Čebotarev theorem. ☐

1. Artin, E., Tate, J.: Class Field Theory. W. A. Benjamin Inc, New York (1968)
2. Berhuy, G., Frings, C., Tignol, J.-P.: Galois cohomology of the classical groups over imperfect fields. J. Pure Appl. Algebra **211**(2), 307–341 (2007)
3. Borovoi, M.: Abelian Galois Cohomology of Reductive Groups. Memoirs of the American Mathematical Society, vol. 132 no. 626. American Mathematical Society, Providence (1998)
4. Bourbaki, N.: Algebra. Chapter II. Springer, Berlin (1998)
5. Bourbaki, N.: Commutative Algebra. Chapter VII. Springer, Berlin (1998)
6. Bourbaki, N.: General Topology. Chapter I. Springer, Berlin (1998)
7. Bourbaki, N.: Algebra, Chapter V. Springer, Berlin (2003)
8. Bourbaki, N.: Algebra. Chapter VII. Springer, Berlin (2003)
9. Cassels, J.W.S., Fröhlich, A. (eds.), Algebraic Number Theory. London Mathematical Society, London: Papers from the Conference Held at the University of Sussex, Brighton, September 1–17, 1965, p. 3618860. Including a list of errata, MR (2010)
10. Česnavičius, K.: Poitou-Tate without restrictions on the order. Math. Res. Lett. 22(6), 1621–1666 (2015). MR 3507254
11. Chenevier, G., Clozel, L.: Corps de nombres peu ramifiés et formes automorphes autoduales. J. Am. Math. Soc. **22**(2), 467–519 (2009)
12. Conrad, B.: Grothendieck Duality and Base Change. Lecture Notes in Mathematics, vol. 1750. Springer, Berlin (2000)
13. Cox, D.A.: Primes of the form $x^2 + ny^2$. In: Fermat, Class Field Theory and Complex Multiplication. A Wiley-Interscience Publication, Wiley Inc, New York (1989). MR 1028322
14. Fein, B., Kantor, W.M., Schacher, M.: Relative Brauer groups. II. J. reine angew. Math. **328**, 39–57 (1981)
15. Fried, M.D., Jarden, M.: Field Arithmetic, Ergeb. Math. Grenzgeb. (3), 3rd ed., vol. 11. Springer, Berlin (2008)
16. Gille, P., Szamuely, T.: Central Simple Algebras and Galois Cohomology. Cambridge Studies in Advanced Mathematics, vol. 101. Cambridge University Press, Cambridge (2006)
17. González-Avilés, C.D.: Arithmetic duality theorems for 1-motives over function fields. J. reine angew. Math. **632**, 203–231 (2009)
18. Grothendieck, A.: Technique de descente et théorèmes d'existence en géométrie algébrique. II. Le théorème d'existence en théorie formelle des modules, Séminaire Bourbaki, vol. 5, Société Mathématique de France, Paris (1995). Exp. No. 195, pp. 369–390
19. Harari, D., Scheiderer, C., Szamuely, T.: Weak approximation for tori over p-adic function fields. Internat. Math. Res. Notices **10**, 2751–2783 (2015)
20. Harari, D., Szamuely, T.: Arithmetic duality theorems for 1-motives. J. reine angew. Math. **578**, 93–128 (2005)
21. Corrigenda, Ibid. 632, 233–236 (2009)

22. Hartshorne, R.: Algebraic Geometry. Graduate Texts in Mathematics, vol. 52. Springer, Berlin (1977)
23. Hewitt, E., Ross, K.A.: Abstract Harmonic Analysis. Vol. I. Grundlehren der Mathematischen Wissenschaften, vol. 115, Springer, Berlin (1963)
24. Iyanaga, S. (ed.): The Theory of Numbers, North-Holland Mathematical Library, vol. 8. North-Holland Publishing Company, Amsterdam (1975)
25. Iyanaga, S. (ed.), Travaux de Claude Chevalley sur la théorie du corps de classes: introductioñ. Jpn. J. Math. 1(1), 25–85 (2006)
26. Izquierdo, D.: Le théorème de Poitou-Tate à partir du théorème d'Artin-Verdier, Mémoire de M2, Université Paris-Sud, Orsay (2013). http://www.math.ens.fr/~izquierdo/Poitou-Tate.pdf
27. Janusz, G.J.: Algebraic Number Fields. Graduate Texts in Mathematics, 2nd edn., vol. 7. American Mathematical Society, Providence (1996)
28. Kato, K.: Galois cohomology of complete discrete valuation fields. Algebraic K-Theory, Part II (Oberwolfach, 1980). Lecture Notes in Mathematics, vol. 967, pp. 215–238. Springer, Berlin (1982)
29. Kato, K., Kurokawa, N., Saito, T.: Number Theory. Introduction to Class Field Theory. Transl. of Math. Monographs, vol. 240, American Mathematical Society, Providence (2011)
30. Koch, H.: Algebraic Number Theory, Encyclopaedia of Mathematical Sciences, vol. 62. Springer, Berlin (1997)
31. Lang, S.: On quasi algebraic closure. Ann. Math. 2(55), 373–390 (1952)
32. Lang, S.: Algebraic groups over finite fields. Am. J. Math. 78, 555–563 (1956)
33. Lang, S.: Algebra, 3rd edn. Addison-Wesley, Reading (1993)
34. Lang, S.: Algebraic Number Theory, 2nd ed., Graduate Texts in Mathematics, vol. 110, Springer, New York (1994)
35. Lang, S., Weil, A.: Number of points of varieties in finite fields. Am. J. Math. 76, 819–827 (1954)
36. Mac Lane, S.: Categories for the Working Mathematician. Graduate Texts in Mathematics, 2nd edn., vol. 5, Springer, New York (1998)
37. Milne, J.S.: Étale Cohomology. Princeton Mathematical Series, vol. 33, Princeton University Press, Princeton (1980)
38. Milne, J.S., Abelian varieties. In: Arithmetic Geometry (Storrs, Conn., 1984), pp. 103–150. Springer. New York (1986)
39. Milne, J.S., Jacobian varieties. In: Arithmetic Geometry (Storrs, Conn., 1984), pp. 167–212. Springer. New York (1986)
40. Milne, J.S.: Arithmetic Duality Theorems, 2nd edn. BookSurge, LLC, Charleston, SC (2006)
41. Neukirch, J.: Class Field Theory. Grundlehren der Mathematischen Wissenschaften, vol. 280, Springer, Berlin (1986)
42. Neukirch, J.: Algebraic Number Theory. Grundlehren der Mathematischen Wissenschaften, vol. 322, Springer, Berlin (1999)
43. Neukirch, J., Schmidt, A., Wingberg, K.: Cohomology of Number Fields, 2nd ed., Grundlehren der Mathematischen Wissenschaften, vol. 323. Springer, Berlin (2008)
44. Raskind, W.: Abelian class field theory of arithmetic schemes. K-Theory and Algebraic Geometry: Connections with Quadratic Forms and Division Algebras (Santa Barbara, CA, 1992). Proceedings of Symposia in Pure Mathematics and vol. 58, pp. 85–187. American Mathematical Society, Providence (1995)
45. Roquette, P.: The Brauer-Hasse-Noether theorem in historical perspective. Schriften der Mathematisch-Naturwissenschaftlichen Klasse der Heidelberger Akademie der Wissenschaften, vol. 15. Springer, Berlin (2005)
46. Sansuc, J.-J.: Groupe de Brauer et arithmétique des groupes algébriques linéaires sur un corps de nombres. J. reine angew. Math. 327, 12–80 (1981)
47. Serre, J.-P.: A course in Arithmetic. Graduate Texts in Mathematics, vol. 7. Springer, New York (1973)
48. Serre, J.-P.: Local Fields. Graduate Texts in Mathematics, vol. 67. Springer, New York (1979)

49. Serre, J.-P.: Algebraic Groups and Class Fields. Graduate Texts in Mathematics, vol. 117. Springer, New York (1988)
50. Serre, J.-P.: Galois Cohomology. Springer Monographs in Mathematics. Springer, Berlin (2002)
51. Serre, J.-P.: Cohomologie galoisienne: progrès et problèmes, Séminaire Bourbaki, Vol. 1993/94, Astérisque, vol. 227, Société Mathématique de France, Paris (1995). Exp. No. 783, pp. 229–257
52. Shatz, S.S.: Profinite Groups, Arithmetic, and Geometry. Annals of Mathematics Studies, vol. 67, Princeton University Press, Princeton (1972)
53. Skorobogatov, A.: Torsors and Rational Points, Cambridge Tracts in Mathematics, vol. 144. Cambridge University Press, Cambridge (2001)
54. Szamuely, T.: Corps de classes des schémas arithmétiques, Séminaire Bourbaki. Vol. 2008/2009, Astérisque, vol. 332, Société Mathématique de France, Paris (2010). Exp. No. 1006, pp. 257–286
55. Weibel, C.A.: An Introduction to Homological Algebra. Cambridge Studies in Advanced Mathematics, vol. 38. Cambridge University Press, Cambridge (1994)
56. Weil, A.: Basic Number Theory. Grundlehren der Mathematischen Wissenschaften, vol. 144, Springer, New York Inc, New York (1967)

Index

A
Abelian variety, 66
Adèle, 167

B
Brauer–Hasse–Noether (th. of), 204

C
Category, 291
  abelian, 6, 296
  additive, 296
  anti-equivalence of -, 295
  artinian, 292
  equivalence of -, 295
  noetherian, 292
  opposite, 293
  small, 292
  sub-, 293
    full, 293
  with enough injectives, 9, 301
  with enough projectives, 10, 301
Cebotarev (th. of), 279
Character, 249
  co-, 249
  Dirichlet, 317
Class field, 212
  Hilbert, 220
  axiom, 110, 191, 194
  ray, 220
Class formation, 235
  P-, 250
Class group
  ideal, 162
  idèle, 166, 179
  ray, 219

Coboundary, 11, 16, 303, 306
Cochain, 16
Cocycle, 16
Cohomological dimension
  of a field, 91
  of a profinite group, 79
  p-, 79
  strict, 81
Cohomological functor, 9
Cohomology
  groups, 10, 71
  modified, 31
  of a complex, 303
Cokernel, 293
Completion
  p-, 69
  profinite, 69
Complex, 303
  double, 307
  homotopic, 304
  tensor product-, 307
  total, 307
Conductor, 220
Coproduct, 294
Corestriction, 25
Cup-product, 43
Cycle, 219

D
Degree
  absolute, 280, 322
  residual, 102
Density (Dirichlet), 279, 319
Dirichlet series, 313
Divisible abelian group, ix
Divisor, 172

© Springer Nature Switzerland AG 2020
D. Harari, *Galois Cohomology and Class Field Theory*, Universitext,
https://doi.org/10.1007/978-3-030-43901-9

Homology (groups), 30
Homomorphism
  augmentation, 29
  crossed, 16

I
Ideal
  augmentation, 29
  fractional, 161
  maximal, x
  prime, x
  principal, 161
  product, 161
Idèle, 165
  finite, 166
  principal, 166
  S-, 169
Index
  of a closed subgroup, 67
  ramification, 102
Inflation, 19

J
Jacobian, 172

K
Kernel, 293
Kronecker–Weber (th. of), 174

L
L-function, 317

M
Mittag-Leffler (condition), 300
Module
  dualising, 131
  flat, 298
  G-, 4
  unramified, 138
Morphism
  epi-, 292
  Frobenius, 103, 165
  mono-, 292
  of cohomological functors, 18
  of complexes, 303
  of functors, 294
  of G-modules, 4
  of strict, x

N
Norm , 24, 29, 253
  absolute, 280
Norm Hasse principle, 191
Number
  p-adic, 100
  supernatural, 67

O
Object
  acyclic, 10, 306
  final, 292
  initial, 292
  injective, 301
  noetherian, 292
  of a category, 291
  projective, 301
  sub-, 292
  zero (or null), 292
Order (of a profinite group), 67

P
Place, 162
  p-group, ix
  archimedean, 162
  dividing another place, 163
  finite, 162
  inert, 184
  split, 173
  totally split, 165, 173
  unramified, 164
Poitou–Tate
  exact sequence, 265
  theorem, 264
p-primary abelian group, ix
p-primary component, ix
Principal ideal (th.), 222
Product
  in a category, 293
  restricted, 166, 261
Proper (map), 125

R
Reciprocity
  homomorphism,, 210
  isomorphism, 120, 210
  map, 121
Reciprocity law
  global, 202
  quadratic, 204
Resolution

Printed in the United States
By Bookmasters